Housing policy
and economic power

Michael Ball

Housing policy and economic power

The political economy of owner occupation

METHUEN *London & New York*

First published in 1983 by
Methuen & Co. Ltd
11 New Fetter Lane, London EC4P 4EE

Published in the USA by
Methuen & Co.
in association with Methuen, Inc.
733 Third Avenue, New York, NY 10017

© *1983 Michael Ball*

Typeset in Great Britain by
Scarborough Typesetting Services
and printed at the
University Press, Cambridge

British Library Cataloguing in
Publication Data

Ball, Michael
Housing policy and economic power.
1. Home ownership – Great Britain –
Social aspects
2. Home ownership – Great Britain –
Political aspects
I. Title
333.3'23'0941   HD7287.8.G7

ISBN 0–416–35280–4

Library of Congress Cataloging in
Publication Data

Ball, Michael.
Housing policy and economic power.
Includes bibliographical references and
indexes.
1. Housing policy – Great Britain.
2. Home ownership – Government policy
– Great Britain.   I. Title.
HD733.A3B25   1983   363.5'8'0941
83–13070

ISBN 0–416–35280–4 (pbk.)

Contents

List of
abbreviations

BSA	Building Societies Association
CIPFA	Chartered Institute of Public Finance and Accountancy
CLNNTC	Coventry, Liverpool, Newcastle and North Tyneside Trades Council
CNC	Cost of New Construction (Index)
CPRE	Council for the Protection of Rural England
CSO	Central Statistical Office
DLT	Development Land Tax
DoE	Department of the Environment
EIU	Economist Intelligence Unit
GMC	Greater Manchester Council
HBF	Housebuilders Federation
HCS	Housing and Construction Statistics
HCEC	House of Commons Environment Committee
HPR	*Housing Policy Review*, 1977
HPCD	*Housing Policy – a consultative document*
HPTV	
I–III	Technical volumes I to III of HPCD
JURUE	Joint Unit for Research on the Urban Environment
LBC	London Brick Company
LOSC	Labour only subcontracting

MHLG	Ministry of Housing and Local Government
MLR	Minimum Lending Rate
NHBC	National Housebuilders Council
NFBTE	National Federation of Building Trades Employers
NSB	Non-speculative builder
OPI	Output Price Index
PEHW	Political Economy of Housing Workshop (Conference of Socialist Economists)
RPI	Retail Prices Index
SHAC	Scottish Housing Advisory Committee
TRG	Training Research Group
WNHC	Workmen's National Housing Council

List of tables

List of
figures

Preface

Owner occupation increasingly dominates housing provision and housing policy in most advanced capitalist countries. Although this book is specifically about owner occupation in Britain, the original impetus to start the research came from an interest in explaining such tenure shifts and their consequences in general terms. Traditional housing research places emphasis on the situation of housing consumers in different tenures and on state policy towards them. For a number of years I have felt such a narrow focus to be insufficient to explain the enormous changes in housing provision that have taken place over the past thirty years. Its inadequacies are indicated by the slow drying up of housing research and the increasing fragmentation and specialization of what remains during years when housing problems are growing in many countries.

Other researchers have recognized the unnecessary confines of the traditional approach. Two main alternatives have been tried. The first is a comparative one where the situation in different countries is examined over time. This approach has the advantage of demolishing certain accepted wisdoms by highlighting empirical situations in other countries where they do not hold. Yet ultimately, unless they question the consumer-orientation of traditional perspectives, cross-country comparisons just reproduce the early difficulties on a global scale. The problems are even compounded by having more data and questions to answer with no theoretically adequate means of dealing with them.

The second and, I think, more adequate approach is to recognize that housing provision involves more than a relation between the state and consumer households. This has led to studies of financial institutions associated with housing or of the construction industry and landownership. Again, however, there are problems, as it is tempting to focus on one agency alone without integrating it into an overall analysis of developments in housing provision. In this way, each aspect of housing provision ends up in its own isolated box. Analysis becomes a series of distinct chapters, each of which does not depend on the others in a meaningful way. Alternatively, the primacy of one aspect may be asserted, like production, with little or no attempt to justify the assumed primacy.

When looking at owner occupation in Britain I have tried to avoid these pitfalls by concentrating on the dynamic processes of change in housing provision and trying to see the role of different types of social agent in them. Particular relations of economic power in housing provision are argued to be key, but neglected, elements in the understanding of housing issues. A brief outline of the approach is given in chapter 1. Because I wanted to get across an interpretation of a whole series of empirical events associated with owner occupation, this initial theoretical exposition is kept to a minimum. Practical demonstration of its usefulness seemed more important. Instead theoretical issues, of which there are many, are confronted in the text when the need arises.

No attempt is made to deal comprehensively with all issues associated with owner occupation. It is impossible to cover every empirical issue associated with a housing tenure; some selection of topics must be done. Here specific questions are asked related to issues of economic and political change as they are felt to be central to the analysis of changes in housing policy. Similarly emphasis has been put on the more neglected aspects of housing provision to avoid repeating material that already is well known. The most neglected aspects I feel are housebuilding, land development and land-use planning. In doing so the book goes beyond the normal confines of housing studies. I hope people who are interested from a non-housing perspective in those topics will also find the relevant chapters useful and interesting.

Like all written material dealing with social issues, this book is a product of its time. The data and arguments presented in it are influenced by the information available when the final manuscript was written in the second half of 1982. Anyone might reasonably ask what the relevance is for later time periods. A number of points need to be made in reply.

The most obvious point is that the re-election in 1983 of the Thatcher Conservative administration for another five years has reinforced the relevance of the critique of current housing policies. No analysis of housing provision can ignore the disastrous effects of the Thatcher government in the housing sphere and in the related areas of infrastructure expenditure and land-use planning. The immediate effects of that government's policies on housing are discussed in chapter 1 and the consequences for planning and the built environment in chapter 8. Yet, sad to say, the policies of the Thatcher government in these spheres have not been unique. They principally continued and exacerbated trends that had developed throughout the 1970s, as that government's ministers delighted in reminding the Labour opposition in Parliament, most of whose senior members had been ministers in the previous 1974–9 Labour administration.

On the political horizon there is little indication, at present, of potential future alternative governments, such as Labour or the SDP/Liberal Alliance, having radically different policies. All that is likely to change is a reversal of the cuts in state housebuilding and public works programmes when a government finally decides that a little reflation via public investment is desirable. Less likely, there might also be a reform of housing finance, including fiscal changes for owner occupiers such as the abolition of mortgage interest tax relief. Yet one of the key conclusions of this book is that reform of housing finance alone will not solve any of the fundamental problems of owner occupation in Britain. The description given of housing policy towards owner occupation as being essentially passive and aimed, at least nominally, at maintaining the status quo is likely to remain relevant for a number of years to come.

The continuity of trends in housing policy in the midst of differences in detail and party political rhetoric is not accidental. It is the product of a lack of interest by successive governments in altering the social relations of housing provision. Contradictions arising from those social relations as a result have continued to exert paramount influence on the housing policy of the state and have severely constrained the options open to any government. Often those contradictions will appear empirically in the guise of difficulties over, say, housing finance or levels of output. Lack of adequate analysis has led to a misconception of the causes of housing problems, because attention is concentrated solely on those forms in which immediate difficulties appear. The pattern of housing policy, therefore, is not simply the product of the ineffectiveness of political reformism but also of inadequate theoretical analysis of the nature of housing provision, as later chapters will show.

Chapter 1 presents the most historically specific information as it outlines the growing housing crisis in Britain in the early 1980s. This is done to counteract the complacency which pervades so much mainstream discussion of housing issues. Housing problems frequently are treated as isolated residual difficulties faced by an unfortunate minority. Yet the housing crisis affects everyone because it is a consequence of the way in which all housing gets provided. Data on the rates of new housebuilding, delapidation and improvement are used in chapter 1 as barometers indicating the extent of the housing crisis in the 1980s. In many respects, however, they are only weak indicators of housing trends as they are prone to fluctuate with short-term variations in the economy, thereby obscuring the more fundamental nature of the crisis. Private housing output at the end of 1982 and in early 1983, for example, picked up sharply from the record 30-year low it had reached previously. The upturn, however, does not look as impressive in absolute terms as it does proportionately and was petering out again by April 1983. Output is still well below the levels of the late 1960s and early 1970s and of any realistic assessment of housing need. Also, justification for the rundown of council housing has been based partially on a supposed switch of new output from council housing to private building for owner occupation. To reverse the growing housing shortage and to justify policies towards owner occupation, private housing output has to be *higher* than the post-war peak levels it reached in the 1960s. There seems little prospect of that happening.

Some people have argued that the slump conditions of the early 1980s have made a new owner-occupied housing boom possible, similar in nature to the one which helped to lift the economy in the 1930s out of the depths of the inter-war depression (see chapter 2). Such a position is contrary to the argument presented here. Some similarities to the inter-war period do exist in the 1980s. There are regional and class differences in the severity of the slump. They help to create both a market amongst the more affluent for new housing and a cheap and docile workforce to build it from out of the ranks of the underemployed and unemployed. There is also a partial return to the passive planning system of the 1930s. In addition the house price average earnings ratio has dropped to a historically low level, whilst interest rates have begun to fall (although nowhere near to the level of the 2 per cent cheap money era of post-1931). The scenario of such a future owner-occupied housing boom, however, not only ignores the likely escalation of building costs if there is a substantial increase in activity, but also forgets to consider the tenure in which the better-off households live. In the inter-war period owner occupation as a housing tenure expanded rapidly from a low level. Many richer households became homeowners for the first time. Now most of them are already owner occupiers. To buy a new house they have to sell their existing one. The market for owner occupation as a whole can expand only by drawing in lower income households from the working class. Failure to do this brought the 1930s' boom to a halt. As the early 1980s' slump has hit those social groups the worst, it hardly seems likely to happen now.

The general point being made in this book is that the problems of housing provision and owner occupation in Britain are structural. They arise from the contradictory effects of the relations between social agents involved in owner-occupied housing provision. So even though there might be a whole series of changes in the contemporary state of the market or possibly even in the characteristics of some of the agencies involved (for example, I increasingly suspect that some building societies and clearing banks would merge if legislation made it possible), the basic structure of owner-occupied housing provision will remain the same. If no attempt is made to deal with the structural problems of owner occupation, along the lines suggested in the final chapter, any future for housing provision in Britain will involve enormous economic and social costs.

One final point to make is that an attempt was made to provide data on the most important aspects of the housing market until the end of 1982. Publication dates of the data did not always make this possible, although information was added until the proof stage of the text. Obviously some of this latest data is not commented on in the text.

Much of the material on present-day speculative housebuilding in this book was collected during 1979 and 1980 as part of a wider project on the economics of housebuilding undertaken at Birkbeck College. I should like to acknowledge with thanks the financial support for that project of the Social Science Research Council under grant HR5181. Andrew Cullen worked with me on the project during those years. My ideas were greatly improved by discussions with him, and without his efforts this book could not have been written. Originally we had hoped to write it together but the temporary employment conditions faced by university research staff made this a practical impossibility for him. I should like to acknowledge with thanks the importance of his contributions to the production of this book. Other friends and colleagues also commented on drafts, offered much needed advice and support, and helped with its final production. In particular, I should like to thank

Ben Fine, Diana Gilhespy, Michael Harloe, Maartje Martens, Doreen Massey, Ron Smith and Marjorie Turton.

★ ★ ★

The author and publisher would like to thank the following for permission to reproduce copyright material: The Building Societies Association for data from the *BSA Bulletin* in Tables 10.1, 10.2, 10.3, 10.4, 10.5, 11.2, 11.4 and Figures 2.1, 4.7, 11.1, 11.3, 11.5; the Controller of Her Majesty's Stationery Office for data in Tables 1.1, 1.3, 3.1, 3.2, 4.1, 4.2, 5.1, 6.1, 8.1, 8.2, 9.1, 9.2, 11.3 and Figures 1.1, 2.1, 2.2, 3.2, 4.2, 4.3, 4.4, 4.5, 4.6, 4.8, 4.9, 6.1, 8.1, 9.2, 10.1; *The House-builder* and W. F. Cunter, COFI, London, for Figure 6.3; Ellen Leopold, Bartlett School of Architecture and Planning, University College London, for Table 5.2; the National Federation of Building Trades Employers for data in Figure 6.2; the National House-Building Council for data in Tables 3.3, 3.5 and Figure 3.3; the Nationwide Building Society for data in Tables 4.3, 11.1 and Figure 4.1; the Royal Town Planning Institute for data from *The Planner* in Table 8.1; *Town and Country Planning* for data in Table 8.1; the United Nations for data from the United Nations Statistical Yearbook in Table 1.2.

1

Housing problems and owner occupation

From complacency to crisis

Since 1945 state involvement in housing provision has been enormous. Housing has been one of three areas of general personal consumption, along with health and education, to which governments were committed by the post-war political consensus. There have obviously been variations in policy over the past forty years but housing has consistently been a key concern of successive governments. Housing provision can also be said to demonstrate the philosophy of the 'mixed economy' cherished by most post-war governments. For, unlike education and health which principally have been provided on the basis of need without direct payment, households still pay for their housing directly via rental, mortgage, maintenance and repairs payments which the state subsidizes in various ways. Ideally this is said to enable subsidies to be selectively directed towards those in need, and to enable households to choose the tenure in which they want to live at reasonable cost. The reality, however, has been somewhat different.

Despite the market-orientated approach to housing provision, state financial involvement has until recently been of a roughly similar magnitude to that in health and education. In 1976, a peak housing expenditure year, state investment and subsidy expenditure on housing was just over £5 billion whilst another £1 billion was 'spent' on tax relief to owner occupiers. This is about the same as the £6 billion spent on the

National Health Service and not that much less than the £7 billion going to education in that year, and it does not include much of the ancillary state expenditure on the built environment (roads, sewers, etc.) that goes along with new housing provision.[1]

In a number of respects post-war housing policy can be said to have been a notable success. The physical housing situation in Britain has improved dramatically and, although state involvement has not been the sole cause, successive governments have been quick to point out the improvement. Since the earlier 1950s through to 1980 there has been a net gain of 200–250,000 dwellings each year. Millions of slums have been demolished whilst thousands of other dwellings have been renovated to modern standards. Housing conditions have continued to improve into the 1980s so that under 5 per cent of dwellings are now overcrowded or lack the sole use of a bath/shower or have no inside toilet. And most of the dwellings still without basic amenities are concentrated in the relatively minor privately rented sector (particularly in furnished accommodation). Added to these quality improvements successive governments have made great play of an apparent 'crude surplus' of dwellings over households that has grown since the early 1960s.[2]

Whilst these physical changes in the stock have been taking place, there has been an even more dramatic change in the tenures in which people live. Most households are now owner occupiers, whereas few were prior to 1914. Then virtually everyone rented from a private landlord but that tenure has declined and been replaced by owner occupation as the majority tenure, whilst council housing grew from virtually nothing in 1914 to constitute a third of the housing stock in the early 1980s (table 1.1). Now over 55 per cent of households are owner occupiers and surveys indicate that it is by far the most popular tenure (although their

Table 1.1 Housing stock and tenure, 1914–81

	1914	1938	1960	1971	1980	1981
Percentage of						
privately rented[1]	90	58	32	19	13	13
local authority	negligible	10	25	29	32	31
owner occupied	10	32	44	53	55	56
out of a total stock of:						
(million)	7.9	11.4	14.6	17.1	21.0	21.1

Source: HPTV I and HCS

[1] Includes other minor tenures

wording helps to produce this answer). So the tenure switch in post-war Britain has been notched up as another policy success.

By the mid-1970s a widespread complacency pervaded political discussion over housing, particularly from successive housing ministers. The growth of both council housing and owner occupation apparently demonstrated the even-handed approach of post-war state housing policy. There might be minor problems over certain aspects of housing policy, but basically the formulas adopted at various stages since the war have been claimed to have worked well. As the 1977 *Housing Policy Review*[3] concluded, minor tidying up of housing policy alone was required. Peter Shore, the then Secretary of State for the Environment, confidently wrote in the foreword to the *Review*:

> We are better housed as a nation than ever before; and our standards of housing seem to compare well with those of similar and more prosperous countries. This should give pause to critics who start with an assumption that present arrangements have served and are serving us badly. (HPCD, p. iv)

And he pinned his faith firmly on the consensus view of maintaining the *status quo*, which essentially meant that the taxation arrangements for owner occupiers would remain the same:

> we certainly do not believe that the household budgets of millions of families – which have been planned in good faith in the reasonable expectation that present arrangements would broadly continue – should be overturned, in pursuit of some theoretical or academic dogma. (ibid., p. iv)

The 1977 *Housing Policy Review* was the high point of post-war complacency over housing provision. By the early 1980s that complacency had disappeared and there was much talk about a growing housing crisis. As two academic commentators noted in 1982:

> Given the present outlook, it seems beyond question that by the mid-1980s some of the improvement in British housing conditions, brought about between 1945 and 1976, will have been undone – perhaps the first major reversal in this sector since the war. (Fleming and Nellis 1982)

This shift, moreover, could not simply be blamed on the onset of economic slump nor on cuts in state expenditure. Whilst both have contributed to the growing crisis there is also a structural malaise in the British housing system.

To an extent the official complacency of the 1970s was misplaced. Housing output had been declining from the late 1960s and the criteria used to measure standards were of the simple nineteenth-century public health variety which ignore many housing attributes important for modern life. Dwelling floor space standards, for example, fell during the 1970s, particularly for owner-occupied housing (see chapter 5), whereas the quality of new construction and conversion left much to be desired in terms of insulation and structural standards. In this respect, local authority industrialized-system flats have had the greatest publicity over condensation and design problems: a number of blocks have already been demolished only a decade after they were built. But such problems are not solely located there.

Britain's post-war housing record has been poor compared with other West European countries. All experienced a housing boom from the late 1950s to the early 1970s, but relative to population size Britain tended to trail behind other countries' housebuilding rates, particularly those with a similar 'welfare state' social democratic tradition (table 1.2). Alarm bells should also have been ringing about indications of the rising cost of providing housing in the post-war era. Additional investment in housing was producing proportionately less and less new housing. Twice as many new dwellings were built for the same real amount of capital investment in 1953–5 compared to 1981, as table 1.3 shows. In part the difference is accounted for by an increase in improvement expenditure and possibly by better unit quality, but the discrepancy is too great to be solely a product of those two factors.

Historically high interest rates, in addition, began to have an effect on housing finance from the early 1970s onwards. In combination with general price inflation and the escalation in building costs they were creating havoc with the traditional mechanisms of housing finance. The upward drift of costs is illustrated by council housing cost rents as a

Table 1.2   Cross-country comparison of number of houses completed per 1000 population, 1955–78

	1955	1960	1965	1970	1978
UK	6.1	6.0	7.0	6.6	5.1
France	4.8	7.0	8.7	9.0	8.4
Italy	4.2	5.8	7.8	7.0	2.8
Sweden	8.0	9.0	12.1	13.5	6.5
W. Germany	10.7	10.1	10.0	7.5	5.7

Source: United Nations Statistical Yearbook 1978

proportion of working-class incomes. Cost rents are the rents that would have to be charged if all the costs of the current system of council housing finance had to be paid out of rents (rather than from rents and subsidies). In 1947 cost rents were 18 per cent of gross adult male manual earnings, yet by 1970 they had risen to 24 per cent and by 1976 to as much as 36 per cent (HPTV I, p. 46). The percentage of post-tax earnings obviously is greater.

Since 1977 the situation has become far worse. The rate of new housebuilding is one of the best barometers of changes in the state of housing provision. It cuts through all the rhetoric of things-are-not-as-bad-as-they-seem, 'crude housing surpluses' and the fallibility of population projections. In periods when housebuilding is expanding, housing conditions for most people will tend to be getting better; when it slumps the problems begin to mount up. Like all barometers the rate of new housebuilding is only an indicator of change, not a fundamental cause. It highlights, none the less, the fact that things were getting worse throughout the 1970s and accelerating rapidly into crisis by the early 1980s (see figure 1.1). In 1969, almost 400,000 dwellings were completed; by 1981 output was down to 198,000, less than half the output twelve years previously. 1981 was a trough year in the building cycle, but data for 1982 showed little sign of a major upturn in the near future.

Table 1.3 Gross fixed investment in dwellings: United Kingdom: annual averages over selected years, 1899–1981

	Average annual number of houses completed (thousands)[1]	*Gross fixed investment in dwellings (£m) at 1970 prices*	*As % of total gross fixed capital formation*[2]	*As % of Gross Domestic Product*[2]	*'Unit cost' at 1970 prices (£)*[3]
1899–1901	145	370	22	1.9	2552
1911–13	60	180	13	0.8	3000
1925–7	217	700	34	3.4	3225
1934–6	351	970	36	4.0	2764
1953–5	328	985	25	4.0	3003
1966–8	401	1750	20	4.3	4364
1973–5	292	1750	19	4.3	5993
1981	198	1210	13	2.5	6111

Sources: HPTV I, HCS and *National Income and Expenditure*

[1] Great Britain
[2] At current prices
[3] 'Unit cost' is a crude measure derived from dividing column 2 by column 1. It takes no account of quality changes nor of the improvement expenditure included in the data (important after 1970). The pre-1945 data are particularly poor

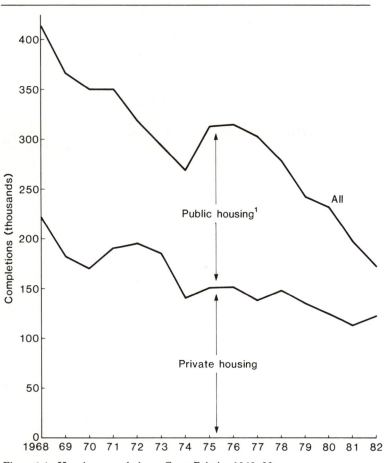

Figure 1.1 Housing completions, Great Britain, 1968–82

Source: HCS

[1] Public housing includes local authorities, new towns and housing associations

When this output is compared with requirements, the extent of growing shortfall is clear. Data on housing need are bound to be rough, given
the difficulties of population forecasting and of estimating needs con-
cealed within the current housing system. But, in what is still regarded
as the best estimate of crude housebuilding requirements (given in the
1977 *Housing Policy Review*), at least 300,000 new houses per year are
needed. That figure has been met in only three years since 1973. The
House of Commons Environment Committee in 1980 forecast a shortfall

of half a million dwellings by the mid-1980s (HCEC 1980) and the collapse of output since then has exacerbated the shortage. Meanwhile the existing housing stock is ageing: over 1 million houses are unfit or require over £3000 repairs, another 1.5 million households' dwellings lack a basic amenity, whilst the number of unfit houses increased by a dramatic 42 per cent between 1979 and 1980. And, on top of these aggregate problems, there is also a severe mismatch between the location and type of dwellings and households.[4]

What has happened?

Economic crises and political responses to those crises have had a dramatic effect on housing programmes. Post-war housing policies have rested on the assumption that both households and the state could continually spend more on housing provision with rising national income. Mass unemployment and declining living standards in the early 1980s sapped households' abilities to pay for rising housing costs, whilst the reaction of governments since the mid-1970s to economic crisis has been to cut public expenditure, and the cuts have fallen disproportionately on housing programmes. One way of illustrating the severity of the cuts is to compare the changes in public expenditure on housing, education and health since the 1976 data given earlier. All three sectors have seen sharp reductions in physical levels of provision but housing has had by far the largest financial pruning. Whilst the three services had roughly similar levels of expenditure in 1976, by 1981 expenditure on housing (including mortgage interest tax relief) was only 60 per cent of the other two, and it is planned to be even less in the future.[5] Council housebuilding as a result has plummeted and, far from being compensated by increased private housebuilding as hoped for, private housebuilding has fallen as well (figure 1.1), although there have been cyclical fluctuations.

In the midst of public expenditure cuts and economic crisis there none the less has been a marked shift in the emphasis of housing policy and state expenditure away from council housing towards owner occupation. This can be seen by looking at the expenditure cuts in more detail. It is worth examining the council housing cuts in terms of their effect on capital and current expenditure separately, as this accounting distinction highlights the two crucial uses of housing finance. Capital expenditure is principally money spent on new and renovated housing by the public sector (councils, new towns and housing associations). Current expenditure is the cost of running the existing stock (paying off loans borrowed in earlier years to finance housebuilding, and undertaking maintenance, repairs and administration). Both these categories of expenditure are political products rather than simple consequences of accounting convention. The division into capital and current expenditure helps none the less to highlight the recent history of council housing cuts.

Capital expenditure on public housing has fallen in real terms every year since 1976. Tory government cuts after 1979 therefore continued a trend that had already been well entrenched during the later years of the previous Labour administration. Real capital expenditure on dwellings by local authorities in 1981 was a mere *15 per cent* of that in 1976.[6] Not surprisingly only 20,600 new council dwellings were started in 1981 compared with 107,600 in 1976.

Up to 1980/1 subsidies on current expenditure slowly rose; pushed up by, amongst other things, rising interest rates. Yet during the 1981/2 financial year the real value of general subsidies was cut by almost 40 per cent and the government did the same again during the 1982/3 financial year.[7] This is done principally by increasing council rents, although rises in rent rebates and allowances will partially offset that cut. In 1980/1 average rent increases of 32 per cent were sought, in 1981/2 they were 39 per cent; and from spring 1982 they rose by another £2.50 a week. With this scale of increase average unrebated rents more than doubled in three years from April 1979. One consequence of these massive rent rises is that some councils have started to make profits on their housing stock. Fifty local authorities managed to achieve this feat in 1981 according to the public accountants' institute, CIPFA; Colchester topped the list with a profit of £1.1m. Within the remaining state subsidies to council housing the proportion financed from local taxes has also been pushed up considerably from a fifth in 1979 to over a third in 1982. In other words, it is central government in particular that is trying to disengage from council housing.

Legislation introduced in 1980, in addition, gave most council tenants the 'right-to-buy' their dwellings at substantial discounts irrespective of whether the local authority wished to sell. Over 300,000 local authority dwellings were sold into owner occupation in England and Wales in two and a half years from 1979 to mid-1982. One hundred thousand were sold in the first six months of 1982 alone at a discount of over 40 per cent of their estimated value. So the disposal programme is accelerating and the best parts of the council housing stock are now the main source of expansion for owner occupation.

The cuts in state subsidies to local authority housing, however, coincided with increases in subsidies to owner occupation. The extent of state subsidy to owner occupation is far from clear. Again it would be useful to make an initial division between subsidies to capital and current expenditure but it is impossible to work out in detail. Capital subsidies take the form of the preferential (zero) tax treatment of wealth tied up in owner-occupied housing, improvement grants, general expenditure on the built environment and the provision of windfall land gains to housebuilders. The latter two are difficult to identify although chapter 8 will show that general state construction-related expenditure has fallen sharply over the past decade whereas the state has improved conditions for builders to make windfall land gains.

On the current expenditure side, the most important subsidy to owner occupiers' housing expenditure is mortgage interest tax relief. These tax reliefs are not shown as state expenditure, yet constitute a loss of state revenue and so are equivalent to extra expenditure. Whilst general subsidies to council housing have been falling, tax reliefs to owner occupiers have been rising. Once the conditions for eligibility for tax relief are determined the state has little control over the magnitude of the subsidy implied as it depends on the level of house prices, the rate of personal income tax, and mortgage interest rates. The conditions of eligibility have essentially not changed since 1974 when a mortgage ceiling of £25,000 was fixed (raised to £30,000 in 1983). The fiscal subsidies to owner occupation in other words have been a long-term trend rather than a product of recent policy.

The growth of the mortgage interest tax relief subsidy has been substantial. In 1980/1 it was in real terms five and a half times higher than it had been in 1970/1.[8] As the number of owner occupiers with mortgages and public sector tenants is roughly similar, comparison with general public sector subsidies is enlightening. Both subsidies grew rapidly throughout the 1970s and for most of the time the overall level of subsidy to each sector was similar. The relative variation in the growth of subsidies was determined primarily by the two tenures' different systems of housing finance. So in the years 1972–4, with an owner-occupied house price boom and rising mortgage interest rates, the owner-occupier subsidy rose fastest. Then council subsidies rose faster for a couple of years owing to interest rate effects and the holding down of council rents as part of an incomes policy (HPR, figure 3). The years 1977 and 1978 again saw big rises in public sector subsidies for similar reasons: then in the early 1980s there was a sharp rise in the subsidies to owner occupiers.

The cuts in council housing subsidies from 1980 onwards totally changed the pattern of a long-term mutual increase in current expenditure subsidies. For England and Wales in 1981/2 the House of Commons Environment Committee calculated the total costs of mortgage tax relief to be £1300m. compared with only £1000m. for the public sector: so by then state subsidies to owner occupation were 30 per cent higher than to public housing. In future years on current policies, as public sector subsidies fall, this ratio will swing further and further in favour of owner occupation. There is nothing to stop the continued increase of the mortgage tax relief subsidy unless the maximum eligible mortgage sum stays fixed whilst inflation erodes the real value of the relief away. The effectiveness of that strategy, however, even if it were adopted, is extremely long-term, so in the forseeable future unless there is a major change in housing policy owner occupation will be the tenure to receive enormous state subsidies.

The housing crisis can be seen to be more than just a physical one of a shortage of new dwellings and a deteriorating existing stock. It is one of the cost of housing as well, and of who pays that cost and, as the economic crisis persists, who can afford to pay that cost. The growing physical deterioration and shortage of housing results from the increasing difficulty of financing the two main existing forms of housing provision, council housing and owner occupation. Throughout the 1970s state subsidies rose dramatically in the face of escalating housing costs. Charitably it could be argued that governments have simply tried to keep down the direct housing costs of individual households. More realistically the subsidies have made it politically possible to avoid fundamental reform of the housing system in Britain.

The increasing cost of council housing provision illustrates the extent of the problem of rising costs. Even after taking account of general price inflation, annual current expenditure on local authority housing rose by an incredible 50 per cent between 1971 and 1980. This additional expenditure on supervision and management, repairs and outstanding debt had to be financed either out of increased rents or subsidies. Yet the council housing stock had risen by a mere 14 per cent.[9] So despite all the outstanding debt that had been paid off in previous years and the effect of inflation in eroding the real value of the debt remaining, council housing was in a financial mess. The organization of local authority housing and its finance was ripe for reform. The balance of political forces in the early 1980s, however, led to its decline to the residual role of a welfare net for those who could not afford to buy.

The cuts on state subsidies to council housing were justified on the general back-to-the-market, reduce-state-expenditure homespun philosophies of the post-1979 Conservative government. Owner occupation was to solve most people's housing problems, an idea that also had been firmly entrenched in the minds of previous governments. State intervention into other housing tenures could then be concentrated on those in 'real need'. But owner occupation, as it currently exists, is also in severe difficulties. New housebuilding is at historically low levels; housing costs, although variable, are still high for many owners; and the state faces a limitless commitment to subsidize it, to mention just three issues.

The crisis of housing provision is also a political one because many people recognize that something must be done about reforming housing policy but no viable alternatives seem to exist. The problem is often posed as being a product of the political power of owner occupiers. Any reform of housing finance, it is argued, must involve increased housing costs for owner occupiers through tax reforms, and any government that introduces such reforms is risking political suicide as over half the population are owner occupiers. As no change in owner-occupied finance means no fundamental shift in housing policy, there is consequently a political stalemate. This view seems borne out in the housing policies of both the Labour and Conservative parties which in varying ways argue for more-of-the-same-old-medicine.

It is undeniably true that the current form of owner occupation has considerable political support and that there is an impasse over changing house provision, but to suggest that individual owner occupiers control, via the ballot box, the limits of housing policy is to take notions of economic interest and representative parliamentary democracy to simplistic extremes. It would imply that governments never undertake actions against the short-term economic interests of the majority of people. Macroeconomic policy, tax increases of any sort, price rises by any nationalized industry, and a whole host of state actions would similarly be 'politically impossible'. The current political stalemate over housing policy must have reasons but deeper analysis is required than simply seeing whether owner occupation has passed the magical 50 per cent of voters mark. Politics must be interlinked with the other strands of the housing crisis before an adequate understanding of the current state of housing policy is clear.

This book suggests a different explanation for the housing crisis and its interrelated economic and political dimensions. It argues that the nature of owner-occupied housing provision has to be looked at in detail first; then it is possible to evaluate why this form of housing provision has gained such broad political support. Having approached the issue in this way, it is then possible to make some comments on the cohesiveness of that political support. In particular, it can be shown that mounting internal problems within owner-occupied housing provision are generating a weakening of political support for it which is creating a space on the political agenda for new forms of housing provision.

The details of and justifications for these arguments are given in the following chapters. The essential points being made are that owner occupation as a form of private market housing provision is not only failing to meet the basic aspirations of most people for decent, cheap housing but also is coming into conflict with the economic interests of most groups in society. The only beneficiaries of the current form of owner occupation are those private interests who dominate it: landowners, building societies and speculative housebuilders (plus a few of the more wealthy owner occupiers). And even they are facing increasing problems because of the workings of the owner-occupied housing market. An increasing lack of correspondence with economic interests is the cause of the weakening of the coherence of political support for the tenure in its current form. Yet because of the nature of the politics of housing provision, reform of the housing system must maintain the existing main housing tenures, including owner occupation. Instead the tenures need to be taken out of the market framework by removing from them the private agencies that currently dominate housing provision. Programmes of land and building industry nationalization are suggested as necessary pre-conditions for successful change to forms of housing provision which can meet the housing needs of the mass of the population.

Analysing housing provision

The arguments presented in the following chapters differ quite substantially from those of many other housing studies, as the comments above indicate. It is perhaps worthwhile at this stage, therefore, to contrast the approach with more well-known ones and to elaborate the broad threads of the argument, as the directions in which they go and the particular juxtapositions of theory and empirical analysis might at first sight seem a little bewildering to those used to sharper divisions between theory and fact and between housing consumption, housing production and the land market.

Most analyses of housing concentrate almost entirely on the consumption effects of housing. Their emphasis is perhaps not surprising because when looking at housing tenures and making comparisons between them the most obvious place to start is where the problems look most pressing and where they seem open to remedial action by the state. The most pressing housing problems concern the living needs of households and the housing costs they have to bear which the state can influence by legislative reform and, in particular, through housing subsidies. Yet presenting housing problems in this way is, at the same time, the specification of a particular approach to looking at them. There are two basic conceptual categories in the approach, households and the state; households have limited means by which to satisfy their housing needs, and the state has the power and the means to deny or satisfy those needs via its housing policies. The linkage between these two conceptual pillars is, therefore, the effect of state policies on households' consumption of housing. This conceptualization of housing can consequently be termed the consumption-orientated approach to housing.

The traditional consumption-orientated way of looking at housing is to treat household needs as given and to examine housing tenures as means of allocating housing. Housing provision is seen, in other words, principally in terms of individual households' consumption and costs which are compared with those of other households on a distributional basis. Housing policy or state subsidies to households' housing costs can then be examined to see whether they are fair and equitable, or correspond to other assumed goals in a check list of desirable ends (such as 'economic efficiency', 'equality of opportunity', etc.). Housing provision, therefore, is seen essentially as a distributional issue.

One much-publicized recent study within the perspective argues that the mortgage costs of owner occupation are more than offset by rising money gains from house price inflation. Owner occupiers make an investment in housing and, over the past fifteen years or more, they have made a very good return (King and Atkinson 1980). Such studies have the advantage of showing that state housing policy unduly favours better-off households (although the task of evaluating distributional implications is not so theoretically straightforward as often is made out). But they cannot explain why state policies take the form they do, and they present a one-sided picture related to who bears the burden of costs instead of examining why those costs exist in the first place.

The politics of housing policy in this approach is a question of correcting past mistakes through a process of political advocacy using the results of the distributional analysis (cf. Cullingworth 1979a). There is, in fact, widespread belief amongst academic experts that not only are the taxation/subsidy arrangements for housing unfair between households and between tenures but also that housing as a whole gets subsidized too much. This argument has been regularly put for many years (cf. Odling-Smee 1975, Clarke 1977, Cullingworth 1979a and Foster 1980). So, according to it, the present housing crisis is one of a misallocation of resources: a failure of the state to get its subsidy/taxation policies towards housing consumption right.

Central to treatments based on such notions of the market mechanism and state intervention is the claim that there are no fundamental divisions within society based on the ownership of economic wealth and control over the productive uses to which that wealth is put. It is possible for households to differ in their economic circumstances, and for conflict to exist, but society cannot be grouped into mutually antagonistic social classes. The isolated position of individual economic agents is maintained theoretically either by assumption (as in neoclassical economics) or by the infinite refinement of households' social classification (as in Weberian sociology). Political issues, moreover, are similarly compartmentalized and isolated as some individuals may group together over one issue (for example, in neoclassical economics over the provision of a public good like a transportation facility), only to break up into new formations for others. Only markets and other autonomous social institutions link society together: hence, analysis is focused upon them. In housing, this means that the price mechanism, a particular bureaucracy or a protest group action over one issue become the isolated objects of study and housing tenures the means by which those topics are approached.

In recent years, a body of housing analysis has grown up which rejects this compartmentalization of housing issues and tries instead to relate changes in housing provision to wider developments within society. Based broadly in the Marxist problematic of historical materialism, this approach accepts the importance of class and class struggle in understanding social conflict and change. Classes are constituted by their relation to the dominant mode of production. Under capitalism, for example, the two major classes are those of the working class and of capital and its functionaries. Those two classes are in conflict as capital's ownership of the means of production and control over their use enables surplus value to be expropriated from commodity-producing wage workers. That surplus value is the source of the profit for capital and so without its existence accumulation is impossible. Class struggle, therefore, necessarily exists within capitalism and will alter in its form and intensity over time.

This approach leads to a reconstitution of the links between households and the state as specified in the consumption-orientated approach to housing. It generates a reformulation of how both the state and the household are conceived as social entities. Particular forms of them are seen as likely to exist in a capitalist society. The state has to help create the economic conditions for accumulation and to contain class struggle in forms that enable capitalist society to continue through coercion or ideological and political manipulation. Households are also products of the societies in which they exist, rather than natural entities with invariant needs. Particular household forms help to reproduce a stable capitalist society, for instance by encouraging conformity and respect for authority. Household forms also influence the cost of reproducing a labour force. In the patriarchal, nuclear family, in particular, subordination according to gender reinforces class exploitation because the domestic labour of the wife/woman reduces the cost of reproducing a workforce, and it helps to create a segment of the labour force that is especially vulnerable and, hence, cheap to employ: married women. The Women's Movement has highlighted the social pressures that lead to this reproduction of the dominance of the patriachal family, and writers on state social policy have highlighted the state's role in that reproduction (Wilson 1977 and Ginsburg 1979).

This aspect of social policy is clearly seen in council housing in the definition of the household needs that 'deserve' housing and in housing management procedures: for instance, families with children have priority of access, even if frequently to the worst parts of the stock, as do the elderly. The relation of households to owner occupation is different. There are no criteria for selecting types of household for this tenure beyond ability to pay. As a market transaction anyone can buy, yet only particular types of household can afford owner occupation in its current form, so inevitably the economically strongest type of household will prevail. Social conformity of household type is therefore reproduced by the pressure of economic necessity. And later it will be argued that state expenditure on housing has switched to owner occupation partly because the economically strongest households are there.

One obvious implication of the Marxist approach is that there is not a unitary housing question with a single answer. Housing provision for the working class necessitates the large-scale expenditure of economic resources financed either out of wages or directly by the state. So, in one way or another, housing affects the accumulation process and becomes an area of class struggle; one that will change over time. The analysis of housing, therefore, cannot avoid the issue of class relations. Putting this conclusion into practice in the study of housing issues is not easy, however. It has raised a great deal of controversy and the interpretation of what is meant by it fundamentally affects how owner-occupied housing provision is examined.

The conclusion about the role of class relations was reached by making some general (and schematic) statements about an abstract mode of production. Yet issues over housing provision exist in particular societies at historically specific points in time, so obviously a far greater theoretical apparatus is required to examine the empirical material thrown up by those issues and to consider their political implications. Perhaps the most difficult and controversial area is the explanation of the nature of class struggle in actual societies, the ways in which it works and its usefulness as a means of understanding social change at any point in time. Deductions from the concept of mode of production only highlight the existence of class struggle and certain tendential forms it may take. That, in itself, is of major importance as it points to the need to go beyond examining any event simply in terms of the immediate way in which it presents itself, yet it is not enough.

Capitalist societies do not function simply to improve the conditions for accumulation. Accumulation might be a dominant influence but that dominance continually has to be reproduced, as aspects of the social relations and institutional forms of any capitalist society will continually tend to threaten it. A number of advanced capitalist societies have state structures and political movements that could produce and have produced social and economic change to the detriment of capital. State policy is a site of class struggle. Yet it is in the political sphere that the problems of class analysis seem greatest as political battles are rarely fought on straightforward class lines. Classes do not enter the political arena as united fronts and there is no reason to think that they will ever automatically do so. Many Marxists, in fact, argue that one of the major roles of the capitalist state is to ensure that the working class remains fragmented at the political level so that capitalism is never threatened by a socialist revolution.

In housing analysis the problems of class relations and politics have led to attempts to simplify the issues by specifying underlying functions of state housing policy. Once these functions are specified attempts are made to classify political groupings around conflicting interests with respect to those functions (see Gough 1979, Ginsburg 1979 and PEHW 1975 and 1976). This has had the unfortunate consequence of reviving the consumption-orientated approach to housing: the functions are effects on households as housing consumers and state policy is the means by which they are created. With regard to owner occupation, for instance, it can be said that it is a housing tenure that ideologically incorporates the working class into the dominant value system or, alternatively, creates a social group of house owners with a clear and distinctive economic interest in housing. As later chapters will show, it is not possible to ascribe invariant functions to housing tenures and this limits the usefulness of the insights that can be gained from this broad theoretical approach.

By placing emphasis on the functional aspects of housing tenures in wider social struggles, most work in this area is forced, once again, to see housing problems and housing policy in distributional terms. The 'level' of the distributional analysis has moved from that of the individual to that of social groupings based on class or sub-categories of classes. But the analysis is still orientated towards the point of consumption alone, which makes it one-sided as forms of housing production and their effects within class societies cannot be considered.

Breaking with the limitations of consumption-orientated approaches necessitates seeing housing provision at any point in time as involving particular social relations. Housing provision via a specific tenure form is the product of particular, historically determined social relations associated with the physical processes of land development, building production, the transfer of the completed dwelling to its final user and its subsequent use. They can be defined as 'structures of housing provision'.[10]

Owner-occupied housing as the name implies simply means that the household owns the dwelling in which it lives. Such personal ownership, however, has become synonymous with housing provision via the private market. Governments may try to regulate this housing market, and they may even generously subsidize it, but they can never control it; the uncoordinated actions of thousands of consumers and a large number of suppliers determine what goes on there. Instead, therefore, of simply being one way in which housing may be consumed, owner occupation has become associated with a particular way in which housing is provided; and with all the forms of landownership, building, finance and market exchange that exist there. There is consequently a particular set of social relations involved in the current structure of owner-occupied housing provision.

Owner-occupied housing provision in Britain consists of interrelations between private landowners, capitalist housebuilders, building workers, financial organizations, specialist professionals involved in house buying and selling, the final owner occupiers, and the state. Each group intervenes at one or more points in the physical process of housing provision; the nature of its intervention and the responses by others to that intervention are the key determinants of present-day owner-occupied housing provision and its problems. The actions of each category of social agent therefore affect the behaviour of the others, frequently in unintended ways; ones that none the less still produce sharp conflicts of interest. Specification of a structure of housing provision therefore is also the elaboration of power struggles that go on over its nature via economic and political processes. Those struggles are dominated by the accumulation of capital both within owner-occupied housing provision itself and within the wider economy.

Many of the problems currently associated with owner occupation arise from contradictions within its present structure of provision, and between aspects of that structure and the wider social context within which it exists. A number of features of owner occupation in Britain, such as periods of rapid house price inflation, have their origins in consequences generated by the attempts of housebuilders to raise their profits and to minimize the conversion of land development gain into land rent appropriated by others. Those attempts by individual housebuilders to improve their individual profitability have had particularly widespread repercussions, a number of which have led to somewhat ironical results. One result in particular has been to create a crisis in production which has generated escalating production costs. Because of its importance, analysis of this profitability crisis in the British housebuilding industry will be of central concern in the arguments that follow. Yet attempts to improve profitability by housebuilders can themselves only be adequately understood by examining the nature of the market in which the housebuilder's product is sold and the means by which households finance their house purchases. Land development, production, exchange and consumption as aspects of owner-occupied housing provision must therefore be examined as a unity rather than as isolated components of a rather complicated jigsaw puzzle.

The structure of owner-occupied housing provision is a series of relations between social agents that are familiar to almost anyone living in Britain. The agents consist of those involved in consuming and providing owner-occupied housing: the owner occupiers themselves, the building societies, estate agents, housebuilding firms, building workers, planners, landowners and yet others participating in the construction and selling of owner-occupied housing. The immediate relations between them are schematically described in figure 1.2. As the broad features of each of these groups are well known, there is little point at this stage in describing them in detail. The relevant characteristics of each will be discussed at the appropriate places in later chapters. Owner-occupied housing provision obviously does not exist in a vacuum. Again, the broader influences on it are well known. The state, for instance, has specific housing policies, whilst particular government economic strategies have indirect and often undesired effects on housing provision. Other economic and social forces also have an effect, like the level of wages and salaries from which housing costs must be paid, or competition in the money markets and the associated movements of interest rates.

Whilst there is general descriptive awareness of the nature of housing provision, there is little understanding of the dynamic effects generated by a structure of provision. So each of the main social relations in owner-occupied housing provision are examined in the following chapters to elaborate their role in producing change. Emphasis is placed on the interconnections between the various components of owner-occupied housing provision.

The next chapter examines how the structure of owner-occupied housing provision in Britain developed during the inter-war years. Chapter 3 then examines in more detail the nature of the speculative housebuilding industry, considering the types of firm operating within it and why there has been such a dramatic shift towards large producers over the past fifteen years. Chapter 4 examines the housing market in which builders sell owner-occupied houses. Emphasis is placed on the economic processes affecting aggregate supply in order to begin understanding why output has declined so dramatically over the past decade.

Chapters 5 and 6 then consider land development and housing production in order to explain why production costs have risen so sharply. Chapter 5 deals with the development process and land banking by speculative housebuilders. Central to an understanding of the development process in owner-occupied housebuilding, it is suggested, is the nature of the struggle between landowners and housebuilders over the conversion of development gain into land rent and the role played by the planning system within that struggle. Chapter 6 then examines the effects of housebuilders' actions with respect to land development on

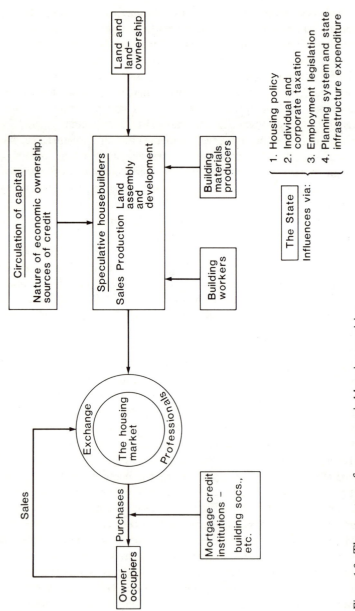

Figure 1.2 The structure of owner-occupied housing provision

the production methods used, on the relation between housebuilding firms and the workers they employ, and on the building materials industries.

Housebuilders are productive enterprises because of their building activities. They are merchants as well, buying and selling in a variety of markets where they try to get the most profitable terms. This is clearest with the selling of the completed product, where each firm has specific marketing strategies. But land purchase and the hiring of building workers are also key market exchanges for a housebuilder. Speculative housebuilders in general terms are no different in this respect from any capitalist enterprise. All firms have to buy the inputs necessary for production and to sell the produced output. Speculative housebuilders, however, are peculiarly susceptible to the potentially opposing pulls of production and exchange because of the importance to them of land development profit. Speculative housebuilders, therefore, should be seen as merchant producers rather than simply as one form of productive capital. The term merchant producer directs analysis towards seeing how speculative housebuilders try to come to terms with the contradictions forced on them by having simultaneously to deal in a number of distinct markets and to produce houses at the same time.

Chapters 7 and 8 consider the much neglected intervention by the state into housing provision through the land-use planning process. The first examines how the planning system arose and the second why it is currently under threat. The role of land-use planning in owner-occupied housing provision can only be understood if the land values question is dealt with. This requires a break with the Ricardian theory of land rent and the associated problem of land value compensation and betterment. The Report of the Uthwatt Committee on compensation and betterment published in 1942 is considered in detail, as failure to implement their recommendations in post-war British legislation is frequently cited as a major impediment to effective land-use planning.

Chapter 9 examines the economic position of owner occupiers and their status as a political force. The following chapter undertakes a similar analysis for building societies, concentrating on their changing economic position. The discussion of the role of the various social agents involved in housing provision then enables the operation of the housing market to be considered. Chapter 11 on the housing market completes the study of the internal workings of owner-occupied housing provision. Chapter 12 relates the analysis to the issues raised earlier in this chapter: namely, why is there a growing housing crisis in Britain? The final chapter then suggests some preconditions that are required for a progressive transformation of housing policy.

This book is concerned with the current structure of owner-occupied housing provision in Britain and its relationship to wider economic and political forces. Much space is devoted to the land market and to the housebuilding industry because they are both vital to an understanding of what is going on within owner-occupied housing provision. It is in these two areas, in particular, that reforms must be made if the housing policy of the state toward this tenure is to change in ways that benefit the broad mass of the population. Such measures, it is argued, enable a fundamental change in the social relations involved in owner-occupied housing provision. In this way the whole nature of the tenure and the social role it plays can be altered without alienating existing owner occupiers. Such proposals are in contrast with treating, as is generally the case, necessary reform as being limited to taxation measures and the supply of credit; such reforms alone do not require any change in social relations and so, it is argued later, are doomed to failure.

There is a danger when talking about housing tenures of assuming that the households living within that tenure automatically constitute a unified social category. This mistake has been made in the past by the present author as well as by many others. This view of a unitary social category (frequently called a housing class) is strongly criticized in chapter 9. It is not expected therefore that the proposals for reform presented in the final chapter will be supported by all owner occupiers. They none the less do represent gains to many current and potential future owners, and to the workers that build those houses. Moreover, they try to embody concepts of public accountability and democratic control rather than the present operation of blind market forces working ultimately in the interests of a few landowners and capitalist enterprises. Most importantly the proposals can help to raise the political consciousness of a significant group of people about the forms of housing provision that are possible within a future socialist society and the need to change the present forms.

2

The origins of mass home ownership

Introduction

Owner occupation became a major housing tenure in the years between the First and Second World Wars. Prior to 1914 less than 10 per cent of households owned their houses; by 1938 the number had risen to almost a third. The main impetus to this growth was a housebuilding boom of unprecedented dimensions. Millions of houses were built at prices large sections of the population could afford, even though mortgage finance was expensive because mortgage interest rates were high compared to other contemporary interest rates and the relative burden of paying off mortgage debt was rising owing to the general fall in prices over most of the period. During these years many existing houses were also sold for owner occupation: 1.1 million of them in fact, 14 per cent of the 1914 housing stock (according to the 1977 *Housing Policy Review*). But these second-hand houses were sold at prices and under conditions primarily set by the new housing market, so it is on that sector that attention must be concentrated.

Private housebuilding was mainly for owner occupation during the inter-war years. Its output exhibited the pattern of a classic building 'long wave' as figure 2.1 shows, rising from next to nothing in 1920 to almost 300,000 houses a year in 1934, a rate never since reached, only to fall off rapidly a few years later as economic conditions turned against the continued expansion of owner occupation.

What is most important about this period is the existence of a unique combination of social and economic circumstances which enabled a

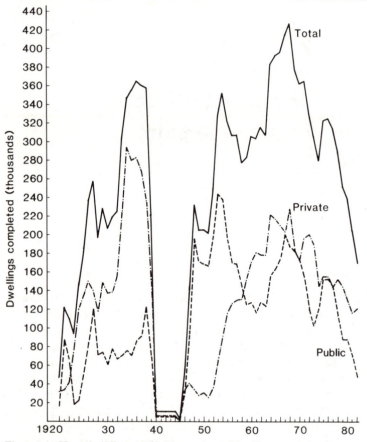

Figure 2.1   Housebuilding, 1920–82: number of dwellings completed

Sources: 1920–38, Richardson and Aldcroft (1968); 1938–81, *Annual Abstract of Statistics*; 1982, *BSA Bulletin*

Note: Public housing includes new towns, the Scottish Special Housing Association, the Northern Ireland Housing Trust and the Northern Ireland Housing Executive, but excludes other housing associations. Private housing is principally built for owner occupation. Total output includes other housing tenures not included in the public or private housing categories.

particular set of social relations to develop in owner-occupied housing provision. Those social relations determined both the success and the limitations of owner occupation during the period; in doing so they influenced government policies towards other tenures. They also developed an internal coherence and social strength to an extent that owner occupation has become synonymous with their existence ever since. The current nature of owner occupation consequently can only be understood in the context of this crucial period.

The decline of private rental provision

Before the First World War most people rented their homes from a private landlord. This was as true for the rich as for the working class. In fact the social status of the relatively few owner occupiers that existed at the turn of the century was broad, ranging from landed magnates in their country mansions through the traditional professions and other petty bourgeois, like lawyers and shopkeepers, to skilled and semi-skilled working-class households such as coal miners. There was also quite a wide variation in the amount of owner occupation between regions, and amongst the working class it tended to be concentrated into a relatively few manufacturing and mining districts. Institutional arrangements for the channelling and availability of mortgage finance and the ability to obtain finance for the initial deposits then, as now, were crucial in facilitating ownership.

Large sectors of the population obviously were completely excluded from the possibility of owner occupation, especially through poverty and casual employment – states which made it impossible to sustain the large regular payments necessary for house purchase. But that economic condition also frequently excluded them from other housing tenures as well, and possibly from any form of housing at all in the real sense of the term. Private landlords would want regular payments as house rents and charged high prices for anything that would be accepted as a reasonable home. Periods of destitution, living in overcrowded slums, and 'moonlight flits' to avoid the rent collector were the lot of the poor and in some areas, like London, quite a substantial proportion of the working class as a whole.

It is difficult to explain the growth of owner occupation simply as a product of increasing affluence and natural inclination on the part of individuals as is frequently claimed. For most households prior to the First World War the problem was one of finding a decent home within their means, immaterial of its tenure. Ownership, for those who could afford it, did not necessarily offer any financial advantage over renting and little or no social stigma was attached to non-ownership. After 1920, however, the advantages swung strongly in favour of owner occupation. It became relatively cheaper, provided new comparatively spacious suburban housing attuned to the consumer goods that were beginning to come onto the market, and as all of those households that could afford to buy flocked into owner occupation it rapidly developed a social kudos. Conversely non-ownership began to become synonymous with the negative social status of being unable to afford house purchase.

Economically consuming housing as an owner occupier using mortgage finance has one important common characteristic with private renting. Both involve the investment of capital not owned by the household in the dwelling in which they live. This makes it possible for households to live in houses whose prices are too high to make outright purchase feasible. With owner occupation, mortgage repayments spread housing costs over time until the occupier becomes the outright owner. With private renting, rental repayments also spread the cost over time for the consumer, although unlike mortgage repayments rents never stop. Landlords are obviously out to make a profit and they will usually have a mortgage loan as well, so the level and flow of housing costs in the two tenure forms are likely to be very different. Yet logically there is no reason why either form of tenure-based finance should produce higher or lower costs for the ultimate consumer. Whether they do depends on the historical circumstances surrounding the form of tenure provision in which they exist. The growth of owner occupation, therefore, can only be explained in terms of the collapse of the structure of provision associated with late nineteenth-century private renting and the growth of institutional forms associated with present-day home ownership.

The collapse of private rented housing provision in the years prior to the First World War was quite spectacular. Housing output fell by two-thirds in the decade prior to 1913 to the lowest level for over fifty years (Cairncross 1955 and Parry Lewis 1965). The unprofitability of building and letting houses on the suburban fringes led to this collapse, whilst the resultant housing shortage gradually forced up rents on existing dwellings and heightened housing distress. A lack of profitability in the letting of new housing therefore increased social tensions between tenants and landlords. Those tensions exploded during the First World War as rents in the munitions districts reached unprecedented heights. This working-class agitation forced a reluctant government to introduce a rents freeze in 1915 and indirectly led to central government subsidy of council housing after 1919.

One problem with private renting is that the rents of new and existing houses are directly related, so the new housing market cannot be segmented off from the existing housing stock. No one will rent a new house if an equivalent existing one is cheaper. Similarly any housing shortage forces up the rents of all houses rather than just those of the vacant dwellings on the market. A temporary over-supply of houses at the prevailing level of rents, on the other hand, does not lead to a general fall in rents, as landlords are reluctant to reduce rents and tenants face high costs in moving to cheaper vacant dwellings. Eviction is costless for the landlord but when the market swings in the tenants' favour they have to pay a high price to take advantage of it and the costs incurred might be more than the rent saving gained. This helps to explain the well-known ratchet effect of nineteenth-century housing rents which rose in times of shortage whilst stagnating rather than falling during periods of glut.

Housebuilding prior to 1914 in light of this relationship was, not surprisingly, related to changes in rent levels. There was a close link between periods of rising rents and increases in housebuilding, and of stagnant rents and falls in housebuilding. Over time, consequently, rents rose at rates unmatched by other prices. Singer's index of rent, for example, was 85 per cent higher in 1910 than in 1845 although increases in unit quality might have accounted for some of the rise (Parry Lewis 1965, p. 156). The impact on working-class incomes over time is obvious. With an expanding population a high economic price had to be paid to house it and, in particular, real working-class incomes had to rise to avoid severe economic distress caused by high housing rents. Only sustained long-term falls in the price of new housing purchased by landlords could have broken the rent ratchet, but that required long-term falls in building costs. As building productivity was at best static (Ball 1978), building costs simply depended on building input prices and so oscillated in phase with the housebuilding cycle. During slumps they fell as wages and materials prices were depressed by unemployment and excess capacity; in upturns shortages raised them again, whilst there was a slow upward trend over time as real wages gradually rose (Maywald 1954).

Owner occupation breaks these links in a number of interrelated ways. One is on the demand side where the current market price is only paid by moving households − not all households − in the tenure. Owners who do not move pay costs based on the time they moved and current interest rates. Their housing costs, therefore, need not bear any relationship to contemporary market trends. During periods of house price inflation they pay less than contemporary purchasers whilst during periods of deflation they pay more. The ratchet effect of private renting is broken, as the market is only for vacant (or vacated) dwellings so prices go up or down depending on the economic conditions there alone. The fact that current market costs are not passed on to non-moving owner occupiers produces a second break in the link: the ability to purchase new houses at the price they are offered for sale depends only on the incomes of their actual purchasers. Activity in the owner-occupied housing market therefore depends only on the economic strength of potential purchasers, not all owners. This means builders have a more reliable market with owner occupation as long as they are prepared to change their product and location according to varying household circumstances. Owner occupation is a segmentable market whereas private renting was not.

The difference between private renting and owner occupation as markets in which to sell housing creates a crucial precondition for the revolution in the organization of the building industry that took place during the inter-war years. Owner occupation made housebuilding and selling the most important parts of the building process, whereas before land dealing and land development had been the most profitable activities. Prior to the growth of owner occupation speculative housebuilders were generally small, undercapitalized enterprises with a comparatively short life span. Land developers made the most profit, buying up greenfield sites, servicing them and leasing them out on an individual plot basis to small builders who would sell off the completed houses to housing landlords (Ball 1981). Sometimes the development and building functions would be combined under one firm which might even let most of the houses directly to tenants, as with the well-documented case of the firm of E. Yates in South London at the turn of the century (Dyos 1961). But this was rare.

The problem with housing production was the slow turnover of productive capital associated with selling to private landlords. This did not matter too much to the land developer who would have little money tied up in undeveloped plots but it did matter to the builder with partially erected dwellings financed on short mortgages. Selling to landlords placed an intermediary, the landlord, between the producer and the final consumer who was solely interested in the investment potential of housing. Landlords' demands would be lumpy (two or three houses, say, at a time), restricted to periods when rents were rising, and there would be only a limited number of potential landlords in any one district. The result was an erratic market, fluctuating between sharp peaks when landlords saw renting in a district as a good investment followed by long troughs of sluggish activity. It is not surprising, therefore, that Dyos found in the development of Camberwell, South London 'a good number of roads in which up to forty or fifty years elapsed between the filling of the first and last building plots'. (Dyos 1961, p. 126). The largest builders in London in the 1870s and 1880s had to spread their activities across a wide area, yet still they could only manage to build thirty houses a year. Twenty years before, ten a year had been the maximum (Dyos, ibid.).

Owner occupation changed all that as builders sold direct to consumers who needed the houses to live in. So their demand was less fickle. It was not restricted to times when housing was a good investment. Customers, moreover, could be attracted from a wide area through advertising and selling drives. The conditions for a mass market were created. By the 1930s output rates per builder of 5000 houses or more a year were common. Once they achieved a rapid turnover of productive capital, speculative builders could gain economies by combining the spheres of land development and housebuilding. Independent land developers consequently were squeezed out during the inter-war years and builders made spectacular profits undreamt of in housebuilding in earlier years.

The inter-war owner-occupied housing boom

The collapse of private rental new housing provision, and certain intrinsic advantages of owner occupation for speculative housebuilders, were important preconditions for the advent of mass owner occupation. Yet by themselves they only created the potential for this new form of housing provision; positive action was needed to create its actual existence. Certain pre-1914 developments helped to produce that action in the 1920s. Owner occupation, for instance, gradually grew in the years prior to 1914, whilst building societies were beginning to look more favourably on this form of housing, switching funds towards it away from lending to housing landlords and speculative ventures (Cleary 1965). Furthermore by 1920 there was an acute housing shortage for virtually all classes in society.

During the 1920s also the barriers to the easy transfer of freehold land were finally dismantled because of a breakdown in the resistance of solicitors to changes in property law. Their interest in maintaining a monopoly over conveyancing had held up much needed reform of the property laws for decades. Restrictive covenants and the sheer impossibility of transferring some types of land and other real property between individuals acted as a barrier to mass owner-occupied housing provision. Owner occupation needs the free, rapid and unquestionable transfer of small land plots and single dwellings without which it was exceedingly difficult to borrow on the security of a property. Solicitors only stopped their resistance to legislation, finally produced in a series of Property

Acts from 1910 to 1925, once a compromise had been reached which did not attack their conveyancing interests (see Offer 1977).

Yet what really created the inter-war housing boom was the unique state of class relations existing in Britain during those years. As is well known, the inter-war period was one of deep crisis for British capitalism. Mass unemployment existed from the early 1920s, associated first with the collapse of the staple export industries and then the world slump of the 1930s. New industries did grow up but in the Midlands and the South East rather than the traditional industrial areas. During the 1920s and 1930s many prices were falling, so real wages for those in employment rose gradually. Speculative housebuilders were in a unique position to take advantage of this combination of events. Strong demand for new housing existed in the more prosperous regions of the country from households who, in the circumstances, could afford to become owner occupiers. Yet mass unemployment provided a bottomless pool of potential building labour power which could be put to work at low wages and at a stringent, almost physically impossible, pace of work. Unemployed workers from the depressed regions flocked to find work in the Midlands and South East, many with skills useful in craft building work, and a building site would be their first, or more likely their last desperate, port of call. The dual nature of the British economy associated with the fundamental restructuring of capital going on at the time, therefore, made housebuilding highly profitable.

The late 1920s and early 1930s were also times when other building inputs were cheap: primary commodity prices dropped by two-thirds between 1925 and 1934, and technical innovations in brickmaking, cement and roof tiles led to falls in their prices, despite increased monopolization of those industries (Richardson and Aldcroft 1968, and Issacharoff 1977). The position of landed property was also exceptionally bleak, which depressed land prices. Prior to the protectionist measures of the early 1930s, agriculture was subject to the collapse in world commodity prices with obvious implications for rent rolls. Incapable of investing capital in agricultural improvement, and subject to death duties and legislation which increasingly favoured the tenant farmer, many landowners were eager to sell out. The weak position of the landowner is illustrated in the fact that agricultural rents did not reach their 1870 money values again until the early 1950s despite intervening price inflation (Self and Storing 1971). Selling land for residential purposes was an outlet that landowners were keen to pursue in these conditions. Speculative builders could acquire cheaply whole farms or even substantial landed estates at the edge of urban areas. Large housing developments of 150 to 300 acres consequently were common, and the practice developed during the thirties of the largest speculative builders holding land banks of three years or more so as to be able to take advantage of such potential large-scale acquisitions when they became available (Bundock 1974). Suburban land prices consequently were exceptionally low compared in real terms to their levels either at the end of the nineteenth century or after the Second World War (see figure 2.2).[1]

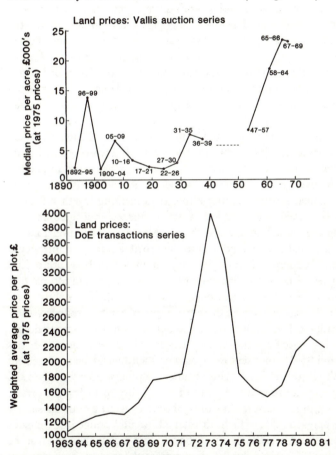

Figure 2.2   Changes in residential land prices, 1890–1981[1]

Sources: Vallis (1972), Evans (1974), Feinstein (1972) and *National Income Accounts Blue Book*, HMSO

[1] Deflated by GDP at factor cost price index

State regulation of private housebuilding during the period was minimal. Conformity to planning and building regulations was easy for the suburban builder. Neither, moreover, stopped some remarkably shoddy building, examples of which led to the Building Societies Act of 1939 which precluded building societies from having formal lending links with housebuilders, who had used those links as advertising material for the 'quality' of their houses. State expenditure also encouraged the owner-occupier boom; initially by giving cash grants to builders to induce them to build. These subsidies were important in the 1920s but they virtually ceased by 1930. Perhaps more important was the pliant and encouraging attitude of state agencies to the provision of infrastructure facilities, especially via local authority provision of basic services and the activities of the London Transport Executive which constructed underground lines into the greenfields north of London, a major area of activity, in advance of speculative development (Jackson 1973).

These circumstances made building for owner occupation a particularly profitable activity in the inter-war period. Two types of capital were involved in owner occupation: one was loan capital and the other the productive capital of speculative housebuilding firms themselves. They remained separate entities rather than merging together. Housebuilders borrowed funds and loan agencies financed house purchasers. Differing economic interests and the contemporary institutional framework maintained this separation. Building societies, for example, could not enter production as formally they could not make a profit (and setting up some non-profit-making production enterprise in competition with speculative builders was out of the question). The only general attempt at a linkage between these two types of capital produced disastrous results and legislation to ban its continuation. Builders would put finance into specific building societies and in return those societies would lend 95–100 per cent mortgages on their houses; an arrangement called 'the Builders' Pool'. Unscrupulous builders, however, used the mortgage guarantee as a guarantee of the quality of their dwellings to prospective purchasers, leading to legal action against a building society for misrepresentation which threatened to undermine the tenuous creditability and hence viability of the building society movement at the end of the 1930s. But, whilst they remained separate, both types of capital in housebuilding had distinctive characteristics compared with capital in other spheres of investment.

The general position of loan capital in owner-occupied housing provision in the inter-war period has to be understood in terms of the wider economy and the potential investment opportunities available to money capital. The scope for investment in manufacturing industry in Britain was limited from the early 1920s onwards due to the crisis of profitability experienced by most industrial sectors. After 1929–30 overseas investment collapsed as did the yields on UK government bonds. A large-scale property market did not exist[2] so there was a surfeit of money capital looking for a profitable return. The 1930s was the era of cheap money (Nevin 1955). The economic conditions in the owner-occupied housing market made it an ideal haven for loan capital. Banks lent to builders on overdrafts and mortgages but, most importantly, direct mortgage lending to owner occupiers boomed. One building society alone, the Woolwich Equitable, saw its assets grow from under £4m. in 1925 to almost £17m. in 1930 and over £30m. by 1935 (Jackson 1973). The return to building society investors was considerably higher than other alternatives, at least until the later years of the 1930s, and the interest rate building societies charged on mortgages made mortgage lending attractive for other types of capital (like insurance companies). Table 2.1 summarizes some of the statistics. Building societies were able to maintain such attractive rates to investors by means of their cartel, the Building Societies Association, which fixed mortgage and investment interest rates for the bulk of the movement. Given the societies' dominance

Table 2.1   Building societies, housing costs and weekly cost of buying, 1930s

	Yield on building society shares (%)	As % yield from Consols	Building society new mortgage advances (£m.)	Average rate of interest charged (%)	Average cost of houses being built (£)	Estimate weekly buying cost[1] s. d.
1930	4.65	103.8	88.8	5.82	411	9 9
1931	4.62	105.2	90.3	5.87	404	9 4
1932	4.52	120.9	82.1	5.87	375	8 7
1933	3.95	116.5	103.2	5.57	362	7 8
1934	3.80	122.6	124.6	5.44	361	7 10
1935	3.64	126.0	130.9	5.21	371	7 8
1936	3.45	117.3	140.3	4.97	384	7 11
1937	3.38	103.0	136.9	4.87	427	8 10
1938	3.37	99.7	137.0	4.83		
1939	3.41	91.7	94.5	4.80		

Source: Compiled from Nevin (1955), ch. VIII

[1] Calculated for 70% mortgage for 20-year term on the 'house price' of previous column in shillings and pence

over housing finance there was little competitive pressure to bring housing finance into line with current general market rates of interest, at least until the closing years of the boom.

Speculative builders also had distinctive characteristics as capital. In the main they were neither established large building firms nor other similar enterprises entering construction. Instead, they generally started off as small firms, often on the basis of capital advanced by the proprietors themselves. Such family firms were generally founded by members of the traditional petty bourgeoisie and usually there was some prior link with housebuilding in one way or another. Few had been active in speculative building prior to the 1920s but by the 1930s some had managed to establish themselves as giant concerns. A few examples of firms that are still well known today illustrates their characteristics.

The largest of the 1930s housebuilders, New Ideal Homesteads, was set up by an ex-local authority surveyor in the late 1920s. Laing and Costain had been medium and small building firms respectively in the north of England before both moved south to London and prospered on the inter-war private housebuilding boom. Taylor Woodrow was set up after the unintended sale of two houses by a 16-year-old greengrocer's son and his solicitor uncle. Wates was a small family concern, two members of which owned a furniture shop in south London. Wimpey was a small stone masons and road building firm until acquired in 1919 by an ex-army major with the support of his quarrying merchant father (see Bundock 1974). Four out of the five largest present-day building and contracting firms became established large-scale enterprises on the basis of their inter-war speculative housebuilding operations (namely Wimpey, Laing, Costain and Taylor Woodrow). Interestingly, however, their long-term expansion was achieved by diversifying out of speculative housebuilding as the boom began to falter. By expanding into general contracting and civil engineering in the latter half of the 1930s, these firms were placed in a good position for the profitable cost-plus government contracts of the subsequent war years. And they stayed in general contracting and civil engineering after the war, so that with the exception of Wimpey speculative housebuilding never became a substantial part of their operations again.

Little change in the production process of housebuilding was generated by the inter-war boom (Richardson and Aldcroft 1968). As was shown above, firms profited from the unique circumstances that gave them enhanced power over the purchase of their major inputs. Costs, therefore, could be reduced without altering production techniques.

There presumably were some scale economies but often they would be related to minimizing the idle time between tasks of a demoralized work-force. In 1931, firms could expect bricklayers to lay 1000 bricks per day and many workers hired worked at even faster rates than that (Jackson 1973). Organizing such rapid flows of work, so that trades dovetailed together on site and were transferred from site to site as required, produced substantial reductions in costs to the firm. Only one major speculative housebuilder, New Ideal Homesteads, tried to integrate vertically across housebuilding operations, haulage and building materials (Fox 1934). This was done in order 'to carry out practically every operation necessary for the complete development of building estates, and a complete organization for the mass production of houses' (company statement quoted in Fox 1934). Such an integration can be a precondition for substantial changes in production methods. However, it meant that Ideal did not have the flexibility to respond to the changed circumstances of the late 1930s, so, from being by far the largest specu-lative housebuilder in the 1930s, Ideal slumped into relative obscurity with the outbreak of war.

One overall result of all these different factors was that throughout the inter-war period deflation in the owner-occupied housing market was the norm. Even when construction costs started to rise after 1934 they only reached their 1929 level by 1938, and were still far lower than at any earlier post-First World War date. Housebuyers therefore bought a com-modity whose price was falling, or at best static, and housebuilders had to cut costs and achieve a rapid turnover of output to sustain profits. Neither could hope that inflation would reduce the effective cost of their debt or in the case of builders encourage them to hold land in order to profit from rises in its price.

The inter-war owner-occupied boom can be summarized as a cycle of speculative housebuilding profitability in a specific historic era. The determinants of the profits of speculative housebuilders were the factors described above, factors that are essentially relations of differing economic power between social agents in housing provision. The differ-ence between house prices and construction costs that fixed the gross profit to be made from housebuilding and the subsequent share of that residual profit going in land prices are the cold statistical summary of those rather complex relations. Their consequences in terms of the profitability of housebuilding can be seen clearly in the number of houses built, especially during the 1930s. Between 1928 and 1936 private housebuilding soared from 118,000 units per year to peak in the latter year at 293,000 new homes, but after 1936 completions started to fall, first by 15,000 units in 1937 and then by 31,000 in 1938. Private housebuilding was rapidly accelerating into a slump; then the start of the war brought housebuilding to a virtual halt. The cause of the initial downturn, however, was relatively simple: the market had been saturated. Lower and lower income groups had to be induced into house purchase, so house prices were being squeezed at a time when building costs were rising. Not surprisingly, builders cut back their output.

The level of wages paid to the working class proved to be the final stumbling block to the ability of speculative builders to sustain their housebuilding operations. This relationship between wage levels and structures of housing provision is a key element in explaining the pattern of housing provision since the ascendance of capitalism in the seventeenth and eighteenth centuries, and also in understanding the contradictions and class struggles which have moulded the housing policy of the state.

Owner occupiers and suburbanization in the inter-war period

It is not the purpose of this chapter to provide a history of housing in the inter-war years but to explain how a particular structure of provision associated with owner occupation arose in Britain during the period. No attempt will be made, therefore, to provide an in-depth analysis. It is important at this stage, however, to place certain features of the period in context as they throw light on certain aspects of the wider social role of the structure of owner-occupied provision that arose then, and on potential reasons for its existence.

Perhaps the first thing that should be made clear is that owner occupation was never solely a middle-class tenure. The strata of the population entering owner occupation were relatively similar to today: that is, the 'middle class' and the upper echelons of the working class (non-manual workers, such as clerks, and skilled and sometimes semi-skilled, manual workers). By 1938, for example, nearly 20 per cent of what were officially classified as working-class households were owner occupiers (HPTV I, p. 38). Even so, it was still the case that most of the working class were excluded from owner occupation. Most were also excluded from the council housing programmes of the inter-war years. Because of some important parallels with the situation now, it is useful to look at inter-war council housing in a little more detail.

During the 1920s many working-class households could not get access to council housing because the level of state subsidies did not enable the rents of the comparatively good-quality houses built to be lowered enough.

> The market for local authority houses was largely confined to a limited range of income groups, that is, in practice, the better-off families, the small clerks, the artisans, the better-off semi-skilled workers with small families and fairly safe jobs.   (Bowley 1944, p. 129)

As far as working-class housing was concerned this was just the same sort of market as that catered for by speculative builders later in the 1930s owner-occupied boom (ibid., p. 178).

The switch in the 1930s from general needs subsidies for council housing to a policy limited only to slum clearance, however, altered the social composition of council tenants. Households from the areas selected for redevelopment joined the earlier, more prosperous tenants. A change in national policy towards council housing consequently created social divisions within council housing that were as significant as those between the tenure as a whole and owner occupation. Distinct strata of the working class were now tenants differentiated by their social characteristics and their ability to pay. The higher strata were generally housed in Unwin and Parker style garden-suburb cottage estates, whilst in contrast the others were more likely to live in the poorer accommodation of cheap, walk-up flats on inner city redevelopment sites (some of the worst of which themselves were the subject of clearance programmes only forty years later).

The social and political consequences of this division within council housing at the time are unclear.[3] But its existence is reflected in the housing policies of both the Conservative and Labour parties during the inter-war years. In retrospect, for example, Wheatley's much vaunted Housing Act of 1923 during the first minority Labour government can be seen as aimed directly at only the upper strata of the working class. Similarly Labour as well as Conservative local authorities in the 1930s presided over the implicit discrimination of allocation policies influenced by the principle of the-best-goes-to-those-that-pay. Officials of the Labour movement and Labour politicians would have both been drawn principally from the same social backgrounds as those favoured by council housing policies, the upper strata of the working class and the lower middle class. So this particular effect of Labourism is perhaps not surprising. Its political importance may also explain the reluctance of many councils to pool the housing subsidies they had received from the various Housing Acts since 1919 as they were allowed to do after 1936. Pooling would have had the effect of raising rents on the 1920s estates and lowering them on later ones in line with tenants' capacity to pay. Despite central government encouragement such rent pooling did not come into general practice until the changed economic circumstances after 1945.

No moral condemnation of councils who failed to rent pool is implied here. Cross-subsidies between tenants within council housing anyway may simply deflect the need for more government subsidy. The point is to note the differential effect of housing policies on types of household within a tenure, and so the existence of a material basis for variations in the political support for any particular housing policy from households in one tenure. There is no stage in history when a simple correspondence exists between housing tenures and social class interests. A crude economism of identifying separate tenures in which different social classes live can neither explain the development of housing tenures in Britain nor the housing policy of the state.

The year 1919 nevertheless does mark the start of significant housing tenure divisions within British society. From then on the growth of owner occupation and council housing gradually eclipsed private renting which, shorn of a continual major new building input, slowly declined as a rump of the worst parts of the existing housing stock.

A number of ideologies about life-styles and housing tenures began to gain widespread currency. The use of owner occupation as a means of ideologically incorporating important sections of the working class into the dominant value system of British capitalism has for example been since the 1920s a recurrent theme of discussions of the effects of owner occupation by commentators on both the political Left and Right. Some remarkably silly things were said in the inter-war years about what home ownership did to the psyches of purchasers by commentators with a strong vested interest in encouraging owner occupation. Adequate discussion of the ideology of owner occupation, however, must be left until chapter 9.

One prevalent ideology that speculative builders were quick to take advantage of was the persistent myth of a rural tranquillity somewhere beyond the edge of the big cities and industrial towns in which most people were forced to live. This myth had been common throughout the years of industrialization during the nineteenth century (Wiener 1981), but the owner-occupied boom of the inter-war years seemed to bring its chimera closer to reality for those who could afford to buy. Status and fulfilment were stressed by housebuilders in their advertisements, as in the following: 'Novean Homes are offered to families of good breeding who wish to acquire a house to be proud of at a cost of less than £1 a week.' Another blurb reached higher planes of ecstasy over their estate where 'the green grass is not banished from the sidewalks; where in spring the trill of the lark may accompany the worker as he walks to the station; where the air is clean and fresh and nature's own life-giving decoction'. Aping the gentry, many estates were also named 'such-and-such Park'.[4]

The reality was much different and helped to engender the political opprobrium that increasingly fell on the speculative builder during the 1930s. Instead of Merry Olde England, suburbia brought only consumerism to even more rigidly patriarchal, nuclear family households. Wider family and friendship networks of earlier city dwelling were lost, whilst the classic sexual division of labour encouraged by the suburban life-style gave husbands long commuting journeys to work and left wives at home to the drudgery and boredom of the soulless estate. One local paper in 1935 commented

> These people have been lured here by tempting advertisements about living in the country. By coming here in their thousands they have defeated their own object – the country has disappeared – a new suburb has been built, and the true gainers are the builders, estate agents and multiple-shop owners.   (Jackson 1973, p. 289)

Suburban owner occupation in the inter-war years, furthermore, was the only way to get reasonable housing for most of its purchasers, so the extent to which it was based on free consumer choice is open to question. As a chairman of the Building Societies' Association in the 1930s said with remarkable candour, 'The new owner occupiers were so by necessity rather than choice. . . . The main source of the increased demand for building society mortgages came from people forced to become owner occupiers because there were no houses to let' (Quoted in Boddy 1980, p. 15).

Widespread dislike of 1930s speculative developments helped to create broad popular support for planning controls in the 1940s. Yet the attempt to equate suburban owner occupation with a rural idyllic myth did not end with the closing of the inter-war years. Throughout the post-1945 period and increasingly over the past decade this association . continues to be made, as a brief glance at any volume builders' housing brochures shows. The contradiction between the pleasures of the countryside and mass home ownership have intensified instead of diminished. In fact, as chapter 8 will show, it is leading once again to an absence of effective planning control.

Those characteristics of inter-war owner occupation, none the less, were the product of the structure of provision associated with it, not the tenure's inherent nature. Builders had to produce on the suburban fringes to make a development profit, whilst house designs, estate lay-outs, building methods and the provision of collective facilities were similarly constrained by profit requirements. Housebuilders found that by playing on people's unrealizable rural fantasies they could sell their houses. Consumer taste could be moulded at the cost of few compro-mises to their building operations. Catchy jingles, a bit of mock Tudor or other 'ancient' covering to the house shell, plus the occasional Indian prince or radio star to open the show house, were felt to satisfy a market where more than anything the cost of the house mattered.

State policies and the inter-war housing boom

Little has been said so far about the role of the state in the development of mass owner-occupied housing provision. This is not surprising as the state did little to create it. The take-off of owner occupation in the mid-1920s was helped for a few years by generous building subsidies introduced by the Chamberlain Act of 1923. But this effect was fortu-itous as they were a response to Conservative doctrinal aversion to council housing rather than specifically to support the growth of owner occupation. If council housing had to be subsidized private house-building should receive equal aid, so the logic went. The drafters of the legislation must have been most surprised that middle-class owner-occupied housing, rather than the hoped-for working-class rental housing, received the private enterprise subsidy (Richardson and Aldcroft 1968). A lot of political noise was made about the desirability of owner occupation but intervention was limited; policy was restricted to one of a reluctant guardian angel. The only time that role led to forceful action was at the end of the period when state judicial and legislative action staved off a threat to the Building Societies Movement caused by imprudent recommendations and investments by them. The 1939 Building Societies Act made sure it would not happen again (Boddy 1980). Not even mortgage interest tax relief was of much significance as few people paid income tax prior to the Second World War. The import-ance of the 1925 Property Act and of infrastructural expenditure has been already mentioned, but in terms of housing policy in isolation the inter-war years were the only time when the state has not had to get closely financially involved in the development of this tenure.

The existence of owner occupation, and the extent of the new housing boom associated with it, on the other hand, was used by successive governments as partial justifications for successive rollbacks from the limited intervention in working-class housing initiated in the years after the First World War. These justifications were exceedingly feeble, hardly hiding the real concern with cuts in public expenditure that underlay them. But they did have the political effect of diffusing opposition by throwing up a smokescreen of ideological debate over tenures (see Bowley 1944). This tactic, of course, continues today.

What is politically most important about the inter-war period is the universal acceptance of the structure of provision that emerged with owner occupation. Owner occupation became synonymous with building societies and speculative housebuilding, and with expansion on greenfield sites and urban decentralization. One consequence was that the political unpopularity of speculative housebuilding led to a cessation of the expansion of the tenure for fourteen years after 1939 until post-war building controls were relaxed by a market-orientated Conservative administration in 1953. The holding back of owner occupation in the years of austerity after the Second World War was primarily justified in Bevan's (the Minister responsible for housing) famous words: 'the speculative builder, by his very nature, is not a plannable instrument' (Foot 1975, p. 71). No attempt, however, was made in the years after 1945 to alter the structures of provision created for owner occupation and council housing (except for the introduction of land-use planning controls). The moulds for these two tenures had been formed in the inter-war years and during them the social relations associated with each tenure had set. They would colour political debate for the next four decades, leading in the 1980s to the current political impasse over housing policy.

Conclusion

The growth of mass owner occupation in the inter-war years was associated with the development of a structure of housing provision which has no exact parallel in any other country where owner occupation is a significant tenure. This structure of provision did not grow by chance. It had its roots in earlier forms of housing provision but prospered because of the peculiar economic and social circumstances of the inter-war years. By the end of the period, however, owner occupation was unthinkable without that structure of provision's existence.

The discussion of the structure of owner-occupied provision in this chapter has centred on certain key economic relations between social agents: landowners, speculative builders, building workers, mortgage merchants and housing consumers were the principal actors. By being placed in a market context their relations to each other seem harmonious enough. Yet from the description given of the position of building workers, for example, it can be seen that such cold economic relations are the outward form of intense and unequal social struggles. To understand a structure of housing provision, therefore, it is necessary to clarify the nature of those social struggles within it, and how they relate to wider social forces. They create a dynamic which leads to change. It will be the purpose of the following chapters to examine the present-day situation. For the inter-war period it was clear that speculative builders played a pivotal role, so attention will be focused on them for a while.

3

The modern speculative housebuilding industry

Although the basic structure of provision associated with owner occupation was created in the inter-war years, economic and social change since then has altered the relations between its social components. Taken together these changes explain the problems of owner occupation in recent decades. This chapter will start the process by looking at the types of firm currently operating in speculative housebuilding. The most notable feature is the extent to which the types of firm operating in the industry have changed over the past twenty years. The reasons for this dramatic restructuring are a major theme of the chapter.

An important watershed for firms in the industry was the crisis year of 1973 when housing starts fell to only half of the previous year, a record drop. Many firms collapsed, others survived only through the ingenuity of their accountants or in name only as the subsidiaries of firms that took them over. The industry has never been the same since; one result was that an output crisis in 1980/1 did not bring a repetition of the bankruptcies of the earlier slump. What had changed was not a greater prudence when making investment decisions but the type of capital operating in the industry. The crisis years after 1973 sped up a process of centralization of ownership of the largest housebuilders and with it a sharp growth in their market share. Firm ownership has changed substantially over the past decade and can be classified along the interconnected dimensions of long-term investment capital and family firm. The latter, as will be seen, is not limited to the smallest producers; the two largest housebuilders are still essentially family firms: Laurie Barratt is the driving force behind the largest producer, Barratt Developments, and the Godfrey Mitchell family still exerts a large but unknown influence on George Wimpey Ltd through personal connections and their large shareholding (held in trust).

The preponderance of family interests within the ranks of the top twenty private housebuilders suggests that a particular type of capital finds accumulation in the industry attractive. The reason is not some sort of hangover from nineteenth-century industrial ownership patterns but the present-day economic characteristics of housebuilding. Such firms are able to take advantage of the booms and slumps in the industry in ways not available to others. Much of this chapter, therefore, will document and explain the reasons for the growth of the large firms.

One theoretical problem has to be dealt with before examination of the industry can start. What sort of characteristics of firms are relevant to the discussion? Of interest are the factors that might influence firms' investment decisions because that affects how they operate in the land market, the production methods they use and ultimately the price of owner-occupied houses. All speculative builders are industrial capital. This categorization, however, explains only their function within the circuit of capital; the way in which they actually operate is influenced by who owns and controls them. The problem is that with twentieth-century firms there is usually a separation of ownership and control, the nature of which is difficult to unravel. Shareholders legally own firms but management has an influence on the operations of the enterprise which may predominate over that formal ownership. The potential boundaries of this division must be sorted out in order to be able to identify the influences on the investment decisions of housebuilding firms.

In order to distinguish the nature of firm control a division will be made between legal and economic ownership. Economic ownership is the

> real economic control of the means of production, i.e. the power to assign the means of production to given uses and so to dispose of the products obtained. . . . This ownership is to be understood as real economic ownership, control of the means of production, to be distinguished from legal ownership, which is sanctioned by law. . . . The law, of course, generally ratifies economic ownership, but it is possible for the forms of legal ownership not to coincide with real ownership.
> (Poulantzas 1975, pp. 18–19; see also Wright 1978)

In this definition economic ownership of capital depends on the relation between top management and shareholders. Where shareholders are diffuse and passive economic ownership resides with management. The founders of a company for instance whose shareholdings have shrunk to small proportions through share sales and dilution still have economic ownership of the company; their effective ownership is especially strong if there is no large and active shareholder. They may, for example, have more power with this ownership structure than if they had kept complete legal ownership by retaining all the shares and borrowing the additional capital from a bank. Where there are large personal shareholdings (common in housebuilding firms) economic ownership is divided between those shareholders and management; the greater the involvement of shareholders in the running of the company the greater is their economic ownership. (Note: those shareholders only have to dominate shareholdings, they do not have to have an absolute majority of the shares.) Similar differentiations can occur for companies taken over by other ones. With 100 per cent legal ownership of one company by another, there is complete integration so economic ownership will depend on the characteristics of the acquiring firm. But transfer of economic ownership also occurs with majority shareholdings, and when a large minority share interest is taken. All of these potential variations crop up in the discussion of the major housebuilders.

In this chapter use will be made of information derived from an interview survey of senior management in housebuilding firms and from a separate analysis of housebuilding firms' published company accounts. To avoid an excessive weight of empirical discussion brief details of these surveys and the sampling frameworks used are given in an appendix at the end of the book. The reader should be warned, however, that the data are not foolproof for various availability and computational reasons. The data are meant to give only a general picture of the industry, one however which is felt to be correct.

The distinctiveness of speculative housebuilding

Before considering the types of firms involved in speculative housebuilding it is necessary to place the industry within the context of the overall UK construction sector. It is unfortunately very difficult to do this as little national data on speculative housebuilding are published separately from the construction industry as a whole. No such thing as a minimum list heading, common for virtually every other industry, exists for construction. Some disaggregated pieces of information by trade of firm and by type of work are presented but no composite picture of sectors of the industry can be derived.

There are two principal sources of data on the construction industry: an annual *Private Contractors' Construction Census*[1] undertaken by the DoE and the annual *Census of Production* of the CSO. Within them little attempt is made to distinguish between the controllers of the production process, the main contractors and the speculative housebuilders, and the subcontractors employed by them. So the numbers of speculative builders building at one time, the number of volume builders or small builders, the amount of direct employment or subcontracting, and temporal and regional differences in them cannot be discovered. Neither source, furthermore, has accurate data on a major part of the industry, that is, the employment and output of large numbers of self-employed workers (and a much smaller number of employed workers) not included in declarations made to government enquiries. Significant proportions of activities undertaken by subcontractors, in particular, are absent from the data, especially with respect to labour-only subcontracting (LOSC). Estimates of LOSC vary from 20 to 40 per cent of the construction workforce so the omission is substantial, especially for private housing where the practice is most prevalent (Phelps-Brown 1968). Some information can be gleaned however about output and building trades used.

As a proportion of total construction work, private housing is comparatively small, averaging 16 per cent for the country as a whole in 1978 and rising to just over a fifth of work in some regions, although sectoral

Table 3.1   Private housing by size of firm, 1978

Size of firm (by no. employed)	Private housing		Public housing %	Other new work %	All repair and maintenance %
	Value (£m.)	%			
0–7	101.6	20	5	4	24
8–24	103.5	20	12	10	26
25–59	67.1	13	14	11	15
60–114	35.6	7	12	9	9
115–599	107.5	21	30	27	15
600–1199	67.4	5	13	14	5
1200+	24.8	13	14	23	6
Total	507.5	100	100	100	100

Source: Private Contractors' Construction Census 1978

variations in construction output mean that this proportion varies. Within private housebuilding, moreover, the characteristics of firms involved differ from the rest of construction. The marked importance of small firms employing less than twenty-five workers, for example, is clearly brought out in table 3.1, although whether as main contractor or subcontractor is unknown; 40 per cent of private housing work was done by such small builders compared with only 17 per cent for public housing and 14 per cent for other new work. There is a corresponding lack of larger builders, especially in the medium-sized categories. The trade of firms is shown in table 3.2 where it can be seen that private housing is the domain of the general builder in a way that no other sector is.

Table 3.2   Private housing: use of trades, 1978[1]

	Private housing		Public housing %	Other new work %	All repair and maintenance %
	Value (£m.)	%			
Main trades:					
General builders	284.3	56.0	43.9	17.7	34.3
Building and civil eng. contractors	98.3	19.3	27.4	28.1	7.8
Civil engineers	18.7	3.7	2.4	16.8	6.0
Total main trades	401.3	79.1	73.7	62.7	48.2
Specialist trades:					
Plumbers	15.6	3.1	2.9	0.8	4.6
Carpenters and joiners	6.1	1.2	1.2	0.9	3.3
Painters	5.5	1.1	1.8	0.8	9.6
Roofers	10.9	2.1	1.2	2.0	4.0
Plasterers	9.9	2.0	3.0	0.5	0.9
Glaziers	5.4	1.1	0.6	0.4	1.8
Heating and ventilating eng.	11.7	2.3	3.4	7.5	6.3
Electrical contractors	7.3	1.4	2.4	6.9	6.5
Plant hirers	19.5	3.8	4.9	4.7	2.4
Other	14.4	2.8	4.7	12.9	12.3
Total specialists	106.2	20.9	26.3	37.3	51.8
Total all firms (£m.)	507.5	100.0	400.5	1606.7	757.4

Source: Private Contractors' Construction Census 1978.

[1] Firms are classified by trade on the basis of the activity which forms the most significant part of their turnover

Similarly, less specialist firms are employed, especially when compared with non-housing new work and repairs and maintenance. Of the specialist trades used in private housebuilding, the less technically demanding predominate; plant hire, plumbers, plasterers and roofers constitute over half the specialist trades.

The data on type and size of firm are indicators of the work required in private housebuilding. Repetitive, fairly simple, traditional building tasks are undertaken on greenfield sites that themselves need only limited preparation and groundwork. This work can easily be done by the workforce of a traditional small builder (which is not to say that it is usually done by such firms), or subcontracted to non-specialist trades. The difference in the nature of the building work required when compared to other sectors is generally one of degree only, but it does make it easy to organize the building process in ways that make it highly flexible to variations in workloads. This characteristic is of considerable advantage to the speculative builder. The physical nature of the building process is itself dependent on the economic characteristics of private housebuilding. And its economics are so distinct from those of the rest of the construction industry that speculative housebuilding can be regarded as virtually a separate industry.

Building firms specialize and concentrate their activities in specific sectors of the construction industry. To an extent this arises for technical reasons and the economies of scale associated with specialization, for instance in the cost advantages derived from having high utilization rates with specialist plant and machinery. Even the largest firms, working across a number of sectors, use fairly autonomous divisions to undertake different types of building work. Speculative housebuilding is no different; most firms in this sector build only owner-occupied housing or have diversified from it solely into the speculative building of factories and offices instead of houses. Few combine speculative housebuilding with other forms of general building work; even the largest building contractors (like Wimpey, Laing and Tarmac) set up autonomous private housing divisions. With a limited number of exceptions, therefore, speculative housebuilding firms are independent entities or parts of larger firms where the overall linkage with the rest of the enterprise is through financial control and long-term corporate strategies.

The interview survey, for example, showed surprisingly few firms switching between council and private housebuilding depending on the relative profitability of the two sectors. This does not mean that building firms never move between sectors of the industry. It means instead that when such movement occurs in and out of private housebuilding, it is based on long-term corporate strategies rather than as one option within the short-term adjustment of a portfolio of contracts.[2] The same feature is also true of the scale of involvement in private housing. It is not possible to expand output rapidly in private housebuilding; the most obvious limitation being the acquisition of a land bank. Large building firms, therefore, cannot use private housing as an output regulator for variations in their other workloads by taking advantage of possible differences in its short-run cycle and those cycles of other construction activities. The gestation period involved in such sectoral switching is longer than the amplitudes of the cycles, so by the time the relative switch has been made the attractiveness of making the switch is likely to have long passed.

The major non-financial linkage between private housebuilding and the rest of construction is in the use of similar building inputs. Private housebuilding, therefore, is in competition with the rest of construction for them. High output in one sector transmits adverse effects on input supply and price to other sectors; an effect which is especially crucial with certain commonly used skilled trades (like bricklaying) and widely used materials (like plumbing and heating equipment and the ubiquitous LBC brick). Simultaneous booms in all sectors of construction can cause considerable bottlenecks and price escalations, as happened in 1972. Start-ups after an idle period caused by prolonged bad weather can produce a similar effect as occurred during the winter of 1979–80.

The links between speculative housebuilding and the rest of the construction industry consequently are, on the one hand, financial and, on the other, as users of construction inputs. There is little joint production with other sectors, either on one site or by shifting between sites and types of work. This autonomy cannot be explained by the technical nature of private housebuilding which is one of the simpler building processes. It exists because speculative housebuilding involves a different type of capital investment and, therefore, a distinct form of management expertise and control. This can be seen by describing the three main types of building firms: jobbing, contracting and speculative.

Jobbing builders are the classic small-time builders akin to the centuries-old, much-romanticized master craftsmen. Only a few workers are directly employed, others being hired on a casual basis, and some tasks are also subcontracted out to specialists. These firms turn their attention to most forms of building work, moving from job to job in a limited locality. The work they undertake includes some speculative ventures, such as house conversions and modernizations or new building when land can be obtained. This type of builder is not a major force in most sectors of the construction industry. In private housebuilding, a jobbing builder is very unlikely to build more than ten houses a year, usually far less. The one-to-ten-house category of private housebuilder as a whole built only 14 per cent of new output in 1978 and it consists mostly of small-time speculative builders rather than jobbing firms.

The small-time jobbing builder sits on the divide between capitalist and non-capitalist forms of enterprise. The proprietor often does part of the work, yet others are employed to increase the firm's profits and the overall objective is to provide a steady and sometimes good income for the proprietor. Such firms cannot compete with larger, capitalist enterprises and operate in areas which are unprofitable for the latter.

Contractors, as the name implies, build to contract. A client wants some building work done, a bill of quantities is drawn up, and construction firms tender for the work specified in the bill on a competitive or selected basis.[3] Usually there is a main contractor who appoints subcontractors to undertake all or part of actual building work (some subcontractors may be nominated by the client's architect). With the exception of private housebuilding and some industrial and commercial development, building to contract dominates the British construction industry and has produced what is often called the *contracting system*. Profits are made by keeping costs below the revenue received from the client through payment of the tender price and successful additional claims.

Competition between firms takes place at the tendering stage so that profit margins on tender prices can be low. Profits on a contractor's capital, however, can be much greater through minimizing working capital and overheads. This means of making profit places emphasis on a rapid turnover of capital and on achieving a profit-maximizing portfolio of contracts that does not leave capital idle or in too many risky ventures (see Ball 1983). Management, therefore, has to concentrate on acquiring a good mix of contracts; on being able to evaluate highly uncertain building costs; on being able to assess a client's willingness to pay and the current state of the market so that the best tender can be submitted; and on ensuring that a project's working capital and its demands on the firm's overheads are kept to a minimum.

In *speculative building* the profit-making process is different and in it development profit predominates. Development profit is achieved by a judicious purchase of land and conceiving of the appropriate residential scheme for the site. The predominant risks to assess are the market-ability of houses built on a site, the price paid for the land, the timing of development, and the overall scale of housebuilding appropriate for the firm's resources in the context of the current state of the market. This is a very different management function from that of the contractor.

These distinct management tasks exist because of differences in invest-ment. In speculative building the turnover time of capital, for instance, is much longer. After six months most contracting projects are self-financing as the client makes monthly progress payments (Ball 1980), yet in speculative development capital has to be invested in land years in advance of building. Furthermore, at the start of development more capital has to be sunk into groundwork and infrastructure, and then into the houses as they are built, but no revenue is forthcoming until the houses are sold. Competition between firms is also different. It takes place at the time of land purchase and house sale, rather than just prior to the start of building as in contracting. These crucial economic features of speculative housebuilding distinguish it from the rest of the construction industry, and explain why speculative housebuilding is organizationally distinct from the rest of construction.

Links, nevertheless, do exist between speculative housebuilding and the rest of the industry. Historically, many of today's largest contractors can trace their initial phases of rapid expansion back to the suburban owner-occupied housing boom of the 1920s and 1930s. Most of the largest building contractors still have speculative housebuilding divisions or subsidiaries. In many cases the involvement is small, and the attraction is as likely to be the tax advantages of holding land for house-building rather than any intrinsic enthusiasm for the sector. Large con-tractors like Leonard Fairclough, Norwest Holst, Higgs and Hill and the London and Northern Group, for example, build a few hundred houses a year. Contractors with major housebuilding divisions or subsidiaries have distinctive characteristics as contractors, and the nature of their involvement can best be explained by examining the type of capital that has come to dominate speculative housebuilding in Britain.

Capital and speculative building

Speculative building, as has just been argued, involves a distinct type of accumulation process. So there is no necessary reason why construction firms should have any intrinsic advantage over other firms within this sphere. Expertise, anyhow, can be hired if necessary. What needs to be examined is the type of capital that finds speculative housebuilding an advantageous method of accumulation. This cannot be deduced *a priori* but depends on specific historical circumstances. Some of these circumstances are external to housebuilding, for instance the types of capital generally in existence and potential alternative rates of profit; some are a product of the contemporary situation in housebuilding itself, such as the cycle of boom and slump, house price inflation and the nature of development gain. The types of enterprise involved in the industry and the reasons for their presence consequently vary over time. And there is no theoretical reason why a uniform type of housebuilder should exist at any period.

A parallel can be found in capital accumulation via landownership. Massey and Catalano (1978) argue, for example, that there are particular reasons why insurance companies have increasingly invested in property and land; reasons associated with changing investment opportunities, the time pattern of income they require and the long-term gains derived from such assets. They identify three distinct major categories of landowner (industrial and financial landownership and former landed property), all with special reasons for investing in land during the 1970s. The parallel is made stronger in that speculative housebuilding is a form of land investment and a particular means of realizing gains from landownership. It involves, however, a peculiarity not explored in detail by Massey and Catalano. Gains from land purchase in speculative housebuilding require actual development. Firms in the industry therefore have to be involved in an industrial process (building) rather than being solely holders of an appreciating financial asset (land).

The importance of analysing accumulation within an historically specific framework is brought out clearly when the extent of the changes in the ownership of the housebuilding industry over the past fifteen years is documented. The traditional view of the industry is one of a large number of medium-sized and small producers operating on narrow profit margins and prone to bankruptcy (cf. Whitehead 1974 and Drewett 1973). The discussion of the inter-war period in chapter 2 showed this image to be untrue for that period; many houses were built by quite large firms so the small producer was at most only one segment of the firm structure of the industry. Little evidence exists on the firm structure of the industry during the 1950s and 1960s but it is clear that particular types of large capital, to a great extent new to the industry, began to dominate housing production during the 1970s. To understand why this happened it is important to distinguish types of speculative builder. Some of the smallest, for example, are hardly capitalist enter-prises at all, sitting, like the jobbing builders mentioned earlier, on the divide between capitalist and petty commodity production. Other small producers are not interested in expanded accumulation but in ticking over at low level of output. Even the largest firms can generally be distinguished as a particular type of long-term investment capital.

The relationship of firms to capital accumulation is important in understanding the changes occurring in the industry. Five types of builder will be distinguished, and the division roughly relates to annual levels of output. Output levels, however, are not the cause but a conse-quence of being a type of producer. The five types are petty capitalist, small family capital, non-speculative housebuilding capital, large capital, and long-term land development capital. The last category includes most of the largest producers. The expansion of this group at the expense of small family capital has been the most significant feature of the industry's restructuring.[4]

Petty capitalist housebuilders

The continual turnover of capital within speculative housebuilding is of little concern to petty capitalist builders. Housebuilding instead is a source of infrequent profit or income. Such firms, therefore, build only a small number of houses each year, possibly up to twenty but usually only about two or three. Administration is done by the proprietor, usually from home, and only a general foreman and a couple of part-time clerical staff need be employed. The rest of the work is subcontracted.

Some capital and a knowledge of speculative housebuilding is needed to start production but a clear profit of £2000 to £3000 per house (only 9–14 per cent on the 1979 average house price of £22,000) brings the proprietor a comfortable income. Higher building rates or higher (and perfectly feasible) profits bring correspondingly higher income. It is possible to stop building (and work) for months or years and start up again when more income is required, avoiding in this way higher rates of personal tax.

Petty capitalist builders do not necessarily build more houses when profit margins rise. They could do the reverse: build more when profit margins are low to maintain their level of income. One such builder interviewed cut back on his average twenty houses per year during early 1979 as the extra profit margins associated with the sudden spurt in house prices made his marginal rate of income tax too high. In two years profit per house from the same development had varied from £1000 to £15,000. Such sharp fluctuations in profit margins and a fear of interrupted cash flows during downturns also discourage these builders from holding land banks (and bank managers from financing them). Land purchase for the next development is usually financed from the proceeds of the previous one. Only where the builder has a sizeable amount of personal capital will a land bank be held as a (profitable) place to invest that capital.

The undercapitalized of these petty producers are highly vulnerable to bankruptcy. For them almost all the work is financed by bank overdrafts and trade credit from builders' merchants, so any unforeseen additional cost, miscalculation on the part of the financially inept, or problem in selling the houses, brings financial disaster.[5]

The movement in and out of housebuilding by petty capitalist builders is quite substantial. Of the 21,000 registered private housebuilders in 1978, 12,000 did not build a house at all in that year and many of them were such enterprises (see table 3.3). Another 7000 built an average of less than three houses each in that year and another 1000 an average of only seventeen. Numerically, therefore, speculative housebuilding firms are predominantly petty capitalist ones; 8092 of the 8710 active builders in 1978 fell into this category but they only built a quarter of the output.

Table 3.3 Number of registered housebuilders and average output by size category, 1978

Houses built	Registered builders[1]	Estimated average output	% of total output
Not active	12,048	0	0
1–10	7,067	2.9	14
11–30	1,025	17.4	12
31–50	223	33.3	5
51–100	191	70.1	9
101–250	115	155.2	12
251–500	60	347.0	14
500+	29	1,692.1	33
Total	20,758	17.1	100

Source: NHBC and our own estimates

[1] Registration is with the National Housebuilding Council which is virtually universal amongst private builders

Small family capital housebuilders

Small family firms differ from petty capitalist ones in that their capital must be regularly turned over in housebuilding so they need a steady minimum throughput of housing. This means that usually they build between 25 and 120 houses per annum. Often they have long ties in speculative housebuilding in the locality, relying on an image of quality traditional building to maintain sales and an intimate knowledge of local markets to acquire land. One builder interviewed capitalized on these advantages and increased their effectiveness by doubling as an estate agent. The firm generally is named after the proprietor who undertakes most of the management tasks. Their longevity leads to their family classification, the firm being handed down through the generations of a family.

The need for a regular cash flow leads to attempts to balance certain key parameters of their operations, especially ones relating to borrowing constraints, employed personnel, the rate of housebuilding and the size of the land bank. An output of 50 houses a year, for example, produces a turnover of £2–3m. and 100 houses double that to £4–6m. (at 1980 prices). But 50 houses annually gives a sales rate of only 4 per month which requires a very smooth sales programme if capital is to be turned over adequately. Any problem in legal completion or a break in a chain of sales leading to a prospective purchaser pulling out could disrupt the flow of sales badly. Hold-ups in just two houses locks up £60,000 or more, raising interest costs and more importantly increasing the likelihood of hitting borrowing ceilings, impairing the rest of the firm's activities. An output of much less than 50 houses greatly increases the likelihood of breaks in capital turnover. Higher outputs spread the risk of interruptions but outputs much over 100 houses strain the management resources of the organization, in that the proprietor can no longer cope with the management functions. Higher outputs also require substantially higher land banks and, therefore, much more capital.

Occasionally the proprietor of a small family firm might have the luck to purchase a large plot of land cheaply (perhaps white land for which planning permission was subsequently acquired). The substantial development profits of this site will then form the financial mainstay of the business for a number of years, reducing the problem of credit availability. One small family builder interviewed had been building on a 350-house site since 1962, and in 1979 development was still going on!

Firms of this type incur office overheads through the need for administrative staff to do the accounts, to buy materials, to manage the usually subcontracted building work and, possibly, to deal with the sales. This adds to the need for steady sales. The family nature of ownership and control plus the incentive to maintain a steady cash flow tends to mean that these firms do not grow once they have reached a plateau related to an optimum output determined by administrative size and the effort put in by the proprietor. One variation to this category is the small-time speculator who builds up a small family firm to a particular size and then sells it off to a large housebuilder, sets up another firm and so on, eventually retiring to a tax haven at a young age.

Non-speculative housebuilding capital

This category consists of a wide variety of firms whose only common characteristic is that they have a major involvement in other activities. This means they tend to operate at an intermediate level of output which does not tie up too much capital but which is unviable without the spread of non-housebuilding activities. This fixes them generally to an output of 120 to 300 houses a year. Administrative overheads and the high risk of running a small number of sites with a high throughput or a larger number of sites with sluggish sales makes this level of output unattractive for those solely involved in speculative building. Three distinctive sub-groups can be identified within this category.

First there are the new entrants to speculative housebuilding who are growing fast and just happen to fall within this category for a time. Sometimes they are independent family housebuilders who uncharacteristically want to expand; more often they are new subsidiaries of established large speculative builders that are being built up to a larger size or, alternatively, subsidiaries of large firms from outside the construction industry who want to set up a speculative housebuilding division.

The second sub-group are predominantly medium-sized, privately owned building contractors using property and speculative housebuilding as a sphere of investment for contracting cash. One company interviewed, for example, fell within this category as it was a substantial regional building contractor built up by its proprietor since the war. Investments had been made in property and a medium-sized speculative housebuilding division over the years. It had a huge land bank equivalent to twelve years' output at peak levels, and the firm was still actively looking for new land and also dealing in land.

The third sub-group are what can be called 'jobbing builders grown big'. They combine a range of building activities, working where they can, and an interchange of contracting and speculative building is used to maintain turnover.

Altogether, however, medium-sized firms are comparatively small in number: only around 120 were active in 1978 with a 12 per cent share of the market. Other types of housebuilding firms are simply more viable entities.

Large capital housebuilding firms

Big capital dominates the industry and generally such firms produce more than 300 dwellings per year. Table 3.3 shows that 47 per cent of the market in 1978 was taken by 89 firms producing more than 250 dwellings. Later data would show an even larger share, and it is likely that even the 1978 figures are understated as firms registered as separate builders are frequently subsidiaries of larger firms. Large firms are very different from their smaller competitors in this industry both in terms of their ownership and the way they operate as productive capital.

Public quotation is usually essential as even the smallest producers in this category, with a modest three-year land bank, will require capital of £5m. or more in a risky venture. Borrowing such capital on a fixed interest basis leaves a firm highly vulnerable to failure. A large minimum profit steadily has to be made to meet those interest payments, which is difficult with speculative housebuilding. Equity capital overcomes this gearing problem to an extent, as dividends can be varied during the course of a profit cycle. Public quotation also enables share issues to be used as a means of increasing the capital available to the company. Barratt Developments, for example, which went public in 1968, has issued additional shares at ten separate times since then, usually related to the takeover of another builder, increasing its equity capital fivefold in a decade (Laing & Cruikshank 1980). Going public for a family firm, moreover, does not necessarily mean loss of either legal or economic ownership. The former is avoided when a majority of the shares is kept personally or in trust (cf. Wimpey, Laing and Abbey) but the latter will still remain if a significant share of the equity is kept or management control is maintained. Going public, therefore, can enhance the ability of a family-owned capital to expand.

Most large housebuilders have highly geared financial structures despite the potential threat to long-term survival they represent because fixed interest borrowing has advantages as well as disadvantages. Borrowing fixed interest funds enhances the rate of return on equity capital as long as the use to which the borrowed funds are put yields a return greater than the cost of borrowing. The problem is to get the right gearing mix between fixed interest and equity capital for the state of the market at the particular time. For example, it is best to use borrowed funds to buy land just prior to a market upturn and to sell the land or build it out, so reducing gearing again, before the boom collapses. Many of the most spectacular failures of large firms in the 1974–6 period, like those of Northern Developments, David Charles, and the Greaves Organisation, occurred because these firms were overgeared at a time when house sales collapsed and interest rates rose.

During the boom years of the early 1970s many firms built up large land banks using borrowed funds on the assumption, shared by those who lent to them, that land prices would never fall and that at worst the land always could be sold. With the 1973–4 slump, some of these over-stretched firms went to the wall, some followed the dubious accounting practice common amongst property companies of capitalizing their interest payments to make the current account profit look healthier, and yet others survived only because of the strength of their equity base (especially true for some family firms). Speculative housebuilding is therefore not just about buying land and building houses at the right times but also about using the most advantageous forms of capital and maintaining a minimum positive cash flow commensurate with them. This encourages particular types of firm structure for large firms.

Successful firms tend to be long lasting, even if overall economic ownership of them changes, because of the long gestation period of building up a land bank, the minimum threshold size for existence, and the management expertise required. Of the top twenty leading house-builders listed in table 3.4 at least half can trace some lineage back to firms operating in the inter-war period and most of the others have been active for twenty years or more even if at much smaller sizes.

Large firms also tend to be regional specialists; only the top three firms can claim to be national ones operating across the whole country. For reasons associated with the development process (see the next chapter) the rest are active in a limited number of regions, and even the national firms produce via semi-autonomous regional subsidiaries. The smaller the total output the smaller is the geographical sphere of activity.

Looking at the number of houses built by each firm in table 3.4 some interesting features can be seen. In the first place there are some quite remarkable differences in size. The largest firm builds 11,000 houses a year and the twentieth only 500; if the output of the ninth to the twentieth

Table 3.4 Housing output and ownership of the leading housebuilders, ranked by their 1979 output

Operating companies	Approximate output 1979	Approximate output 1980	Legal ownership	Land bank years at 1980 output[1]	Large shareholdings in parent[2]	Link to inter-war speculative housebuilder	Comments
1 Wimpey Homes	11,000	9,000	Subsidiary of G. Wimpey	5	Mitchell Trustees (minority, just)	Yes	Recently reconstituted subsidiary of the UK-based building contractor with wide building, civil engineering and property interests in UK and overseas.
2 Barratt Developments	10,000	11,000	Independent	3			Formed 1958. Parent company of thirty operating subsidiaries. Main activity: speculative housebuilding with some contracting and property development. Recently diversified into US speculative housebuilding.
3 Tarmac Homes	4,000	4,000	Tarmac	2–3			Parent of twenty-two speculative housebuilding companies. Move into housing has been recent, 1973–4. Contracting and quarrying still main activities plus property holdings in UK and overseas.

(cont. overleaf)

Table 3.4—cont.

Operating companies	Approximate output 1979	Approximate output 1980	Legal ownership	Land bank years at 1980 output[1]	Large shareholdings in parent[2]	Link to inter-war speculative housebuilder	Comments
4 Broseley Estates	3,500	2,200	Guardian Royal Exchange Assurance	2–3		Yes	Broseley was formed out of restructuring of Metropolitan Railway Country Estates acquired in 1973 by Guardian.
5 William Leech	2,500	2,000	Independent	10+	W. Leech Trustees (minority)	Yes	Family firm. Large purchaser of land before and after Second World War.
6 Comben Homes	2,500	1,600	Hawker Siddeley[3]	5		Yes	Doubled in size, after intervention by Hawker Siddeley in 1979, through acquisition of Orme Developments.
7 New Ideal Homes and Willett Homes	2,000	1,500	Trafalgar House	4	Broackes and Matthews families (minority)	Yes	Trafalgar House is a conglomerate with property, construction, publishing and transportation interests, run by its founders, Nigel Broackes and Victor Matthews (Matthews left to manage the hived off publishing interests in 1982).

#	Name							Notes
8	Bovis Homes	2,000	1,600	P&O Ltd	n.k.			Bovis Homes is part of Bovis Construction, a subsidiary of P&O Ltd whose main interest is shipping. Lord Inchcape, grandson of the founder of P&O, currently runs the company.
9	Bryant Homes and Focus	1,600	1,700	Bryant Group	5–6	Yes	Bryant family (minority)	Family firm; Mr A. C. Bryant is chairperson and managing director. Switched from public to private housebuilding during 1970s.
10	Whelmer and Salvesen Homes	1,500	1,000	Christian Salvesen	n.k.		Private company owned by Salvesen family	Salvesen's other interests include transportation, seafood, cold storage and open-cast coal mining in US.
11	Bellway	1,300	1,000	independent	10+		Bell family (minority)	Company split in 1979 into a property company and this housebuilding one. Both quoted.
12	Fairview Estates	1,200	800	Independent	4		Cope family (minority)	Founded by Cope (ex-estate agent) in 1961; has diversified out of housebuilding into property since 1975.

(cont. overleaf)

Table 3.4—cont.

Operating companies	Approximate output 1979	1980	Legal ownership	Land bank years at 1980 output[1]	Large shareholdings in parent[2]	Link to inter-war speculative housebuilder	Comments
13 Wilcon Homes	1,000	1,000	Wilson (Connolly) Holdings	7	Wilson family (majority)	Yes	Primarily speculative housebuilder but building up property portfolio. Wilcon Homes set up as distinct subsidiary in 1976.
14 Westbury Estates	1,000	1,100	Private company	n.k.	Joiner family		Rapid expansion: turnover rose 3½ times 1977–80.
15 Galliford Estates	900	500	Sears Holdings	n.k.	Clore and Stainer families		£1bn+ turnover conglomerate created by Charles Clore; includes Selfridges, William Hill, British Shoe Corp., engineering and vehicle distribution companies.
16 Laing Homes	800	950	John Laing	9	Laing family and trusts (majority)	Yes	Company still managed by members of Laing family.
17 Heron Homesteads	1,000	540	Heron Corporation	3	Ronson family trusts (majority)		Motor and petrol distribution company with £250m.+ turnover built up by Ronson since the war. Diversified in housing, property and consumer products (but still less than 15% of turnover).

No.	Company				Ownership / Family	Quoted	Notes
18	Davis Estates	700	600	Wood Hall Trust	n.k.		Yes	Large British company (£500m. p.a. turnover) whose major interests (80%) are in Australia. Largest woolbrokers and pastoral agent in Australia, also in coal mining and engineering. Major interest in UK in construction: Fairweather (contractor) and residential development companies.
19	Abbey Homesteads and Cape Homes	600	700	Abbey	n.k. (large, valued at £46m. in 1980 accounts)	Gallagher family 70% shareholding		Irish company founded in 1936 with construction, property and engineering interests in Ireland, Cyprus and UK. Rapid growth in 1970s with housebuilding almost half of turnover.
20	Taylor Woodrow	500	500	Taylor Woodrow	n.k.	Taylor family (minority)	Yes	Company founded by Frank Taylor in 1920s. Turnover now over £400m. so private housing is very small part of activity.

Sources: Construction and financial press, stockbroker and company reports (especially annual review of industry by Laing & Cruickshank), Extel and McCarthys

1 Estimates of land banks are very approximate due to poor data and large white land banks of some companies

2 Minority shareholdings by families vary considerably in their importance from almost 50% to less than 10%. 'Family' denotes shareholders with same surname quoted in annual accounts

3 Hawker Siddeley's majority ownership of Comben has varied as share issues and purchases since 1979 have altered its stake, sometimes dropping below 50%. Its economic ownership of Comben however is unquestionable

producers are added together they still do not build as many houses as
the largest producer. There are, in other words, quite sharp steps in the
size profile of producers. Chapter 5 will show that this is closely related
to the extent of each firm's geographical spread and the size of their
operating units. 400 to 500 houses per year is the minimum viable size
for a publicly quoted builder, which means that many firms not listed in
table 3.4 are clustered around the output level of the smallest one shown
there.

Comparison of output levels between the two years 1979 and 1980
shows that outputs vary quite sharply. The smaller the firm's output the
proportionately worse can be a drop in sales; some of the firms producing below 1000 houses per year, for example, had output drops of over
40 per cent between the two years shown. The obvious implication is
that greater size enables the risks of market fluctuations to be minimized
through having a diversified market presence with more sites and more
house types. This advantage of a market spread influences the accumulation strategies that management of different firm sizes adopt. The
largest producers can rely on a more rapid turnover of capital than the
smaller ones. Margins per dwelling unit can be trimmed in the knowledge that the comparative annual return on capital may be greater
because of its quicker turnover than is the case for smaller competitors.
This explains why the larger, faster-turnover firms are prepared when
necessary to pay more for their development land than are their smaller
competitors, at prices which are greeted with incredulity by the smaller
firms (as discovered in interviews). A strategy of growth can also provide
some shielding against market fluctuations (see figure 3.1).

Different strategies by management are nevertheless only relative; all
housebuilding firms face the same economic framework of speculative
housebuilding and have to operate within it. Simply to discuss differences as a result of firm size, moreover, ignores the crucial issues of
finance and economic ownership. Their importance can be seen by looking at the recent company histories of the largest producers.

The major housebuilders

It is quite difficult to discover precisely who are the largest housebuilders in Britain and how their output changes. Speculative housebuilders are often reluctant to reveal details of their operations, in some
instances even the annual number of houses they build. This is partially
because of the 'secrecy' embodied in the mythologies of the development

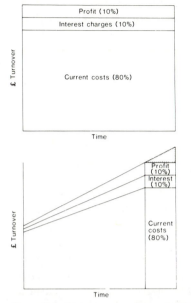

Figure 3.1   Output expansion and debt financing

Note: This figure illustrates a simple example of the advantages of growth for a speculative builder. It assumes that market saturation for a firm's product is not reached (housebuilders can limit their share of any local housing market by growing through movement into new local markets); that land banks are a fixed ratio to turnover; that there are constant returns to scale; and that interest charges relate to current land banks.

If turnover drops the profit of the static firm is immediately affected. Falls of 10 per cent or more threaten viability as outgoings cannot be met. The land bank has to be reduced (possibly forcing land sales) and building work must be curtailed. The firm is forced to contract which creates problems with its debt structure. The expanding firm has more leeway as the fall in demand will be on the planned increase as well as the current level of demand yet costs are a function of current output (planned increases can be shelved). So they are less likely to experience actual falls in output than is the static firm.

Much depends, of course, on the actual temporal incidence of current costs and debt charges and on the ability to alter them. Land purchase is the major reason for borrowing, and shelving expansion plans usually means adding less land to the firm's land bank. The ultimate limit, of course, is market saturation.

and building worlds but also for fear of giving competitors an advantage. Knowledge of the number of houses being built, plans for expansion into new regions, financial strength and, especially, detailed costings (a total secret) can indicate to competitors where, how much, and at what price a housebuilder will bid for land. For this and other reasons public pronouncements are sometimes deliberately inaccurate and are prone to exaggeration.[6] No official published statistics exist and few builders publish the number of houses built in their annual reports, so reliance has to be put on possibly inaccurate press statements by the builders themselves and by financial analysts.

Details of the ownership of what are believed to have been the twenty largest housebuilders in 1979 are given in table 3.4. Changes in the relative outputs of firms occur, so their ranking varies over time. Changes in ownerships are also rapid in some cases. The data presented in table 3.4 and the following discussion of the major firms should be treated as a particular 'snapshot' of the industry. Its importance for the study of owner occupation is in illustrating the types of enterprise involved in speculative housebuilding and the principles governing their actions.

There are differences in ownership characteristics but the twenty major firms can be grouped into three main categories: four of the firms are owned by major independent contractors (Wimpey, Tarmac, Laing and Taylor Woodrow); eight belong to groups of companies under independent ownership whose main activity is or has been speculative housebuilding (Barratt, Leech, Bryant, Bellway, Fairview, Wilcon, Westbury and Abbey); and eight are subsidiaries of large conglomerates which have bought into speculative housebuilding (Broseley, Comben, Ideal, Bovis, Whelmar, Galliford, Heron and Davis).

The ownership of the parent companies of these top twenty firms is also interesting. For at least fourteen of them there are significant shareholdings by family interests, and another, Barratt Developments, although not recorded as having a large family shareholding, is still run by its original founder. Apart from Barratt, all the other seven independent speculative builders have substantial family ownerships of shares; most also are still managed by a member or appointee of that founding family. All the independent firms consequently cannot be taken over without the agreement of the family or individual founder. This perhaps accounts for their continued independence, especially given the wealth of these family shareholders who have little or no incentive to sell up.

The legal and economic ownership of the major housebuilders is unique for such a large-scale industry. This is clearest for the eight independent 'family' firms but is also true for those owned by building contractors and conglomerates spread across a number of industries. All the owners of the major housebuilders in one way or another have the financial means to make the investment strategies of these housebuilders relatively independent of the short-run profits cycle of the industry, and most owners are interested in investments yielding a long-term return rather than necessarily an immediate one.

One aspect of these characteristics can be seen in the investments in industrial and commercial property that some of the firms undertake. Speculative housebuilders are in a good position to act as property developers and investors as the substantial cash flow from house sales can be used to finance such projects, whereas pure property companies generally have to use borrowed funds. A property portfolio and the rental income derived from it, moreover, can provide a steady income for a housebuilder, shielding the firm against the problems of high gearing in a variable-profits industry. Some housebuilders have built up quite large property interests. The independent companies are most notable in this respect, although others are subsidiaries of groups with large property divisions (e.g. Wimpey and Trafalgar House). Fairview and Bellway are now essentially property developers with residential interests; Bellway demerged its housebuilding and property divisions into two separately quoted companies in 1979.

Pointing out the distinctive nature of speculative housebuilding firm ownership, of course, does not explain why it is like that; nor does it explain why this pattern has been reinforced in the 1970s particularly by the entry of non-building firms into the industry through takeover. A brief history of the ten largest firms will help provide an explanation.

To start with the largest housebuilder throughout the 1970s, Wimpey, is perhaps to begin with the exception. But paradoxically it does illustrate some of the key advantages for large capitals of investment in housebuilding. Wimpey is the largest and probably the most diversified British building contractor, with worldwide interests in building contracting, plant hire, civil engineering, property development and property ownership, as well as in speculative housebuilding. Private housebuilding has been a major part of Wimpey's activities since its initial rapid growth in the inter-war years. Unlike many other contractors, Wimpey moved back into private housing in a substantial way after the relaxation of building controls in the mid-1950s. In 1977, for example, private housebuilding provided 29 per cent of Wimpey's total profits (Simon & Coates 1978). Wimpey has always had a general policy of wide diversification with strong central financial control. This strategy undoubtedly reflects the existence and policy of one individual, Godfrey Mitchell, the founder of the modern George Wimpey Ltd. Ownership is similarly centralized. This centralization of ownership has led to a continuity of policy after Mitchell's retirement.[7]

Building contracting can be highly profitable[8] but the rate of profitability fluctuates considerably and, in addition, the inflow of contractual payments is uneven. For these two reasons contractors' incomes can be highly variable. There are often only limited opportunities to reinvest that income back into contracting as new contracts require little capital investment until work actually starts, and even then the investment is often relatively small. (The exception is where large machinery must be bought to be in a position to win contracts, a situation that can occur in civil engineering especially in respect of earth-moving equipment.) Building contractors, therefore, can be in the position of having large incomes that cannot be ploughed back into their main activity. At such times contracting firms could also be vulnerable to substantial profits tax, the precise impact depending on contemporary tax laws.[9]

Residential development and ownership can help increase and smooth out profits, and, at the same time, reduce the tax bill. Peaks in contracting profit can be reduced by investing in land, and troughs in that income can be offset by the additional income flowing from house sales, especially as cycles in the two types of construction activity rarely coincide. Most contractors have opted for property investment which provides similar benefits rather than residential development, but some like Wimpey have done both. Property ownership and private housebuilding are both speculative activities, so neither type of investment guarantees success. Tarmac's property investments in the mid-1970s, for example, were in the main expensive failures. Some contractors who moved into speculative housebuilding in the early 1970s were similarly badly hit by the 1973 slump. To be attractive investment in residential development has to be large and long-term, use established expertise, and be based on money capital looking for a long-term profitable outlet. In this way, the risks of being caught in a downturn or over a misjudged land purchase are minimized.

If housebuilding is used as a sphere of long-term investment advantage can be taken of the secular rise in land prices. Land purchase and development is more profitable than land dealing alone, but changes in land prices illustrate the basic advantage of residential development: namely that residential land prices rise faster than the general rate of inflation. This was even the case through the 1970s, although sharp fluctuations in the price of land meant that in some years short-term speculation could be hazardous. Since 1974 residential development land has had the additional advantage of being liable for tax relief on stock appreciation when held by bona fide housebuilders. In addition stricken housebuilders with large land banks caught by the slump could be acquired at cheap prices. The timing of investment in residential development in general can also primarily be counter-cyclical. Most purchasers of development land will buy during periods of rising prices as that is when credit is most easily available (see figure 3.2). Firms, however, with continual access to money capital can concentrate their land purchases during troughs when land is relatively cheap. The land can then be held and sold (developed or undeveloped) during upturns.

By investing in private housebuilding the value of money capital can be maintained over the long term and a real rate of return achieved in an inflationary context when so many other potential spheres of investment are in economic crisis. It is possible that the attractiveness of residential development (and the property sector as a whole) for a particular type of capital is, in fact, as much a reflection of the general crisis of British capitalism and its consequent low rate of profit than of the profitability of residential development itself.[10]

Contractors who get involved in speculative housebuilding do so consequently because of their need for a particular type of investment outlet rather than because it involves building work. Speculative building might suit some contractors' investment requirements but it is unlikely to suit them all. The extent of Wimpey's involvement, for example, is likely to result from the firm's unique combination of ownership, diversity and size. As one of the largest speculative housebuilders they have the expertise and can spread the risks of housebuilding to a considerable extent. The capital also exists to purchase land on a long-term basis, and at the best times, because of the existence of income from contracting and the nature of the firm's ownership. Wimpey are renowned for their schemes to minimize tax incidence (Hird 1975 gives a number of examples) and the company because of its ownership structure is reluctant to distribute profits to shareholders. To quote stockbrokers Grieveson, Grant & Co. about Wimpey in their investment research review of the construction industry 1981, p. 109:

> The dividend payout ratio has appeared decidedly cautious against this trend in earnings and is obviously a factor of the 49.9% holding in the Group by Grove Charity Management. . . . The low yield and modest dividend payout ratio of George Wimpey has been a feature for several years and there is little prospect of the group adopting a more liberal stance towards the income requirements of its shareholders.

Speculative housebuilding helps both to minimize the firm's tax incidence and to act as a place for the investment of the undistributed profits.

Other contractors are generally not in the same position. Their

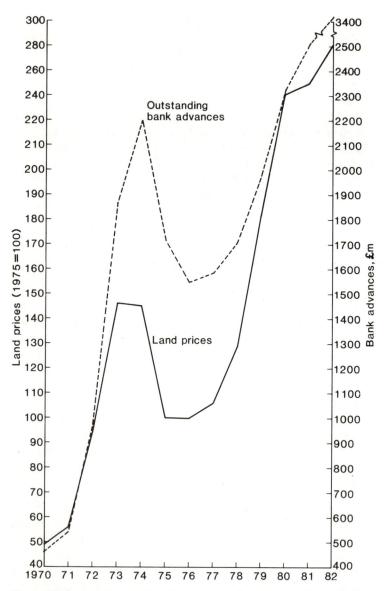

Figure 3.2 Land prices and bank advances to construction firms, 1970–82 (current prices)

Sources: HCS and *Financial Statistics*, HMSO

ownership structures may necessitate a greater distribution of profits to shareholders or their capacity to have large amounts of capital available for long-term investment in residential development may be limited. Even so, some of the other large building contractors are in a position to take some advantage of residential development because they have uneven inflows of cash from contracting and centralized ownership that reduces the incentive to distribute the cash to shareholders. This perhaps explains the involvement of many other contractors in residential development in, at least, a small way. It is interesting to note that perhaps the contractor with the most similar family ownership structure to Wimpey, John Laing, has been rapidly expanding in private housebuilding since the hiving off of their property division as a separate public company in 1978. Turnover in private housebuilding trebled between 1976 and 1979, and the contribution to profits rose almost five times owing to judicious land purchases (Grieveson, Grant & Co., ibid.).

The argument about the specific nature of the capital invested in private housebuilding also explains the ownership of most of the other large housebuilders. Tarmac, the third largest speculative housebuilder, is like Wimpey a building contractor with unique characteristics. Tarmac's traditional main activity is the quarrying and laying of roadstone; it expanded by branching out into other areas of contracting closely associated with roadbuilding. These activities were highly profitable and Tarmac subsidiaries were major participants in the blacktop price rings discovered in the mid-1970s (see Direct Labour Collective 1978). The end of the motorway building boom in the mid-1970s threatened to bring about a severe contraction in Tarmac's workload. The large cash reserves derived from contracting were used to purchase property, to diversify overseas, to purchase a general building contractor (Cubitts) and to acquire a series of speculative housebuilders. The rapid expansion into speculative housebuilding proved to be by far the most successful of Tarmac's diversification attempts. John Maclean, its major housebuilding subsidiary, was acquired cheaply during the slump at the end of 1973; yet by 1979 the housebuilding divisions were contributing a quarter of Tarmac's profits after the division's turnover and profits had both more than doubled in four years.

Hawker Siddeley traditionally has had no involvement in construction, yet its move into private housebuilding was for similar reasons to those of Tarmac. Hawker Siddeley's main aircraft-building activities were nationalized in the mid-1970s, for which it received large amounts of money as compensation; the firm then went on an acquisition spree converting itself into a broad-based multi-national holding company.

Part of the money was used in 1978 to acquire a majority interest in Comben Homes' parent company, Carlton Industries. With the financial backing of the Hawker Siddeley Group, Comben was then able to purchase Orme Developments in the same year, effectively doubling its output (Ball and Cullen 1980). A year later the residential land and housing work-in-progress of Wiggins Construction were bought. The shares issued in association with these acquisitions marginally reduced Hawker's shareholding in Comben to below 50 per cent but the principle of using the purchase of a speculative housebuilder as the best means of investing in land is clear, especially in the continued application of the principle in the acquisitions of Orme's and Wiggin's land banks.

Christian Salvesen, the parent of Whelmar, is another large, profitable firm that entered housebuilding as an investment outlet for money capital that was surplus to the requirements of their other trading activities. The firm is a private company registered in Scotland and is almost totally owned by members of the Salvesen family. Started by an expatriate Norwegian in 1848 as a shipping company, whaling became the firm's major activity after the 1890s. The company recognized the impact of over-fishing in the early 1960s, sold out to the Japanese and so then had plenty of capital and a need to invest it somewhere. Extra capital was generated by their subsequent moves into cold storage and food transportation, as they managed to take advantage of the early stages of the frozen food boom; after which time Whelmar was acquired. More recently the company has recognized the long-term prospects for coal and invested in open-cast coalmining in the USA in the wake of the new economic liberalism of the Reagan administration.

The Salvesen case is interesting again in terms of the timing of its involvement in speculative building. The sector was entered by a series of acquisitions at the end of the 1960s and start of the 1970s, trough years for the housing market and hence a time for cheap acquisitions. Contracting activities were quickly dropped as unprofitable, and the major acquisitions in housebuilding since then have been to obtain building land, again during a downturn in the market in 1973. A housebuilder with large land stocks in Scotland was bought in 1973, Hawker Homes, and also a brickmaker, J. and A. Jackson (Ball and Cullen 1980). Brickmakers have large stocks of land as a future source of clay. However, as builders since the eighteenth century have realized, potential brick land is usually also potential housebuilding land. The realizable asset value of a brickmaker's landholdings consequently may be well above their book value, and well above the cost of acquiring the firm.

The profitability record of the Salvesen divisions since the mid-1970s has not been particularly inspiring; and it has led some financial analysts to suggest that this was the principal reason why Salvesen has not gone public. This is to ignore, however, the differential taxation and cash flow requirements of publicly quoted firms and private ones like Salvesen. Only the Salvesen family, their accountants and tax advisers ultimately can say whether the movement into housebuilding has been a success.[11]

Broseley Estates illustrates the case of another type of capital that can utilize the benefits of residential development: insurance companies. The attitude and interest of insurance companies in property ownership and their emphasis on long-term returns are well known. More recently they moved into property development as well as ownership (Massey and Catalano 1978). The Guardian Royal Exchange Assurance's ownership of Broseley Estates applies a similar logic to residential development.

Broseley is not only unique as the only major housebuilder owned by an insurance company but also in its past. The firm purchased in 1973 by Guardian Royal Exchange, to be restructured with its new Broseley Estates name, was the Metropolitan Railway Country Estates Ltd. This company had been formed in 1919 by the Metropolitan Railway to help create business for its North London commuter services and to derive some of the land increment resulting from the company's investment in new railways. This company, therefore, was the exception to the general restriction on private railway companies deriving land betterment from the creation of a railway by dealing in land adjacent to their routes. In nineteenth-century British legislation betterment was as much the prerogative of the private landowner as it has been for most of the twentieth. Then, however, the fight was not between one type of property ownership and a diverse set of groups pushing for betterment to go to the state but directly between two private groups, landowners and private railway companies. The general inability to acquire any betterment directly contributed to the low profitability of British railway companies in the nineteenth century, and contrasts sharply with the USA where independent land interests generally were almost nonexistent and where the state used land grants to encourage private infrastructural development, particularly of railways.

The Estates Company does not appear to have been nationalized along with the Metropolitan Railway to form part of London Transport in the 1930s. Nor did London Transport itself do any land dealing. Company reports indicate that after the Second World War the Estates Company was mainly active first in Bristol, then in the North West. By the early 1970s it had got into severe difficulty. Real turnover slumped by 70 per cent between 1971 and 1973, and profits were sustained only by large sales of land (£8m. in 1972, almost four times that year's trading turnover). Its acquisition by Guardian Royal consequently was cheap. The years 1974 and 1975 then saw declining profits whilst the company was restructured under the Guardian Royal's ownership. For the next few years both output and profitability rose dramatically. Yet to illustrate how takeover by a large firm does not always guarantee short-term success Broseley was badly hit by the 1980–1 slump with profits dropping to a third of their 1979 level in 1980.

The two final cases to be discussed of conglomerates moving into private housebuilding, the Trafalgar House Group and P&O Ltd, constitute hybrid examples of the reasons given for the other entrants to the sector. Trafalgar House was originally a property company founded by Nigel Broackes in the late 1950s which subsequently branched out into building contracting (e.g. Trollope & Colls and Cementation), transportation (Cunard), and most recently newspapers (Express Newspapers and Morgan Grampian). Trafalgar's ownership of New Ideal Homes can be explained in terms of its property interests and again the nature of the company's ownership.

Trafalgar took over the Ideal Building Corporation, the New Ideal parent company, in 1967. The reason for the acquisition was primarily because of a change in tax laws. Corporation tax was being introduced which particularly adversely affected the profits distributed by property companies to their shareholders. Broackes wanted the advantages of property development and ownership to be combined with trading activities to benefit from the new tax system. Established property companies could borrow more cheaply than virtually any other enterprise, and their shares had high stock market valuation because of the attractiveness of their property assets. Takeovers consequently would be a cheap way of acquiring new assets, and trading companies' investments would qualify for tax relief and investment grants. A list of ailing and hence cheap companies was drawn up, and Ideal Building was the first of a series of acquisitions. The tax laws have now changed but the Trafalgar example highlights the importance of the tax system in determining the attractiveness of investment in housebuilding.[12]

The Peninsular and Oriental Steam Navigation Company (P&O), which snatched Bovis Ltd from the edge of liquidation, shares with

Tarmac, Hawker Siddeley and Salvesen the characteristic of a decline in its main activity and a need to find new investment outlets. In common with building contractors there are also cyclical profit fluctuations in its main activity, shipping. It also shares a characteristic of a number of the larger construction companies: continuity of family control. P&O was founded by the father of the current chairman, Lord Inchcape. The latter is also one of the richest people in Britain, with wealth derived from inheritance and from the success of the Inchcape and Co. trading company. These characteristics perhaps are the most important ones when explaining the company's takeover of Bovis. Since that date, however, the company has had one characteristic shared by no other firm discussed so far: a profits crisis, boardroom rows and a fear of takeover brought about through a series of disastrous investment mistakes (see *Financial Times*, 11 May 1979 and 11 September 1981). P&O underwent a traumatic period of restructuring during the 1970s as not all of its attempts to break out of general cargo shipping, with its cyclical characteristics and increasing foreign competition, proved successful.

Bovis Ltd was acquired in 1974. The latter was heavily involved in speculative housebuilding during the early 1970s boom with almost 50 per cent of its profits coming from this activity in 1973. The mistimed purchase of a secondary bank (Twentieth Century Banking) and over-extended land purchases brought the firm to the edge of liquidation when acquired by P&O. Bovis' subsequent restructuring and success as a contractor and speculative builder have been a mainstay of P&O's profits, to the extent that P&O sold its North Sea interests rather than Bovis when it fell into financial difficulties in 1979. Bovis rapidly increased its housebuilding activities back to the scale existing prior to its mid-1970s debacle. Once again there have been acquisitions of some fairly large housebuilders to gain land banks in new regional markets; for example, B-Vis Construction was acquired in late 1978, a substantial Midlands firm.[13]

The last three remaining firms in the top ten housebuilders are all independent, publicly quoted companies: Barratt, Leech and Bryant. That status has to be qualified, however, as each has strong connections with the founding family. Barratt was formed in 1958 by its present chairman, Mr (now Sir) L. Barratt. W. Leech was formed in 1932 and the octogenarian Mr Leech was still a director of the company in 1979, although management control had passed to a Mr Adamson in 1967 (an accountant) and then to Adamson's son in 1979. In 1976 three-quarters of the company's shares were held by the William Leech Foundation.

This holding was run down to 30 per cent by 1980 but, with directors' holdings of over 10 per cent, the company is invulnerable to takeover. Family interests are also strong in Bryant Holdings as the present chairman and managing director is the founder's grandson and again the directors hold a sizeable (18 per cent) chunk of the equity. All three companies through the nature of their ownership can be seen to have more in common with the other ten major firms than with the traditional notion of the credit-constrained housebuilder. All of them can take a longer-term view of housing development, and each has expanded rapidly in private housebuilding during the 1970s. Thereafter their characteristics diverge, representing an assortment of the reasons for involvement in the sector listed earlier for the other firms.

Bryant Holdings is a well-known West Midlands company. In the post-war period through to the early 1970s it concentrated on local authority housing, especially for Birmingham Corporation where between 1964 and 1970 it was awarded almost half the Corporation's work (Direct Labour Collective 1978). Heavy losses and the long-term decline of the local authority sector led to a switch into private housing which rose to over half of turnover during the 1970s as part of a corporate strategy to change the company from a building contractor to a housing and property developer. Another aspect of the company's involvement in council housing brought the firm to public prominence during 1978 with a corruption case involving three directors of the company who were sent to jail for bribing the chief architect of Birmingham Corporation. This case could only have added to the bleak outlook facing the company in the public sector, so it shares with other entrants into speculative housebuilding a decline in its traditional market as the principal stimulant to expansion there.

Leech is different as it has always been predominantly a speculative housebuilder. It has spread rapidly out of its North East base since 1967. The company illustrates clearly how a private housebuilding operation can be a useful vehicle for investment in residential development land. It has a huge land bank consisting of 10,000 plots with planning permission, and a further 3800 acres without permission (roughly equivalent to 38,000 plots); much of it has been held for years. Expansion into new regions, in addition, has taken place through the acquisition of small local companies and their land banks (the most usual means of geographic expansion for a speculative builder). The company has close links with another North East based family company, Bellway (no. 11 in table 3.4). They jointly developed Cramlington New Town, north of

Newcastle, Britain's only private enterprise new town, a project which arose out of a suggestion by Northumberland County Council in 1958 as a response to complaints about a shortage of building land (Elphick 1964). Bellway and Leech announced a proposed merger in 1981 but it fell through at the last minute.

Barratt Developments is the last detailed case history to be considered. In 1968 when it went public and branched out from the North East (at the same time as Leech) it produced 500 houses; in 1980 it had overtaken Wimpey to become the country's largest housebuilder and has plans to expand well beyond its 1980 11,000 per year output. Rapid growth, aggressive marketing and sectoral and geographical diversification (predominantly within speculative housebuilding) have been hallmarks of the firm's success, and it has frequently been argued within the trade press to have spearheaded the new emphasis on marketing in the industry and to have forced competitors to follow suit. Rapid growth has been the key feature of Barratt's success and, whilst the company clearly has been able to take a long-term view of its investment decisions, its expansion seems to belie what has been said previously, as the nature of the ownership of its capital seems unimportant in comparison to its internal management strategies, whereas what has been argued up to now is that the restructuring of the ownership of capital in the house-building industry has been one of the driving forces behind changes in management strategies.

The apparent contradiction is diminished, however, when it is realized that Barratt is very much a personal product of its founder and of the general restructuring of firm ownership during the 1970s. The relinquishing of direct legal ownership by going public and in subsequent share issues has not lost Barratt economic ownership of the firm with which he is so personally associated; nor obviously does loss of personal legal ownership of a company necessarily mean loss of personal wealth. It is far better to own a limited number of shares in a large successful company than to own completely a small struggling housebuilder. (Barratt's remaining shareholding alone was worth almost £3m. at 1981 share price levels.) In addition, the ability of Barratt to expand to become a national housebuilder was aided by and perhaps dependent on an ability to acquire companies with heavily undervalued land banks during the 1970s. All the top ten companies surveyed so far did this; the cheap acquisition of stricken housebuilders was a means of getting low-cost land banks. So these purchases and their timing were major components of the restructuring of ownership that took place. Cheap land obviously will enhance profits when built on a few years later. But Barratt, in addition, adopted for a while accounting policies which, though permissible, led to some comment in the City. Concern over its reputation led Barratt to run a full-page advertising campaign in the *Financial Times* in October 1979 to convince City institutions of the desirability of investing in its shares. 1981 saw a doubling of its share price, so favourability seems to have been restored. The problem arose from the ambiguities of historic cost accountancy principles in an inflationary environment, as Laing & Cruikshank explain in their 1980 *Construction Industry Review*:

> In so far as Barratt's rapid growth has attracted suspicion from outside the group, it has tended to centre around the acquisition policy and we have thought it right to discuss this in some detail. The accounting principle that concerns us is a simple one. Housebuilders are normally acquired for a price which relates to the value of their land bank – their stock in trade; there may be little difference between paying £1m. for land at auction or £1m. for a company with equivalent net assets. However the Company acquired will be carrying the land at book cost and not market value; if their land (stock) is not revalued to market levels the effect is that the goodwill on acquisition (the difference between book value and market value) is transferred to the profit and loss account as the land is developed. This artificially (and incorrectly) inflates profits in the post-acquisition period. Sooner or later this works its way through the system and to maintain the profits momentum further, and larger, acquisitions are needed. This, it is argued, has been the basis on which Barratt Developments has grown during the seventies. . . . How significant a distortion has this been and should it now be of concern? In the mid-seventies, when published profits were running at the £5–6m. level, the impact (especially from Wardle and Bracken's £6m. [goodwill]) could have been considerable. However, by the time we come to 1979, it is difficult to argue that the tail end of Janes pre-acquisition land bank had played a material part in the overall group total of over £20m. By now, it seems safe to assume that Barratt's profits no longer owe anything to the accounting policies applied to its earlier acquisitions. They have played their part and the Company is now able to institute changes in accounting policy designed to introduce a greater degree of conservatism.
>
> (Laing & Cruikshank 1980, pp. 13–14)

Like previous companies discussed, financial acumen has consequently played an important part in the growth of Barratt. The company has succeeded not so much by going against the tide of trends within the industry but rather by taking full advantage of them.

Tax, finance and private housebuilding

Because of the importance of financial factors it is worth considering them in a little more detail at this stage. The advantages can only generally be achieved by companies rather than by individuals. Many of them concern taxation laws in Britain.

Tax relief for stock appreciation has already been mentioned. It also exists in other industries and most of British manufacturing industry currently pays virtually no tax because of it. Many other industries, however, have been in severe crisis throughout the second half of the 1970s so capital invested in them could become worthless, whereas in housebuilding an appreciating asset, land, is being held. Housebuilding consequently represents a relatively safe investment for companies that might otherwise have a tax problem. The marginal cost to such conglomerate enterprises of failed speculative ventures in housebuilding, moreover, will be net of corporation tax, effectively reducing each pound lost to only 48p.

A good example of such tax effects was the reaction by particular firms to the sharp fall in land values in 1973 and 1974. A review of land banks when land prices fall can lead to large extraordinary losses appearing in the profit and loss account. Firms with non-housebuilding activities can use the tax credits generated by such reassessments to offset the tax payable on their other activities. To cut their net of tax losses, therefore, many firms with non-housing activities annually revalued their land banks during this period (as part of a continuing assessment of their assets). For independent speculative builders, on the other hand, the appearance of such large extraordinary losses could spell disaster, as shareholders and creditors might lose confidence in the firm's viability as a result. These firms consequently tried to hold off admitting such asset value losses and waited for land prices to recover their previous level.

The purchase of land, or the takeover of a company, is an input cost for a firm so they can declare it as a cost for taxation purposes which again reduces the effective cost. Another taxation factor relates to the timing of declared profits. Housing developments frequently take years to complete, and firms have considerable leeway in deciding when they declare profits on a development. The longer the time before they are declared the less is the real incidence of taxation, especially in inflationary times.

Wimpey, for example, are believed to credit profits on housebuilding only after the last house on an estate has been sold (Hird 1975).

In general such taxation benefits of land outweigh any incidence of development land tax (DLT) firms might incur. That tax was designed to lower the profits derived from development gain, but even without the offsetting tax advantages for builders mentioned above, its actual incidence to them is generally very low. The returns to the Exchequer from the tax have to date been derisory, which is perhaps not surprising given the exemptions available to housebuilders. One financial commentator could confidently write an article entitled 'Who's afraid of DLT' (*Construction News Magazine*, April 1981) and in an update article after the Budget a month later could concisely explain the builder's new DLT situation:

> In my article last month called 'Who's afraid of DLT' I suggested that in many cases DLT was not the bugbear often imagined. Since I wrote the article three DLT relaxations have been proposed in the Budget. The first relaxation concerns house-builders only. When they start a project of material development on their building land, there is no DLT liability if the market value of the land is not more than 50 per cent higher than the cost plus, of course, the £50,000 exemption available in any year. In some cases this higher threshold will reduce or eliminate the liability, but I expect that in most cases housebuilders will have been exempt anyway on the grounds that they will have paid a price reflecting the full development potential, and of course they are exempt in such circumstances if they start the development within three years.
>
> (M. Parry-Wingfield in *Construction News Magazine*,
> May 1981, p. 14)

It should also be remembered that tax laws change and with them the advantages of owning a speculative housebuilder. The case of Trafalgar House and the Ideal Building Corporation has already been cited. Similarly, the ownership of a holding company may change which can also alter the position of its housebuilding subsidiaries. The case of Davis Estates, no. 18 in table 3.4's list of major housebuilders, is a useful illustration of this point. Its parent company until 1982 was Wood Hall Trust which also owns Fairweather, a building contractor active in public housebuilding in London and the South East. In terms of Wood Hall's major interests ownership of these two medium-sized building firms at first sight seems odd. Wood Hall was a large (£500m.+ annual turnover) British company whose main interests are in Australia where it is the largest woolbroker and pastoral agent. UK tax laws provide a rationale for this link between Australian sheep and British housebuilding. Profits from abroad are subject to the full rate of corporation tax whereas UK-based profits are subject to lower liability as they can be offset by tax allowances on, for example, investment and stock appreciation. British companies with high overseas earnings therefore have a strong incentive to acquire UK-based operations, financed out of overseas earnings, to incur a much lower corporation tax liability. If Wood Hall had this idea in mind the outcome was less than successful, as its Building Group's performance, especially Fairweather's, has been very poor. In 1981 the Building Group lost £7m. on a turnover of £48m., a loss equal to more than the total capital employed. The whole taxation position was transformed, however, by the takeover in 1982 of Wood Hall Trust by Elders IXL, an Australian Company, who quickly announced their desire to withdraw from the UK building industry.

Taxation is not the only financial advantage of landownership. The benefits of borrowing for companies that are not too highly geared have already been explained. The advantages of borrowing are further enhanced in inflationary periods as the real value of the loan repayments will fall over time. The effect is similar to that of an owner occupier's mortgage repayments. If a housebuilding firm buys land with borrowed capital, it will get (corporation) tax relief for the interest payments, the monetary size of the loan will remain constant, yet the asset will appreciate. Like owner occupation itself, there is a front-loading of real costs whilst the financial benefits rise over time. None of these points, however, should be taken to imply that the longer land is held the greater is necessarily the gain. A rapid turnover of land can also bring benefits. The precise nature of firms' land banking activities will be discussed in chapter 5.

The growth of long-term land development capital in the housebuilding industry

A lot of space has been devoted to the case histories of the largest producers not so much because of their intrinsic interest but to show the changes that have occurred in the leading firms in the industry and in their ownership. Only one firm, Wimpey, has an ownership and relative size that can be traced back more than fifteen years. Large firms have moved in through takeover to occupy leading positions, and their motives related predominantly to long-term, large-scale investment. A new and particular type of capital, therefore, has come to dominate speculative housebuilding. This intervention has meant that the major housebuilders are now relatively immune to the housebuilding cycle. A sudden downturn in demand does not threaten the financial existence of these housebuilders as they now (like Wimpey before) have the financial backing of large corporate enterprises. Creditors need not fear for the security of their loans with such firms as the parent company can give their housebuilding subsidiaries financial backing and might anyway be the main source of credit.

Their actions contrast with the commonly held view that speculative housebuilding is dominated by relatively small firms whose activities are constrained by the availability of credit, so that every time a credit shortage occurs they have to curtail their building and land acquisition. This belief has been reinforced by a number of studies of housing supply based on data from the 1950s and 1960s that argued for this key role of credit availability (cf. Vipond 1969, Whitehead 1974 and Hadjimatheou 1976). Monetary restrictions associated with the stop-go cycle of macroeconomic policy, therefore, would hit both the demand and supply sides of the private housing market by restricting building society mortgages and bank advances to builders.

Whether or not this was the case during that period, the changing structure of ownership has reduced the impact of such credit restrictions. These housebuilders can plan building programmes a number of years in advance. Market downturns might affect turnover temporarily but investment plans do not have to be curtailed. The major investment of speculative builders is in land, so such firms can take a longer-term view of land acquisition, perhaps by holding land for a longer time period but more importantly by being able to acquire and develop land at the best times. In this way, the market position of these firms is gradually strengthened. This has crucial implications for both landowners and for the planning system, because it means that the strength of builders has been considerably enhanced in the struggle over the price and the availability of land.

The change in ownership has altered the dominant influences on accumulation in the industry. Speculative housebuilding is now dominated by firms with accumulation strategies related to the whole structure of their corporate enterprises and the nature of their ultimate owners. As a result, the speculative housebuilding industry is now dominated by long-term development capital whose existence arises from the advantages for specific ownership structures of profits from land dealing and land development.

A large annual output is usually necessary to be successful at land dealing and development, as chapter 5 will show. So the increased market share of the largest producers in the industry is a result of the change in the nature of the dominant type of housebuilding capital. In other words, investment strategies related to market share by the largest producers are products of the particular type of capital they are, instead of the other way round with market share determining the type of capital in question.

What makes it possible to say that long-term land development capital dominates the housebuilding industry is the investment strategies of most of the major producers. Capital invested in housebuilding now increasingly comes from large enterprises whose access to funds does not depend on the state of the housebuilding cycle, nor on the fears of some jittery bank manager, but on the long-term profitability of investment in housebuilding. They do not have to fear the sudden recall of credit or, possibly, even sharp but temporary downturns in the market. Instead they can have a wider strategy of investment aimed at maximizing the profits of their parent companies (or dominant owners), an objective which will influence their housebuilding strategies, development activities and production processes. The development aspects of housebuilding are especially important to them because their ability to take a long-term view enables them to raise their development profit above the norm. They can take advantage of adverse general market conditions, for example buying land when its price is depressed or building on it when the market picks up again, whereas other housebuilders cannot. The financial and taxation aspects of speculative housebuilding moreover attract particular types of capital: ones that want to retain or possibly to avoid declaring profits rather than distribute them to shareholders.

If the changing nature of capital in speculative housebuilding does alter the reaction of the industry to economic pressures, that over time could lead to changes in the nature of those pressures. Even so, long-term land development capital cannot mould the market in its interests. The fundamental economic nature and problems of speculative housebuilding are the framework for accumulation in this industry, no matter what type of capital is trying to accumulate within it. So to talk of this type of capital is not to talk of some rapacious profit-grabbing machine that, say, manipulates prices to its advantage. Nor is it to talk of 'unfair' competition with big monopolies squeezing out 'helpless' small firms.

The demise of the smaller firms has been a product of their increasingly untenable situation in the face of a changing market. A revival of their fortunes (and some can still be made) is not possible therefore by restraining long-term land development capital. The smaller firms produce a more expensive product, so a revival anyhow could be achieved only with much higher house prices and a bottomless pit of loan finance.

The expansion of the new dominant enterprises, nevertheless, has not been a simple exponential process. They tend to take a bigger share of an expanding market and then contract their share during downturns, although, not surprisingly for such a speculative industry, behaviour varies between firms. Also it has been the small family producers who have felt the competitive squeeze the greatest, whereas the petty producers seem to have been comparatively immune, as a look at firm size changes in the 1970s shows.

Firm size changes in the 1970s

The market share during the 1970s of housebuilding firms of different size is given in figure 3.3. The most striking feature of this figure is the marked increase in the share of firms in the 250 houses or more per year categories. In 1969 they had only a quarter of the market, yet by 1979 their share had risen to over half; it therefore doubled in ten years. This increase in market share, however, has been cyclical; increasing during booms and decreasing during slumps. In the crisis years of 1973–4 all categories below 250 houses per year increased their share, whereas the largest 500+ category dropped from having 27 per cent of the market in 1972 to only 15 per cent in 1974. Then after 1974 the 500+ category increased very strongly, more than doubling in the next four years.

Part of the changing importance of each size category simply reflects the fact that builders build more houses during boom years; so individual firms will tend to move between size groups depending on the stage of the cycle. But this feature cannot explain much of the change in market shares, as the variation in each size group's share is not a gradual cascade through the size categories but predominantly a change at the extremes between the largest and the smallest builders. The largest group lost the biggest market share in the slump of 1973–4, the smallest (1–10) gained the most, rising from 11 to 20 per cent in two years, while the intermediate categories just managed to maintain their share. Given the very large absolute fall in completions during 1973–4 the implication in terms of output is that the petty capitalist builders just about managed to

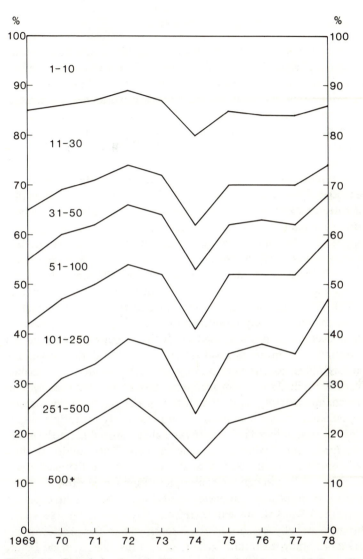

Figure 3.3 Speculative housebuilders: market share by size of firm, Great Britain, 1969–78

Source: NHBC data

Note: The largest firms' market share has continued to grow since 1978. NHBC data for 1982 show the following market shares – 1–10 houses built: 14%, 11–30: 10%, 31–100: 12%, 101–500: 18%, and 500+: 46%. Again, the trend has not been continuous. It depends on the state of the market – the largest firm's share fell back during the early 1980s slump and then rose as the housing market subsequently picked up. The later data are not strictly comparable with the previous information as the NHBC now groups firms with the same holding company together, eliminating the element of double-counting in the earlier data.

maintain their output levels, the small family and non-speculative house-building capital (NSB) firms fell at the rate of the market decline whilst the major producers cut their output the most. In part the drop in output by the big firms as a whole was the result of bankruptcies, yet much of the change reflects the ability of such firms to vary their output depending on the profitability of housebuilding. This ability is not shared to the same extent by smaller firms, although there are important differences between the smaller size groups.

The reaction of the NSB medium-sized firms to market downturns is the most difficult to calculate because of the variety of enterprises within this category. The ones with substantial expansion plans are likely at least to maintain their output, perhaps temporarily accepting lower margins. For the other two sub-groups within this category, their response will probably depend on the relative profitability of speculative housebuilding compared to their other activities. Building contracting, for example, was hit by a similar slump to that of private housebuilding in 1973, so contractors using housebuilding land as an investment might have been forced to rely on realizing income from that investment to sustain overall profits. As to small family firms, they need to maintain a relatively steady turnover, it was argued earlier. But during downturns they are unlikely to start new developments as the capital is not available. What they will do is try to maintain sales on present sites even at reduced, and possibly negative, profit rates. For various reasons, therefore, NSB and small family firms will want to avoid cutting their output even if its profitability has fallen. In general they are unlikely to be in any better (or worse) market position than any other housebuilder, so their overall output will tend to decline at the same rate as the market as a whole, especially through the bankruptcy of individual firms.

Petty producers in contrast were the most successful in terms of output during the 1973–4 slump and, therefore, increased their market share. This was partially because of the insensitivity of their sector of the market (generally small infill sites) to downturns. This demand factor, however, was reinforced by the continuing need of these builders for income, which might even have led them to increase their outputs during the slump period.

In order to understand what happened to the different size categories of firm during the period of recovery after the slump, it is important to realize that there is always a substantial movement of speculative house-building firms between size categories and in and out of the industry. Booms obviously attract more firms to housebuilding and enable those already building to expand output. Slumps do the reverse. The boom and subsequent massive slump in output of the early 1970s not surprisingly exhibited those features. The slump in particular can be seen as a period of restructuring of capital; bankruptcies and mergers remove the weaker firms and the activities of those remaining are often reconstituted in attempts to avoid in the future the worst effects of sharp falls in sales. The strongest of the remaining firms will then be the ones to expand in the next upturn. Comparison of firm losses from the peak in 1972 to the bottom of the slump in 1974, and then the subsequent gains during the upturn to 1978, enable this restructuring process to be examined.

There was a net loss of almost 2000 speculative housebuilding firms between 1972 and 1974 (see table 3.5). As small firms (1–30 houses) outnumber the rest, the greatest absolute loss was not surprisingly amongst the small. The proportionate change, however, shows a very different picture: the greatest proportionate loss was amongst the largest producers (either through ceasing building or by cutting output and so falling into a smaller size category). Whilst the smallest group (1–10 houses), showed hardly any percentage loss at all, the two largest size categories in table 3.5 were decimated; in 1972, 53 firms produced over 500 houses, by 1974 they had fallen to 12.

In the upturn from 1974 to 1978 a different picture emerges. The size

Table 3.5 Speculative housebuilders: losses of firms 1972–4, gains of firms 1974–8, by size of firm

| | No. of firms active | | | Firm losses 1972–4 | | Firm gains 1974–8 | |
	1972	1974	1978	No.	% of 1972 number	No.	% of 1974 number
1–10	7380	7037	7067	343	4.6	30	0.4
11–30	1879	971	1025	908	48.3	54	5.6
31–50	453	225	223	228	50.3	−2	−0.9
51–100	395	167	191	228	57.7	24	14.4
101–250	229	107	115	122	53.3	8	7.5
251–500	82	24	60	58	70.7	36	150.0
500+	53	12	29	41	77.4	17	142.0
Total active	10471	8543	8710	1928	18.4	167	2.0

Source: NHBC data

categories up to 250 dwellings a year showed only modest percentage increases in firm numbers, whereas the two largest categories had dramatic increases of around 150 per cent. This again reflects the ability of larger firms to vary their output but it also shows that the shake-out of smaller firms during the mid-1970s slump was permanent. The numbers of petty producers was hardly affected; pressure instead was on the small family and NSB types of firm.

The largest producers, however, did not just bounce back to their old outputs; a significant reduction in the number of large builders also took place. Only 29 firms built more than 500 houses in 1978, compared with 53 in 1972; the total output of the 500+ category had fallen slightly between these years from 53,000 to 49,000 but the average annual output of each firm had risen steeply from 1000 in 1972 to almost 1700 in 1978. The restructuring of the largest producers, therefore, reinforced the concentration of speculative housebuilding into larger units. It also increased the centralization of ownership of capital in the industry by enabling large firms to take over weakened ones.

The reasons for the decline of the small and medium-sized housebuilders

The 1973–4 slump was a shock from which many small and medium firms never recovered, yet the economic position of those firm types had by the mid-1970s become so poor that during the subsequent upturn few firms of similar size replaced them. The changed nature of owner-occupied housing provision was the principal cause. Rising construction and land costs badly affected the small and medium firms. Each new development requires progessively more capital and, as was argued earlier, more credit to finance it even if the rate of profit is good. These firms, therefore, more quickly reach credit ceilings imposed by their bankers; yet, after the experience of the 1973–4 slump, bankers will take a jaundiced view of lending to such builders. The problem is not one of slow readjustment after one bad slump; the whole market is now more unstable. So the long-term risk of being unable to repay loans on time has increased. Such firms do not have the time horizon by which to take advantage of variations in the market. Only those firms set up as invest-ment outlets for larger firms outside the industry can partially offset this growing problem.

A subsidiary factor, and one most resented by the smaller firms, is the increasing enroachment of larger firms into spheres of the market traditionally left to them. Rising land prices, the house price premium on small, 'non-estate' sites and their desire to move up-market, have encouraged many larger firms to embark on smaller schemes than they would have previously envisaged. The once-sheltered sphere of up-market developments has consequently been subject to greater competition, and so the smaller firms have found it increasingly difficult to acquire this type of land.

The economic difficulties faced by many small family firms, therefore, have been the main cause of their decline, a decline speeded up through the demand for land by large builders. Many such firms with good land banks have been acquired by large firms.[14] The proprietors of the acquired firms are given the chance to get out of an increasingly unprofitable business without necessarily incurring the taxation that would result from winding up their company's activities and selling its land bank. The acquiring firm has the advantage of obtaining land at below market price and possibly being able to roll over substantial tax reliefs.

Corporate profitability and speculative housebuilding

Earlier sections have shown how the housebuilding industry has changed. Domination by long-term land development capital, however, does not guarantee that those firms will be highly successful; they just stand a much better chance of being profitable than do smaller enterprises. Speculative activities by their nature imply that some firms make large profits and some lose. The point is to be able to go against general market trends or, more accurately, to anticipate changes before others do.

To illustrate the financial outcome of speculative housebuilding it is useful to examine the financial performance of a sample of firms over time. In order to do that, obviously, only successful 'survivor' firms can be considered. So there is a minimum impact of market change in the sample that is not representative of the industry as a whole. Although this aspect should be borne in mind, it is perhaps more important to be aware that there is no such thing as a representative firm. No attempt is being made to describe average behaviour. The case histories earlier show, if nothing else, that the average firm does not exist nor is such an abstract construct useful in explaining the behaviour of those that do. What is more important is to explore the structural constraints faced by any individual enterprise. A constraint faced by the research itself in fact determined the nature of the sample, namely the ability to find a consistent set of published accounts from 1971 to 1978 for a sample of firms whose major activity was private housebuilding. (The firms in the sample therefore are all large, publicly quoted ones.)

The published annual accounts of companies are not designed for economic analysis so such an exercise is fraught with difficulties. The problem is made worse in that an examination of individual companies' performance also is a comparison between companies, and there is every reason to expect that different firms with the same economic performance will report dissimilar accounting results. This latter problem arises because of the contradictory functions that accounts are supposed to perform. Accounting information is meant to provide information for four distinct groups: management, shareholders, the Inland Revenue and other concerned parties (e.g. workers, financiers and suppliers). Published company accounts reflect this divergence of interest and the legal framework that has arisen as a result. Most companies publish the minimum required by company law. Within that legal framework some firms, wishing to impress shareholders, will try to present a rosy picture of current profits, whereas others will be more intent on avoiding taxation.

Any analysis of company accounts, therefore, can give only a rough guide to economic performance, and many interesting issues cannot be explored. But it is still useful to look at speculative building firms' accounts because of the sheer severity of the changes that have taken place during the 1970s. Basically, it enables the framework within which speculation occurs to be examined: the framework first, by looking at the general trends in company performance, and then the speculation by looking at performance differences between firms. Each will be dealt with in turn.

The general trend in housebuilders' performance during the period under study, 1971–8, was dominated by the early 1970s' housing boom. For a two-year period, spanning 1971 to 1973, profits reached record levels as unprecedented increases in house prices leapt ahead of rises in construction costs. The crisis of 1973 brought this profits boom to a sharp halt, and the general slide in profitability after 1973 lasted a long time; profits fell for four consecutive years through to 1978. This fall is measured in figure 3.4 by net trading profit as a percentage of turnover. Net trading profit shows the general underlying gross profitability of housing development, that is profit prior to interest charges, having subtracted from revenue the costs attributable to wages, materials, plant

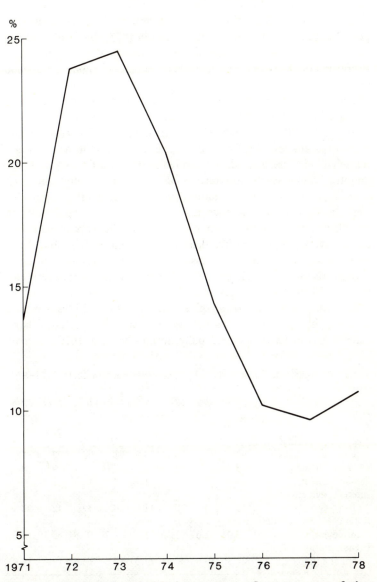

Figure 3.4 Speculative housebuilders: net trading profits as percentage of sales, 1971–8

Sources: Annual company accounts

hire, overheads, depreciation and land. A clear profit cycle is shown; the
peak profit margin on sales was 25 per cent in 1973 which indicates why
housebuilding was such an attractive speculative activity at that time.
Profits fell below 10 per cent in 1977 and initially the house price boom
of 1978–9 brought only a comparatively modest boost in the rate.

These profit data are historic cost based and take no account of
inflation accounting methods. The latter would, of course, show much
lower profits. But as much of the inflation-adjusted decrease would
relate to profits resulting from stock appreciation, and the main source of
housebuilders' profit, land, is classified as part of a firm's stocks, it is
doubtful whether such procedures would give a more accurate, inflation-
proofed picture. Despite the ambiguity about 'real' profit margins, it is
clear that gross margins have fallen substantially in housebuilding.
Whether or not a profit margin of 10 per cent (the approximate 1977
trough average figure) is 'reasonable' is, of course, debatable. Yet it
perhaps helps to explain why firms have cut back on output, waiting for
margins to rise. Above all, every firm will want to halt this general slide
in profitability.

The squeeze on profit margins since the early 1970s has been
paralleled by a rise in the impact of interest charges on company finance.
Interest payments rose substantially during the early 1970s, partially

Table 3.6   Aggregate financial data[1] for speculative housebuilders, 1971–8

	1971	1972	1973	1974	1975	1976	1977	1978
% contribution to aggregate profits[2]								
net trading profit	90	95	96	94	91	90	90	87
rental income	6	3	3	5	7	8	9	8
investment income	3	2	1	1	2	2	1	4
Interest payments as % of aggregate profits	14	13	22	41	39	39	42	30
Ratio of sales to capital employed[3]	2.01	1.61	1.58	1.39	1.70	1.94	2.10	2.01
Pre-tax profits as % of capital employed	22.9	30.8	32.1	17.0	17.0	13.6	12.0	14.9

[1] Historic cost accounting data for a sample of 22 private housebuilders
[2] Aggregate profits are defined as net trading profit plus rental income plus investment
income. Percentages shown therefore sum to 100 in the absence of rounding errors.
Exceptional items are excluded from aggregate profits
[3] Sales relate to total trading activities, and capital employed includes non-trading investments

owing to rising interest rates but primarily because of the sheer volume of funds being borrowed. In the firm sample, there was almost a tenfold increase in aggregate annual interest payments in just four years from 1971 to 1974, and between 1971 and 1974 the proportion of aggregate profits used to pay interest charges rose from 14 to 41 per cent (see table 3.6), indicating the severity of this adverse gearing effect. Many firms spent the mid-1970s trying to run down their borrowings. To an extent they were successful: the aggregate interest charges for firms in the sample dropped by 20 per cent between 1974 and 1978. The relative capital requirements of speculative housebuilders, furthermore, will depend on the stage of the housebuilding cycle and the extent of investment in land, with booms tending to shorten turnover time because of increases in sales. A variation in the ratio of sales to capital employed in the housebuilding cycle was apparent for the sample (table 3.6). It was also noted earlier that firms have attempted to minimize the adverse effects of high gearing during downturns in the housing market by investing in rent-bearing assets. This rising trend for rental income can be seen from the sample in table 3.6. Trading profits, nevertheless, still constituted the major source of profit (around 90 per cent).

Interest payments grew in significance for the firm sample over the 1970s. From 1974 to 1978 they hovered around 40 per cent of aggregate profits and only dropped to 30 per cent in the upturn year of 1978 (see table 3.6). The high interest rates of the early 1980s have reinforced the problem. The simple arithmetic of high interest rates in the face of falling profit margins makes housebuilders far more sensitive to the potentially adverse effects of high gearing. So firms are now far more wary before they commit themselves to substantial borrowing (with obvious implications for the type of capital active in the industry). The variation between firms is substantial, so not too much should be read from simple general trends when considering any particular individual company. Speculation in any sphere is concerned with spotting general trends and avoiding any potential adverse consequences of change, and the winners and losers vary over time. The features discussed above consequently are the average effect of a wide diversity between and within individual firms.

Table 3.7 and figure 3.5 illustrate this variability for just two aspects of company performance: turnover and profit margins. Turnovers of firms rise and fall, frequently against the overall trend and sometimes in surprising circumstances. One firm, for example, managed almost to double its previous turnover in the slump year of 1974! Part of this apparently incongruous variability will be the product of the scale and nature of individual firms' development projects. If a comparatively large number of schemes have just been started or another firm is taken over, turnover will jump, whereas if sites are being completed whilst new schemes are yet to be producing output, turnover could equally slump. It is precisely the timing of such phases relative to the state of the housing market which helps make housebuilding such a speculative activity.

Figure 3.5 shows that there is a wide dispersion of individual firms' profit margins around the average. Even in the peak year of 1973, when one firm had a margin of over 40 per cent, another managed only a desultory 2 per cent. It is interesting to note that there is a wider dispersion of margins when the market is changing, when there is general uncertainty as to the direction of change, than when the pattern of change is more expected (compare 1972–4 with 1975–7). This, of course, is a classic characteristic of a speculative market.

Finally, to reinforce the point about the speculative, and hence risky, nature of private housebuilding even for the largest producers, it is worth just briefly looking at the performance of the ten major firms described in the earlier case histories. For some their housebuilding activities have prospered, but others have slumped during the short-run downturn of 1980–1 (see chapter 4). The biggest contrast is between the two largest firms. Barratt continued to surge ahead: pre-tax profits rose by £6m. in both 1980 and 1981, UK housing output was pushed up

Table 3.7   The variability of speculative housebuilders' financial performance, 1971–8[1]

	1971/2	1972/3	1973/4	1974/5	1975/6	1976/7	1977/8
Change in real turnover[2]							
Firms with rising turnover	16	7	10	17	17	15	14
Firms with falling turnover	6	15	12	5	5	7	7
Aggregate change	+21.3	−10.1	−3.7	+12.7	+23.3	−5.7	+3.5
Range %	+160	+99	+76	+180	+516	+84	+55
	to	to	to	to	to	to	to
	−50	−61	−42	−23	−25	−45	−34
Change in profit margin over previous year							
Firms with rising margins	19	14	5	3	7	9	9
Firms with falling margins	2	8	17	19	15	13	12

[1] Estimates derived from company accounts survey (22 firms)
[2] Real turnover equals actual turnover deflated by house price index (1975 base year)
[3] Profit margin defined as net trading profit as % of firm's actual (undeflated) turnover

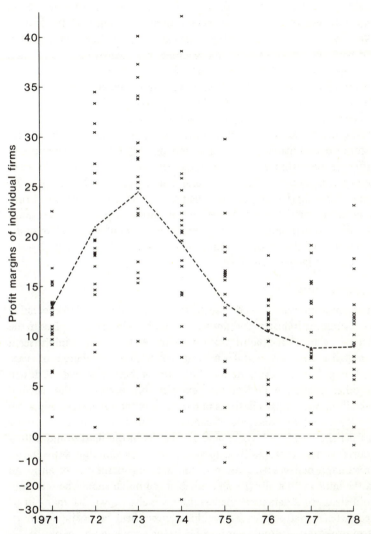

Figure 3.5 Speculative housebuilders: frequency distributions of profit margins, 1971–8[1]

Source: Estimates derived from company accounts survey

[1] Profit margin is net trading profits as a percentage of sales; dotted line is aggregate margin for whole of 22-firm sample

slightly in 1981, two Californian housebuilders were taken over in 1980 and 1981 respectively, and a successful rights issue raised £23m. in May 1981. Wimpey, on the other hand, was badly hit by the downturn. Output between 1979 and 1980, for example, plunged by over 2000 units and the 1981 picture was still bleak. The 20 per cent fall in Wimpey's overall profits between 1980 and 1981 was attributed by the financial press to problems in its housebuilding division. Tarmac, Comben, Leech and Broseley have also seen housing profits squeezed in the early 1980s, whilst Bovis experienced only a short-term drop and Bryant has continued to expand. These variations highlight the consequences of different marketing and investment strategies, in terms of the location and type of developments and of land banking policies. No firms at the time of writing (December 1982) had announced any intentions of running down their housing operations. Wimpey in particular has committed itself to trying to catch up with the turnover of Barratt. The short-run effects of the housebuilding cycle consequently have not clouded these firms' longer-term view.

Conclusions

Substantial changes have occurred in the nature of firms in the speculative housebuilding industry, and since 1973 they have been taking place in the face of a general decline in profit margins. At first sight it might seem paradoxical that new capital is moving into an industry where profits are declining. It is, however, the specific nature of that capital and the dearth of other profitable outlets that explain this paradox. The section describing the changes in firms noted many of the advantages of private housebuilding for this type of capital, and how they try to speculate against the trend by buying land at troughs in the market and selling houses at the peak. Each capital via its housebuilding subsidiaries presumably believes it can be one of the 'winners' in this market, although in the outcome not all will succeed. Not too much should be read into published profit data none the less, as it was also shown that many firms had ownership structures that would not necessarily be looking for short-term profit nor want to record profits in the company's accounts.

There is a clear link between changes in the housing market and changes in the firm structure of the industry. Both changes have led to greater pressures to increase housebuilders' development profit. This has altered the nature of struggles going on within the structure of owner-occupied housing provision. To see how this has come about the next chapter will look at the new housing market in more detail.

4

Building for the new housing market

In order to understand how speculative housebuilders go about building houses some prior knowledge of the housing market is required. There have been major changes in its workings over the past twenty years. This chapter will give a brief review of them by looking at the movement of house prices. But principally it is concerned with aggregate housing supply. The various sources of housing supply in the owner-occupied market will be considered and then attention focused on new house-building. The object of the analysis is to understand aggregate housing supply as a component of the structure of owner-occupied provision, so it will concentrate on the economic processes affecting supply. The discussion supports a very simple argument about housing supply: first, the influence of new supply on the owner-occupied market far outweighs its comparatively small size in total supply and, second, that in fixing output levels housebuilders in aggregate react to the classic capitalist criterion of profitability. Despite much vaunted claims to the contrary, land prices and physical shortages of land and other inputs seem to have little effect on the overall level of output.

A changing housing market

In comparison with the inter-war years the most obvious difference in the present-day owner-occupied market is house price inflation. The years of falling house prices in the 1920s and 1930s have never been repeated; inflation is now endemic. Rising prices are central to the viability of contemporary owner occupation because of their consequences for housing finance. The real value of owners' mortgage debt is gradually eroded away, and owners get money gains from house price rises which they may use to finance improvements in their housing situation through trading-up the market. Inflation, consequently, has shifted market power in the direction of existing house owners. Yet a long-term relative increase in house prices is a comparatively new phenomenon. It is only since the 1960s that house prices have consistently tended to rise faster than the general rate of inflation. As a consequence the gap between house prices and the general price level has widened considerably, as can be seen from a comparison of the two price trends given in figure 4.1 for 1970–81. Short-term bursts of house price inflation widen the gap between house prices and the general rate of inflation, and subsequent periods of stagnant or slowly rising prices bring periods of convergence between the two series. But over the long term the gap between rising house prices and general inflation remains and is widening.

Since the early 1970s house prices have not simply tended to rise faster than the rate of inflation but have also become highly volatile. At the same time the demand for owner-occupied housing has also started to

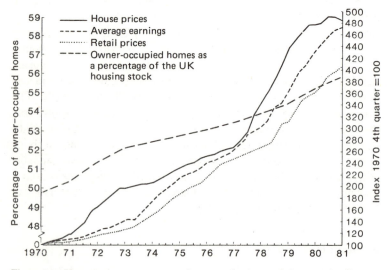

Figure 4.1   House prices, average earnings, retail prices and the growth of owner occupation, 1970–81

Source: Nationwide Building Society, *Housing Trends*

vary sharply over short periods, such as month-to-month, particularly from second-time house buyers. So the UK housing market is now an unstable one. This change post-dated the advent of general real house price rises by more than a decade. Figure 4.2 shows how the rate of increase in house prices varied between 1957 and 1981, both in absolute terms and relative to changes in the retail price index. It can be seen that there were three distinct phases of house price movement, phases which follow closely each change in decade. The 1950s was a decade when prices rose less than the rate of inflation (partially as a consequence of a rapid rise in house prices immediately after the Second World War). In the 1960s house prices rose steadily faster than general prices, and only after that did the pattern change to one of a long-term steep rise in prices punctuated by booms and slumps. This short-term volatility of the housing market is an important new characteristic and its particular significance for this chapter is its effect on new housebuilding (other aspects will be dealt with in chapter 10).

Although house price inflation does not directly correspond to general price inflation, a closer relationship exists between house prices and average earnings: in the 1970s average house price rises shot ahead of

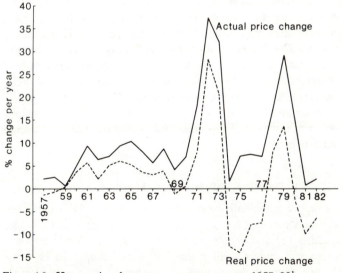

Figure 4.2   House price changes per cent per annum, 1957–82[1]

Source: HCS

[1] Prices at mortgage approval stage. See *BSA Bulletin* 19, p. 21, for details of construction of house price index

earnings during two periods, 1971–3 and 1978–9, and then in both cases they slowed down whilst earnings slowly caught up (figure 4.1). Over the long term, therefore, changes in house prices are linked to changes in incomes. This result is not surprising as rising real incomes imply a greater effective demand for housing. So it can be said that the growth in demand for owner occupation associated with rising incomes has sustained an increase in house prices. As it stands, however, this statement is a truism. It explains nothing as it only states a necessary condition for house price inflation, not an ultimate cause. For instance, with other consumer durables, like cars, audio equipment and video recorders, rapid increases in demand have been matched by falls rather than increases in prices because of concomitant dramatic reductions in the costs of producing them. This has not happened with housing because building costs have risen closely in line with house prices.

The sharp rise in construction costs explains why the output of new owner-occupied housing has slumped so dramatically over the past fifteen years. Starts in the depressed year of 1981, for example, were less than half those of a decade before. Large increases in house prices are now needed to induce any substantial new housing output. The interlinkage between rising house prices, escalating building costs and declining housebuilding is a key indication of a growing crisis in the present structure of owner-occupied housing provision. All three elements feed off each other. Cost increases lower the profitability of housebuilding so builders cut back on their output; over time that increases the shortage of housing which forces up prices; and from the demand side the instability of demand creates production problems for housebuilding which gradually pushes up building costs. There is no single cause for these factors; the problem is a structural one inherent in the present form of owner-occupied provision.

The sources of owner-occupied housing supply

Housing supply can initially be subdivided into new and existing dwellings, although renovation and subdivision of existing structures into flats straddles the divide between existing and new. Completely new houses represent quite a small proportion of total supply. In 1978, for example, new completions were only 15 per cent of the estimated house sales of that year. Supply from the existing stock comes from three major sources. By far the most important is sales by existing owners moving within the tenure; they provided 64 per cent of second-hand supply in

1976 (HPTV I). Another 30 per cent came in the form of houses which previously had been owner-occupied by households that had dissolved or moved to another tenure. Finally about 7 per cent were transfers of dwellings from the rental tenures. Over 80 per cent of house sales in a year, therefore, are likely to be of existing owner-occupied housing. This predominance of existing owner housing has led a number of analyses of owner occupation to treat new housing supply as a relatively subsidiary issue, concentrating instead on the exchange process in the second-hand market. The *Housing Policy Review* of 1977 was the most important of them. Whilst the virtual neglect of new supply helped to justify the Review's support of the *status quo* in owner occupation, its intellectual foundations are poor. What is important in the housing market is not so much the quantitative amount of each source of housing supply but their economic effects, and it is there that the importance of new housebuilding comes to the fore.

A useful way to start examining the economic consequences of different categories of supply is to consider the relation between the amount supplied from each source and the rate of the change of house prices. If the supply of owner-occupied housing was completely fixed obviously only the state of demand would affect price. Supply varies, however, and the different types of supply from new and second-hand sources do so for different reasons. Leaving aside deterioration or improvement of the housing stock, changes in supply depend on the nature of the transaction of which they are part. Sales of existing owner-occupied dwellings and new ones represent different types of transaction. The flow of existing owner dwellings onto the market results from existing owner moves within the tenure and out of it. The selling of houses already in the tenure obviously does not alter the total stock. Their influence as a source of supply consequently depends on what happens to the households that own them. For owners moving within the tenure, the selling of a house is matched by the purchase of another; it is an intermediate link in a chain of sales. Imbalances between supply and demand can be created in this way, however, if the aggregate of moves that owners want to make is not matched by the houses they supply: for example, when more movers want to trade up than to trade down. A similar effect can be generated with the sales of houses of previous owners not buying again. Here the crucial matching is between owners leaving the tenure and those entering it, both numerically and in terms of the types of houses offered for sale and wanted for purchase by new entrants. Adding up the overall effect of the existing owner stock it can be seen that its role as a source of supply is a passive one. Demand plays the active role. House sales influence demand as wealth is realized from those sales. But, even so, the part that existing house sales have in price formation depends on the extent to which they match and influence demand. New additions of housing to the owner-occupied stock, on the other hand, play a more active role in price formation.

New housebuilding provided almost two-thirds of the net increase in the owner-occupied stock between 1914 and 1975, although the proportion has varied over time. In the first two decades after the Second World War, for example, more houses were transferred from the privately rented sector than were built for owner occupation (table 4.1). During the 1960s and 1970s, this position was reversed as the stock of rented housing declined. But, again, what is more important than the proportion is the extent to which their quantity is influenced by the current market price for owner-occupied housing.

The advantage to private landlords of selling off their houses to owner occupiers has been a long-term one (Nevitt 1966). The timing of those sales, therefore, has primarily depended on gaining vacant possession and is only secondarily influenced by the actual state of the owner-occupied market. Throughout the 1970s, therefore, there was a steady transfer of rented dwellings fluctuating between fifty and sixty thousand a year (Boleat 1981). Presumably as the stock declines further this source will gradually dry up. Similarly, the rate of council house sales is dominated by political ideology and by which political party is in national or local power. Council sales, furthermore, are generally to sitting tenants who could not otherwise have become owners, so the transaction is

Table 4.1   Additions to the owner-occupied housing stock, 1914–75

	millions				
	New building	Net addition from other tenures	Demolitions and changes of use	Total additions	New building as % of total additions
1914–38	1.8	1.1	negligible	2.9	62
1938–60	1.3	1.5	−0.1	2.7	48
1960–75	2.6[1]	1.1	−0.2	3.5	74
Total	5.7	3.7	−0.3	9.1	63

Source: HPTV I, table 1.24

[1] Includes conversions

essentially independent of the general market. This means that the only source of new supply which is sensitive to the contemporary state of the owner-occupied market is new housebuilding. So, whilst newly built houses are a relatively small component of annual sales, their influence on price is far higher.

The supply of new housing

Like the market it serves, housebuilding is very volatile. Builders are not interested in holding stocks of houses, then waiting for the next upturn before releasing them. Exorbitant holding costs mean that they build for a quick sale, so housing output closely follows the level of market activity.

The long construction period for housing means that two stages, starts and completions, jointly indicate the level of production. Quarterly data for both series from 1964 to 1982 are given in figure 4.3. The starts series varies more sharply than completions. Once a dwelling is under way it has to be completed eventually, so much of the sensitivity to economic factors is centred on the decision to start building. Completions, none the less, show similar, if smoother and lagged, cycles to starts. Both series in figure 4.3 show fairly regular cycles, averaging three to four years from trough to trough. Since 1963 there have been five cycles (1963–6, 1967–70, 1970–4, 1974–7 and 1977–81) and a downward trend. The downward trend is marked: from a quarterly level of completions averaging around 55,000 in 1964 there has been a gradual decline to around 28,000 per quarter in 1981. Over the period, therefore, private housing output dropped by half. In other words, house prices have not risen nearly fast enough, despite their actual rate of increase, to keep housing output up.

Despite being seasonally adjusted, the data on starts and completions are not entirely weatherproof. Exceptionally bad winters will show dramatic drops in starts and completions (e.g. first quarter, 1979), and good summers will encourage building. The introduction or hoped-for repeal of legislation will similarly affect the series; the sharp rise in starts in early 1967 was an attempt to beat the introduction of the Betterment Levy (Gough 1975a) and, although showing the effects of poor weather, the falls of the first quarter of 1979 were also likely to be in anticipation of the Conservative party winning the 1979 General Election and repealing the Development Land Tax. (It was not repealed, to the disappointment of housebuilders.) Material supply bottlenecks, labour shortages

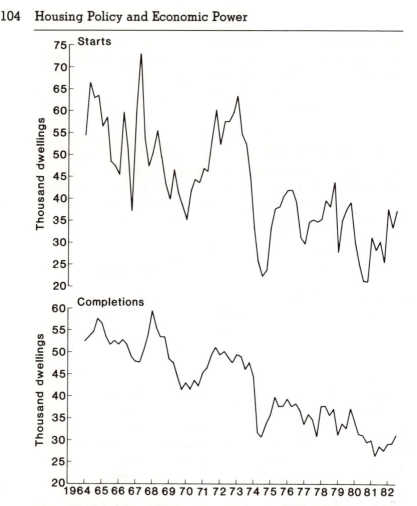

Figure 4.3 Private housing starts and completions, Great Britain 1964–82, quarterly, seasonally adjusted

Sources: Housing Statistics and HCS

and the 1972 building workers' strike will also affect the series (and help explain the peculiar behaviour of starts in 1971 and 1972). Yet the overriding influence on the series will be the immediate economic one of the profitability of housebuilding.

A firm can react in a number of ways to falls in profitability. Reductions in output can take place through cutting back new starts on existing sites, by slowing down the rate of completion of houses already under construction, or by delaying the development of new sites. Each possible option has various time responses and implications for the amount of capital invested. Slowing down completions is a quick short-term response, limited only by the number of houses under way. It does, however, increase the capital tied up in work in progress. Cutting back on starts on existing sites reduces output in the months ahead. Working capital is lowered, yet site overheads still exist. Halting new site development is obviously the longest time response, and could bring into question longer-term expansion plans or even corporate viability. The extent to which each of these responses is adopted, therefore, depends on the expected severity of the downturn and on the firm's financial position.

The sensitivity of output to contemporary changes in profitability appears to be quite marked. The speed with which housebuilders can adjust their output levels means that current profitability is central. Other factors affecting firms' expectations seem only to have a secondary influence on output although presumably a much greater one on land purchases. A comparison of quarterly changes in output and profitability brings this relationship out clearly.

Because either starts, completions or housebuilding-in-progress can be used as measures of activity, an initial choice of output measure has to be made. None in isolation is particularly satisfactory. Even the data problems are controversial. The mid-1970s saw a debate over whether economic factors had any direct effect on construction times; see Duffy (1975) and the reply by Gough (1975b). Duffy argued that once a house was started only physical constraints (weather, supply of materials, etc.) would hold back building. The debate, however, was conducted prior to the slump of the mid-1970s when construction times rocketed, even though substantial excess capacity existed, showing that physical constraints could not be the sole reason.

The problems with the data arise because builders use each stage of the building process to regulate the turnover time of their capital, so no stage in isolation can give an adequate picture of a firm's level of activity. Completions offer a poor indicator of current trends because often they are not closely linked to the time houses are sold. Contracts for purchase may have been exchanged or under negotiation well before the time of completion, especially in recent years as more builders gear their building rates to sales. On the other hand, the completed dwelling may not be sold for a long time, reflecting a previous mis-estimation of market trends. Starts, on the other hand, do not indicate when the dwelling will be completed. A dwelling is officially counted as started as soon as foundation work is begun, yet builders increasingly tend to keep ground floor slabs idle until orders pick up. So an increase in starts may be absorbed in apparently longer construction times. The holding of idle slabs presumably explains some of the dramatic measured increase in construction times over the past twenty-five years. In the mid-1950s the average house took eight months to build, in the 1980s it took twenty months. This indicates either a massive collapse of productive efficiency or, more likely, a weakness in the starts data as a measure of output. On already serviced sites it is possible to build much quicker. One small builder interviewed said he could build a traditional three-bedroomed terraced house from slab in under seven weeks. The new timber-framed methods offer construction times of six to eight weeks from slab. The response at the margin by the industry, therefore, can be rapid.

An improvement on using either starts or completions as the measure of output is to use a combination of both: net starts. Net starts refer to the net addition to houses being constructed, that is total starts minus total completions. It acts as an indicator of new commitments of capital to housebuilding. It is not the same as total investment in building, as additional capital is still required to complete those houses already under construction. Capital, moreover, might also be invested in new plant and equipment or in increasing the size of land banks. Net starts, none the less, show the commitment to new output by firms in the industry. It has the advantage over data on starts alone in that it takes some account of variations in construction times and completions.

A comparison of changes in net starts with the rate of change of house prices indicates the response of housing output to the general state of the housing market. Figure 4.4 shows the situation in the 1970s and early 1980s. The boom of the early 1970s is clearly seen as is the dramatic slump of 1973–4. There is a mini-boom in net starts from summer 1975 to summer 1976, then a sharp drop. The response in the second house price boom of 1978–9 was very limited; when net starts finally picked up they were immediately brought down by the bad winter of 1979 and only revived for another six months. The depths of the downturn were reached in mid-1980, and during 1981 net starts showed a marked revival despite falling house prices. The close fit between changes in net starts and house prices of the early 1970s does not, therefore, appear to survive the mid-1970s slump. This divergence took place because house price rises are only a poor proxy for changes in profitability as they do not take account of building costs.

It is difficult from aggregate data to estimate the actual profitability of

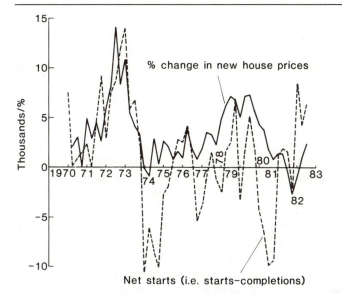

Figure 4.4 Quarterly changes in new house prices and net starts 1970–82
Source: HCS

housebuilding. It is possible, none the less, to get a fairly good indication of how profits are changing. The surplus available for payments for land, builders' interest costs, overheads and profits must come from the difference between the revenue received by builders from house sales and the costs of constructing those houses. This surplus can be defined as gross development profit. So data on the relative change in average house prices and construction costs should indicate whether gross development profit is increasing or decreasing at any point in time. It is impossible to calculate the percentage change in gross development profit as the relative weighting to give to the indices of changes in house prices and construction costs is unknown. Nevertheless an ordinal index indicating the direction of change can be calculated, and this is shown for 1970–82 in figure 4.6. As with housing output this seemingly simple calculation is made complicated by difficulties over construction data. There are no accurate data on how either new house prices or private housing construction costs change in the short to medium terms. Approximations have to be used in both cases – although the resulting index of changes in gross development profitability does seem to be a plausible one. The data problems are discussed in detail in an appendix at the end of this chapter.

Figure 4.5 shows the quarterly variation in the two indices used to

derive gross development profitability. The sustained movement of
house prices has already been discussed but figure 4.5 shows that construction costs also rose rapidly during the 1970s, although they tailed off
in the crisis years of the 1980s. The resulting general picture for gross
development profitability was bleak: short periods of rising profitability
were followed by long falls. One profit slump in the mid-1970s lasted for
three years from 1973 to 1976 and another persisted throughout 1980 and
1981. In fact, there were as many quarters when gross development
profits were falling as when they were rising. The estimate of gross
development profitability is compared with net starts in figure 4.6, where
it can be seen that changes in net starts correspond closely to changes in
gross development profitability. The lag between the two series varies
(three to six months either way). In part this is to be expected: in sustained
upturns the rate of increase in net starts is likely to race ahead of
profitability to avoid missing the boom (non-profit factors seem to have
temporarily choked off the rise in the early 1970s boom); information on
sharp downturns is also likely to take time to filter through the market,
slowing the fall in net starts; and periods of fluctuating profitability are
likely to lead to cautious responses and a closer fit of the two series. The
speculative housebuilding industry, therefore, seems to respond (and be
able to respond) quite quickly to the state of the market.

Figure 4.5 Quarterly changes in new house prices and construction costs,
1970–82

Sources: New house price and private new housing output price indices, HCS

Detailed knowledge of the industry, however, is required to explain all the variations and differences between the two series. The sharp rise in net starts once the rate of decline of profitability started to slow down in 1975, for example, was most probably aided by a readjustment of the mix of supply to demand (houses under construction being related to the housing mix of the previous boom rather than to the contemporary market), and by the large land stocks held by housebuilders (or their creditors) purchased during the earlier boom. Building them out would reduce interest costs. A number of housebuilders interviewed in the survey suggested that many plots of land purchased by builders in the early 1970s could only be built on once prices reached the levels attained in 1978–9.

Changes in gross development profitability should not be treated as the sole factor determining net starts. One reason for caution is that it only considers the main source of profits, development gain, and a substantial cost factor, construction costs. Other sources of profits and costs

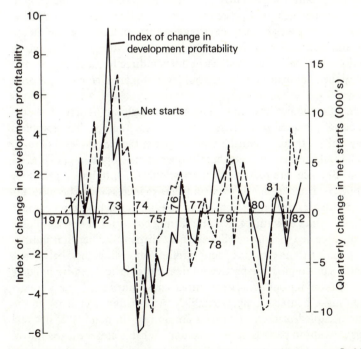

Figure 4.6 Quarterly changes in net starts and gross development profitability, 1970–82

Source: Estimates derived from HCS data

could alter the picture presented by gross development profitability alone. In particular the drain on profits represented by interest charges, taxation including development land tax, and other costs associated with holding and purchasing land have not been considered. The close relationship between gross development profitability and net starts that seems to exist does, nevertheless, tend to discount the claims that have been made that housebuilders have consistently been unable to respond to market conditions during the 1970s because of non-profit-related constraints. Those constraints fall into two main categories: ones associated with demand and ones associated with the supply of building inputs.

The most frequently cited demand constraint is a shortage of building society finance. Most new housing output, however, is covered by special mortgage arrangements between building societies and housebuilders. Most building societies are prepared to 'guarantee' mortgages on housing built by particular builders with whom they have entered into an arrangement, subject to the normal lending rules of the society. In part, these special mortgage arrangements explain why new house sales fluctuate less sharply than existing ones. Mortgage constraints, therefore, are likely only to affect builders indirectly via chains of sales and by their influence on the general level of market activity and price change.

The other major quantitative demand constraint is obviously the number of potential purchasers of new housing output: housebuilders always talk in terms of gearing their output to the current level of sales, which would seem to indicate that the quantitative aspect of sales is overriding. But sales are only made at acceptable (i.e. profitable) prices for the builder. Builders hardly ever cut prices to increase sales because it is not profitable, using instead non-price inducements to encourage demand during downturns. The marketing strategies of housebuilders will be discussed in more detail in the next chapter. But here it is important to note that in practice there is often little difference between a quantity-constrained perspective on demand as articulated by builders ('no-one is buying our houses') or a price-constrained one ('prices are too low to make housebuilding profitable'). As inflation is endemic to the owner-occupied market, price rises are normal (in nominal if not real terms), except in periods of severe slump. What is at issue is the rate of price change. During periods of sluggish demand, prices tend not to rise fast enough to maintain profitability. In downturns, therefore, builders cut the quantity supplied not its price (yet obviously perceive the problem as a quantitative lack of demand), whereas in upturns both prices and quantities tend to rise.

Housebuilders tend to operate a two-stage strategy over fixing the selling prices of their houses. At the time when the decision is made to start a phase of a development, expected selling prices are fed into the calculation of whether it is profitable to start building. Those prices will then tend to be a minimum below which the builder will not go. However, as building proceeds prices may be raised if demand for the houses is strong or if house prices in general are rising. If costs rise, on the other hand, wiping out the profit at the original minimum price whilst the market cannot sustain a compensating rise in price, building will be slowed down or brought to a halt. Over the long term, consequently, builders are dependent on increases in demand to sustain the price rises necessary to ensure a good development profit in the face of rising construction costs.

Comparison between the price of new houses and existing houses is made difficult because of variations in the mix of house types between the two (i.e. the data are not quality constant). Yet some interesting changes have occurred between them over the past ten years. In 1970 the average price of new and existing houses was virtually the same; by 1980 a significant gap had opened up with new house prices approximately 13 per cent higher. As figure 4.7 shows, a major shift took place after the first house price boom of the 1970s. For the three years prior to 1973 the

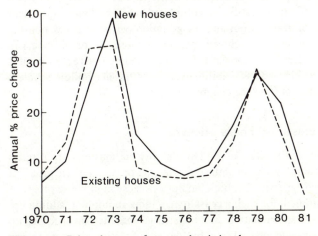

Figure 4.7 Price changes of new and existing houses, per cent per annum, 1970–81

Source: BSA Bulletin

change in new house prices followed that of existing houses; since then new houses have been the market price leader. This change has occurred for two reasons. The first is related to an increase in the quality of new houses with a general move up market by builders, which will be discussed later. Second, there has also been a genuine change in price formation by housebuilders associated with the radical transformation of the industry that occurred in the 1970s. Firms dominating the industry are now market orientated so they are far more responsive to market signals than before, including a more rapid response to conditions favourable for price rises.

The other types of non-profit constraint are associated with the supply of building inputs. Shortages of building materials and labour have constituted important constraints during certain periods: the problems during 1972 have already been referred to. Such shortages in general, however, do not take the form of a quantity constraint; limited availability instead forces up input prices, affecting the profitability of development. The methods of production adopted by speculative housebuilders are a major cause of such shortages as they have generated a crisis of production (see chapter 6).

Land availability is often cited as the most important physical constraint facing housebuilders, caused by insufficient land release by planning authorities and delay in the granting of detailed planning permission. Despite the widespread belief in this role for land availability it is difficult to justify. Most builders hold land banks equivalent to a number of years output and the gestation period of a new housing development is usually eighteen months or more. As price booms in general rarely last much longer and are followed by severe slumps, it is difficult to see how land availability over the course of a building cycle is a major constraint (also see chapter 8).

Do land prices affect house prices?

The question of the effect of land availability on housing output raises the equally thorny problem of the effect of land prices on house prices. It is frequently suggested that high land prices cause high house prices. A common argument for such a causality contains an implicit 'adding up' theory of price determination: in it house prices are the sum of adding up land costs, construction costs and builder's profit. Ricardian rent theory and its modern variants would dispute that conclusion by arguing for a residual view of land prices. Residential land prices, it argues, depend on the profitability of housing development. When profits rise builders bid up the price of land and vice versa, so land prices do not cause house price rises but are a residual consequence of the level of house prices relative to construction costs.

One plausible variant on the Ricardian view of rent goes some way towards the conclusions of the 'adding up' theory. It accepts the broad mechanisms implied by the view of land rent as a residual but, in addition, confers the existence of a monopoly power on landowners. This power enables landowners to restrict the supply of residential land over time, then the housing shortage consequently created forces up house prices and so enables higher land prices to be charged. The question then is whether landowners actually have such effective power.

There are many potential sites that can be developed for housing, so the ability of landowners to limit the supply of land to builders and to continue holding off that supply for long periods of time is essentially an empirical question rather than a theoretically demonstrable one of universal validity. In modern Britain certain conditions may be said to exist to create such a situation. Landowners, for instance, only rarely are in an economic position where they are forced to sell at a particular time. Land usually is a debt-free source of wealth to them, and they are not land-dealing capitalists. So regular returns from land dealing are not generally necessary. Strongly restrictionist views of the planning system can also be added to the argument to reinforce the apparent strength of a limited number of landowners with land designated for development.

Examination of changes in new housing supply suggests that such views are not as well founded as they might seem. The change in the price of residential land is dependent on the change in gross development profitability rather than the other way round. This can be seen by comparing changes in development profitability with changes in land prices, which is given in figure 4.8 for 1970–81. Despite the fact that land price data are available only on a six-monthly basis and that both series are derived on different sampling criteria, there is a remarkably close lagged relationship between changes in development profitability and residential land prices. It actually appears that when their profits rise builders do buy more land, forcing up its price, and when profits fall they buy less, apparently validating the Ricardian residual view. The percentage change for land prices is much greater than for gross development profit, presumably because land represents only a proportion of total house price. If the residual view of land price is correct the proportion of house price which corresponds to current land price should

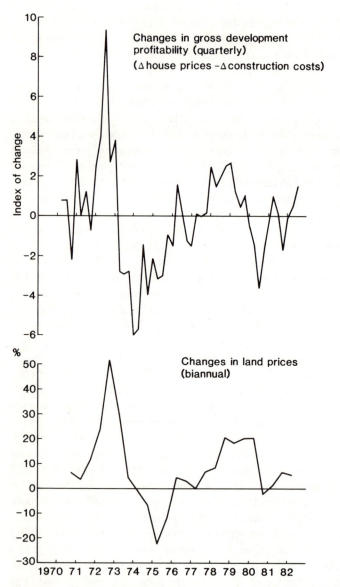

Figure 4.8 Changes in gross development profitability (quarterly) and land prices (biannual), 1970–82

Source: HCS

also vary in line with the building cycle. During periods of rising development profitability the proportion of house price represented by that profit will rise, and vice versa. So the share of house price attributable to land prices should also rise and fall with development profitability booms and slumps. Table 4.2 indicates that this is broadly so for the 1970s and 1980s. But it should be remembered that land price refers to land actually traded during that time period, whereas with land banking virtually all of the land used for building will have been bought in previous years. So the actual cost to the builder of the land built on may have been far less than the shares shown in table 4.2. Finally, the table also shows that land prices have not risen in the long run as a proportion of house prices as would be expected if land shortages were the principal cause of the long-term upward trend in house prices.

Whilst the addition of a severe land restriction rider to the Ricardian theory of rent does not seem to be empirically valid for the modern British context, the unadulterated Ricardian view itself has theoretical weaknesses. Its hypothesis may fit the data presented so far, but it fails to answer the vital question of how housebuilders have managed to avoid landed property exercising a monopoly power over land price. Answering this question destroys the validity of the passive, residual theory of land prices.

Though it may empirically be true that landowners do not have absolute monopoly power over residential land, they are still, none the less, involved in a struggle with housebuilders over the price of development land. The actual market power of landowners, therefore, is contingent on the specific empirical forms taken by the struggle. To conceptualize the struggle in terms of the categories being used here, landowners and

Table 4.2   Land and housing prices, 1970–82

	Average land price per plot (£)	Average land price as % of new house prices
1970	908	17.7
1972	1727	22.0
1974	2663	23.5
1976	1848	13.7
1978	2376	13.4
1980	4460	16.4
1982	5202	20.4

Source: HCS

builders are fighting over the conversion of gross development profit into land price. Landowners try to wait for the best market conditions before selling their land. There have also been periods when they have almost wholesale held off supply, hoping to change market conditions in their favour and at the same time use their undoubted political weight to get political action over their 'plight'. The most well-documented cases of such a collective response are the periods of land legislation associated with successive Labour governments in 1947, 1967 and 1974 (Cullingworth 1979b). Yet at a local level they could easily be more common. Builders, on the other hand, organize their development and production processes in ways which facilitate their retention of development profit at the expense of the landowner (see Chapter 5).

This distributional struggle has wider implications. As production methods affect construction costs, attempts by builders to stop the conversion of development profit into land price may affect production methods and so indirectly raise the price of housing. The result is that the Ricardian theory of rent must theoretically be rejected, since it is invalid to assume, as the theory does in fixing the parameters determining the residual profit, that techniques of production are determined prior to the distributional struggle over rent. They are instead a crucial element of that struggle. This theme will be taken up in more detail in later chapters as theoretical rejection of the passive Ricardian view is central to understanding the housing land development process and the relations in it between housebuilders and the other principal agents: landowners, building workers and the state (in its role as land-use planner).

The movement up market by housebuilders and its consequences for owner occupation

House price inflation has led to a growth in the market power of previous owner occupiers by increasing the significance of owners' money gains from housing. Housebuilders have recognized this shift in purchasing power and altered their product accordingly and moved up market. This is a long-term trend rather than a temporary phenomenon, yet even so there are cyclical effects. The purchasing power of previous owners is greatest during booms as this is the time when the largest amount of money gain is realized, so during the following downturns first-time buyers tend to increase in market importance. If large-scale builders concentrated on only one section of the market they would face severe problems because of such shifts in the composition of demand. The move up market, therefore, has been associated with an increased range of houses built, and changes in emphasis on different parts of that range during the phases of the short-run housebuilding cycle.

The extension of the range and the growing importance of up-market houses is clearly shown in house price data over a ten-year period (figure 4.9). Switches in emphasis with the cycle can also be seen: the slump of the mid-1970s saw a move back towards the first-time buyer as the strongest sector of the market, the upturn of 1978–9 witnessed a marked shift up market and the early 1980s (not shown) a re-emphasis on the lower end. The nature of the existing stock in particular regional markets plays an important part in these changes, but the implications for the industry and the planning system are great. High-quality houses need higher-quality sites. Often this quality cannot be created in the development process but instead puts pressure on 'high status' neighbourhoods and on the green belt.

Some data are available to illustrate the change in the physical characteristics of newly built houses associated with the changes in types of houses built by housebuilders. What is being squeezed by the move up market is the lower to medium type of dwelling, that is, two- to three-bedroomed semi-detached houses. Semis dropped from 35 per cent of new houses built in 1969 to only 24 per cent in 1980; land-hungry bungalows showed a similar percentage fall (27 to 14 per cent); detached, on the other hand, rose rapidly from 22 to 41 per cent of the market. A similar situation has occurred with number of rooms: from less than 4 per cent in the early 1960s, four-bedroomed houses constituted over 20 per cent of the market in 1979 (source: HCS).

Floor-area data bear out these changes whilst highlighting the importance of the increased range of house types. The 1970s saw a significant reversal of the trend of the 1960s towards standardized new house sizes. By 1970, as table 4.3 shows, 61 per cent of all new dwellings had floor areas of between 750 and 1000 square feet. But by 1978, almost 30 per cent of new dwellings were being built with floor areas less than 750 square feet whilst there was also a slight increase in the larger floor-area categories, indicating the construction of a much wider range of properties. Yet total floor areas have not increased substantially despite the trend to up-market developments, so generally rooms have got noticeably smaller.

In terms of the aggregate owner-occupied stock, the cumulative impact of building new houses of lower average floor sizes, combined with a

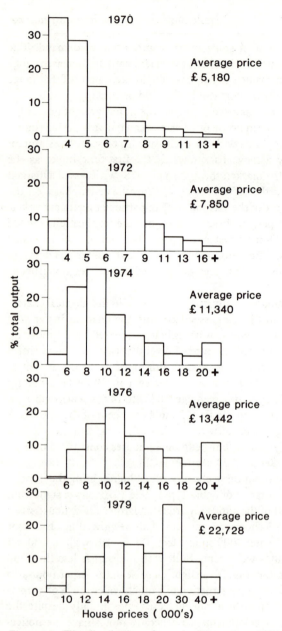

Figure 4.9 The distribution of new dwelling prices, 1970–9
Source: HCS

growth in conversions of existing houses to flats, resulted in an appreciable fall in the average floor size of the owner-occupied stock between 1970 and 1978 (see table 4.3). New entrants to owner occupation have been worst hit by these declining space standards. Because of rising house prices first-time buyers have tended over the years to buy increasingly smaller, lower-quality dwellings. The percentage of new houses, for example, bought by first-time buyers dropped from 63 to 41 per cent between 1970 and 1979. Older and, most probably, lower-quality housing is being bought instead. There was also a marked switch away from the classic three-bedroomed, semi-detached house to smaller dwellings: 42 per cent of first-time buyers bought semis in 1970 but this had slipped to only 32 per cent in 1979.

The downward shift by first-time buyers obviously left more of the higher-quality stock for trading up by previous owners and relative price changes have aided these realignments of households within the stock. Rising house prices do not affect all houses uniformly. During price booms the type of house wanted by those with the greatest increases in purchasing power (i.e. existing owners benefiting from rising money gains) rises fastest, unless relative supply changes match the changing composition of demand which, given the long supply lags, is unlikely. The increasing differentiation during booms might be reversed slightly during downturns when lower-quality structures may have the fastest price increases, as undoubtedly occurred during 1974–6; yet the overall long-term trend is towards increased price differentiation within the stock. So over time the additional housing quality achieved by the same

Table 4.3 Floor areas of owner-occupied dwellings, 1962, 1970 and 1978

	% of totals				
	New dwellings			All dwellings	
Superficial sq. ft	1962	1970	1978	1970	1978
Under 750	26.6	15.9	29.5	17.9	24.5
750–999	53.5	61.0	42.5	53.3	46.7
1000–1249	14.6	15.6	16.3	17.5	16.5
1250–1499	3.4	5.2	7.0	6.3	6.6
1500–1749	1.1	1.5	2.6	2.6	2.6
1750+	0.8	0.8	2.0	2.4	1.9
	100	100	100	100	100

Source: Nationwide Building Society

real extra expenditure has fallen: 'trading-up' is not what it used to be. It should also be noted that builders are responding to a clear and, at least for existing owners of middle- to upper-range houses, worrying market trend. The market power of existing owners is increasing, hence the move up market, but the times when they can use that power are becoming increasingly volatile and uncertain. It depends on upturns in the market. First-time buyer demand is more steady, their bread-and-butter as one volume housebuilder put it; hence the switch back towards it during downturns. Existing owners in these market sectors, therefore, are increasingly likely to be unable to sell their houses (i.e. move) when they want to. Such problems are likely to encourage their interest in housing reform.

Conclusions

New housing output is too small a proportion of the total stock and it takes too long to build for it to have a significant impact on short-run changes in house prices. A lack of new housing output, nevertheless, has helped create the conditions in which demand has pushed up prices in line, and sometimes faster, than increases in incomes. The contemporary housing situation in Britain as a result has certain broad similarities to the situation prevailing before 1914 when private renting was the dominant housing form. Once again there is a reluctance by the state to provide public housing and, in the absence of a political response, living standards are again being held back by problems of new supply in the private housing market.

It has not been profitable for builders to provide more houses. In fact, house prices during the 1970s had to rise by such large percentages to induce the new housing output that was produced. To put it bluntly, the private housebuilding industry has been completely incapable of providing enough new houses to keep prices down. And prices did not rise fast enough to stop output dropping dramatically. Explanations for the lack of housebuilding profitability do not have to be searched for in interest costs or in land prices; central instead has been a growing crisis in housing production, one manifestation of which has been escalating construction costs.

To argue that the lack of new owner-occupied housing is a root cause of problems in the tenure is not to suggest in a moralistic fashion the need to reapportion blame to speculative builders. The crisis of production is a structural effect of the current form of owner-occupied housing provision, not a mysterious inefficiency of supposed unscrupulous speculators. To conclude, furthermore, that there are problems in the building industry is only to open a new starting-point. Justification for such a crisis cannot remain based on the relative movements of a shaky dotted line called the 'estimated change in construction costs'. The workings of the housebuilding industry must be considered in more detail.

The notion of the structure of owner-occupied housing provision as a set of relations between social agents associated with the stages of transferring a greenfield site into an inhabited dwelling gives a method by which to proceed. The relationships between social agents at each stage of the housebuilding process have to be considered. Discussion of land prices and the theory of rent in this chapter also highlighted an issue that must be pursued further. There is a direct struggle over the profits of land development between housebuilders and landowners, and there are subsequent effects on housebuilders' relations with the other agents involved in production, especially workers, and with the state land-use planning system. Each will be dealt with in turn in the following chapters.

Appendix: Estimating changes in gross development profitability

Changes in gross development profitability are calculated by comparing changes in new house prices with changes in private housing construction costs.

The problem with the new house price index (and the same difficulty arises with the all-house price index as well) is that it is not a true price index in which quantity is held constant.[1] 'Quantity' in housing is not simply a function of the numbers of units constructed but also of their quality. 'Quality' is an open-ended, immeasurable term yet it broadly corresponds to house features such as floor area, number of rooms, internal fittings, insulation and so on. These change over time for two reasons: because of longer-term trends in the production of houses and through short-term variations in the composition of houses produced. Overall trends in the quality of housing are not of concern here but short-term output composition is a problem.

There are two problems with the composition of output. First, there are variations in the regional composition of output. House prices differ between regions and, although national trends dominate, there can be significant differences in both regional output and price movements (Neuberger and Nichol 1975). If, for instance, proportionally more housebuilding is done in the higher-price South East region compared to the previous quarter, the unweighted index would mistakenly indicate general house price inflation as a result. Regional variations are unlikely to vary directly with the housebuilding cycle so the distorting effects on the index will not show any consistent pattern. The second problem concerns the types of houses built and this one is related to the housebuilding cycle, as this chapter showed. During upswings in the housing market housebuilders find it more profitable to sell a greater proportion of middle- to upper-market dwellings than during slumps or periods of stagnant activity when more lower end of the market dwellings are produced. As the sample data used to calculate the house price index are not weighted for such quality variations, house price rises during booms and falls during slumps are exaggerated. The size of the overestimate is difficult to quantify but the marked cyclical shifts in output composition described in this chapter (cf. figure 4.10) suggest that it is substantial. The advent of large-scale bank mortgage lending in the early 1980s further complicated the biases of the price index for those years. The house price index is based only on building society mortgages, and banks tend to lend at the upper end of the market, so the decline in prices in 1981 is likely to be exaggerated by the index. Any calculation of changes in gross development profit, because of the problems with the house price index, is an overestimate. These effects, however, are partially offset by differences in the size of development profitability across house types, which fluctuates directly with the housebuilding cycle, as the proportion of total house price represented by development profit is greater for more expensive dwellings (see chapter 5).

Accurate information on construction costs is notoriously difficult because of the heterogeneous nature of building work (Fleming 1966 and Butler 1978). The problem is now worse for private housebuilding than for most other construction sectors. Cost changes to the purchaser of construction products depend on changes in wages, the price of materials, the ratios in which those materials are used, on productivity in the building process and on the size of building firms' profit mark-ups. Until recently official statistics calculated only a global construction cost index, called the Cost of New Construction (CNC). A fixed bundle of labour and materials was assumed and price variations for each component of the bundle fed into the index, and periodic crude adjustments were made for changes in productivity, overheads and profits.

Calculation based on:

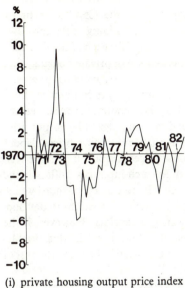

(i) private housing output price index

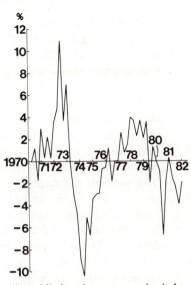

(ii) public housing output price index

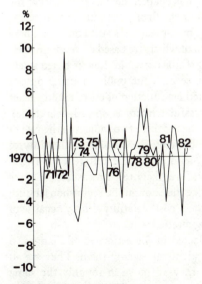

(iii) price index of public sector housebuilding

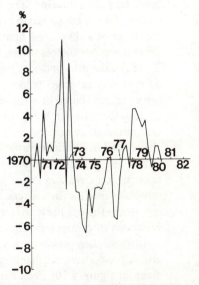

(iv) Cost of New Construction Index (ceased publication 2Q 1980)

Figure 4.10 Alternative estimates of changes in gross development profitability, 1970–82

Source: Social Trends, 1980, 1981

Because of criticisms over the approach of the CNC to technical change and variations in profitability with the building cycle, the CNC was superseded in 1978 by output price indices for each of the five main sub-categories of new work. With the exception of private housing, they are now based on weighted data from tender prices, avoiding objections to the CNC (Butler 1978). Tender prices only apply to contracting, however, not speculative building. For private housebuilding consequently a compromise approach was adopted to derive an index. The output price index (OPI) is a weighted composite of changes in house prices, labour and materials: roughly 50 per cent house prices and 50 per cent the rest. The materials and wages elements try to catch input cost inflation and the house price element changes in profits. It is a very crude index which by including house prices manages to include the variations in development profitability that this chapter wants to consider. However, if the difference between house prices and this OPI is taken this double-counting should balance out and a damped variation between the rest of the index and house prices remain.

The other options to use for an index of construction costs are measures based on public housing. The previous chapter showed that neither the composition of construction tasks nor the firms operating in that sector are the same as in private housing: but at least it is still housing. There are two possible indices to use, both of which are based on tender prices. The first, the price index of public sector housebuilding, is based solely on currently accepted tenders; the second, the public housing output price index, is based on a distributed lag function of current and earlier tenders (because work from successful tenders is spread over time). Unfortunately, when comparing either of these two indices to the rate of change of house prices, variations between the current profitability of public and private housebuilding, as much as the difference between construction costs and house prices, are likely to be measured. This is especially true for the unlagged price index of public sector housebuilding, where there is likely to be marked 'cost' volatility simply because of variations in contractors' tendering strategies.

All these data problems are reflected in the estimates of changes in gross development profitability made using each of them. They are all given in figure 4.10. Fortunately they all move in roughly the same direction. The one based on the price index of public sector housebuilding is the most maverick, but as expected its distinctive behaviour seems to be related to relative sectoral profitabilities. It diverges sharply above the other measures during 1974–5 and also in 1981, reflecting the squeezed profit margins in council housebuilding of those times. (The measure based on the public housing output price index moves in the opposite direction in 1974, presumably because of the weighting in it of high-priced tenders from earlier years.) The two measures based on the private housing output price index and the CNC index generally move closely together. This is not surprising as the major components of change are the same: different weightings of materials and labour costs. The CNC index reflects more closely materials and labour cost changes. This presumably accounts for the sharp divergence in fourth quarter 1972, reflecting the generous wage settlement at the end of the 1972 building workers' strike, and similarly for 1976 when materials prices rose rapidly. (The materials index by itself rose by 23 per cent in 1976.) A reverse weighting effect seems to have occurred in early 1978.

Explanations of the differences between the indices do not seem to require any statements about productivity change. But some of the indices should more accurately reflect productivity increases than others. So to hazard a guess that productivity improvements were non-existent in their effect on construction costs looks reasonable.

Of the four possible measures of development profitability the two based on the CNC index and private housing output prices seem best as they avoid the relative sectoral profitability problem. As the CNC index ceased publication in 1980, the private housing output price estimate was finally settled on.

5

Housing development and land dealing

The development process in speculative housebuilding

The development process involves a number of interrelated, but temporally separate, activities: the initial purchase and assembly of land sites, the conception of housing schemes, the determination of the time of building and finally the selling of completed houses. It consequently wraps itself around the actual activity of housebuilding. This particular division between development and production is unique to modern speculative building. It differs sharply for instance from development in public housing where housebuilding is generally for use rather than for sale, creating its own relation between land assembly, project design and execution. The reason for the unique division, of course, is that speculative builders are part of a specific structure of housing provision.

An immediate concern of housebuilders is to generate at least a minimum positive level of cash flow, because development and production times are long and the market for their product is variable. In reaching their financial objectives firms face conflicting pressures over land acquisition, sales revenue and production costs. The outcome is an overriding concern with land banking and sales rates per site, combined with an attempt to have a diversified market presence and low production rates of fairly standardized house types on individual sites.

At first sight, building firms may seem solely passive reactors to external forces. This is certainly how the problems of speculative builders ideologically are treated within most political and polemical debate. Yet such passivity is misplaced as housebuilders' drive for profit plays a determining role. The way in which each housebuilder arranges its development process determines the firm's individual relation to those external pressures and, in total, firms' actions help to determine the characteristics of the markets in which they operate. So to talk about the development process is not simply to discuss detailed questions of marketing houses, company administrative organization and land banking, but also to describe the concrete forms of a crucial aspect of the structure of owner-occupied housing provision.

There are three particular issues of wider concern about the development process on which this chapter will concentrate. The first is the marketing-orientation of speculative housebuilding. Marketing centres on the conception and selling of housing schemes and is one aspect of the speculative housebuilder's merchanting orientation noted earlier in chapter 1. Another central component of the development process is land acquisition, which again is a merchanting function, this time of land dealing. So the development process is dominated by merchanting activities existing at both its ends as a stage in housing provision. Recent changes in the housing and land markets have heightened the importance of these merchanting activities, which help to explain the transformation in the firm structure of the industry outlined earlier.

One element of marketing obviously is associated with the design of houses. Consideration of design makes it possible to address a further question which constitutes the second major concern of the chapter: to what extent can private housebuilders be said to satisfy the housing needs of their house purchasers? Although only new housing will be examined here, this question more generally is central in discussion of the current structure of owner provision, for it is often claimed that the success of owner occupation means that the sorts of housing existing in the tenure are ideal types of physical house provision. Higher standards than they provide for households of particular income levels, for instance, are often claimed to be an unnecessary luxury. Discussions of market exchange and need can very easily become vacuous and metaphysical so the analysis will be limited mainly to certain empirical aspects of design. None the less some interesting conclusions about needs can be derived.

The final topic of the chapter concerns housebuilders' land banking and land dealing operations. They are the means by which firms ensure that the land development profit made in the rest of the development process is not whittled away in the price they have to pay for land. The social relation between building firms and landowners is embodied in the purchase of residential development land. The two are involved in a struggle over the appropriation of the financial gain from residential development, even though the form of that struggle frequently does not seem one of conflict but rather of harmonious market exchange. The way in which speculative builders organize their development process helps to generate the appearance of harmony whilst ensuring that much of the gain remains as their profit. To talk of the nature of a speculative builder's development process therefore is to talk of that struggle.

It might seem odd that the selling of houses is being classified as part of the development process. It is there because with the control of sales more of the development profit accrues to the firm undertaking house-building rather than its having to be divided between an estate developer and a builder. The total profits from such unified development also are generally higher because the overall scheme has a greater design unity, and hence is more marketable, and can be timed more readily to fit in with variations in the state of the housing market.

The advantages of controlling sales were brought out clearly in the housebuilder interview survey, especially in a number of cases where it was found that banks were directly undertaking housing development. This unusual role for banks arose because of the mid-1970s house-building slump when banks were receiving land as collateral for the unpaid debts of bankrupt builders. Some banks realized that it is better to develop housing schemes rather than sell the land or semi-completed sites. In one instance of this activity discovered in the survey, a bank contracted out the actual design and building work to a speculative housebuilder, but in another case there was more direct involvement with a member of the bank's management running the sixteen site remnants of a failed firm. These apparently drastic measures are extreme cases of a general tendency for merchant and secondary banks to advance loans in return for a share of the development profit instead of on a fixed interest basis (EIU 1975). One firm during the early 1970s, for example, usually paid 75 per cent of a house's gross profit directly to their financiers (interview). As larger firms have increased in importance over the past decade this type of banking activity correspondingly has diminished in importance as fewer builders now face such financing constraints.

The type of development process being discussed in this chapter is typical of larger firms. Small firms do not have identifiably distinct development processes because of their scale of operation. Development and building for them generally occur simultaneously as they tend to build a few houses only and do not hold substantial land banks. Most of a small builder's profit, none the less, is derived from the conversion of land use to housing so their profit is still primarily a development one. But the importance of land development profit does not significantly influence the way in which they organize their activities. It is when firms have land banks that the development process gains its overriding significance.

Marketing and the organization of the development process

The way in which firms structure their development activities primarily depends on the nature of the two markets they have to operate in: the owner-occupied housing market and the land market.

The owner-occupied market can be divided into three broad sectors and this categorization is given in table 5.1. There obviously are important regional variations in house prices (and to an extent in dwelling characteristics) but the definitions do show the main criteria distinguishing market sectors: they are floor space, residential density and house type. Within this basic categorization there are important sub-categories

Table 5.1 Owner-occupied housing major market sectors

Sector:	
Lower	Houses up to 800 sq. ft
	Bungalows up to 700 sq. ft
	Mainly semi-detached, linked or terraced
	Small flats up to 700 sq. ft
	Normal density 12–14 per acre (30–5 per ha.) and upwards; higher for flats
	Selling at less than £15,000 (1978 prices)
Middle	Houses 800–1000 sq. ft
	Bungalows 700–900 sq. ft
	Mainly detached or linked-detached
	Density 10–12 per acre (25–30 per ha.)
	Selling at £15/20,000 (1978 prices)
Upper	Houses of over 1000 sq. ft
	Bungalows of over 900 sq. ft
	Almost invariably detached, often with space for double garage
	Luxury flats and apartments
	Density 4–8 per acre (10–12 per ha.) higher for flats
	Selling at more than £20,000 (1978 prices)

Source: HBF/DoE (1979)

related, for example, to the standard of fittings and finish ('basic' or 'quality') or to the style of design ('traditional' or 'modern').

Each house built by a speculative builder is aimed at one of these sectors. Most large firms do detailed market research of local housing markets to identify gaps in the current supply where the relative price of particular sub-sectors is comparatively high and potential competition low. Such gaps depend partially on contemporary national market trends and partially on local demand and supply imbalances. Land acquisition is undertaken on the basis of these marketing projections. Land purchase is consequently the product of a dual process of market evaluation (market sector identification and estimates of expected house prices and building costs) and site evaluation (land prices, site development costs and site suitability for a particular housing sector). Both aspects influence each other: some sites are ideal only for particular sectors, and the sector in question could be oversupplied; conversely, many sites can be moulded to suit sectors and larger sites can be made to cater for several through careful design and landscaping. The trend towards the 'new vernacular' design style has aided the ability to supply more than one sector from a site by increasing the visual autonomy of small groups of houses. In one scheme visited by the author, for example, a virtual new town had been created over the past twenty years, yet even so the housebuilder was trying to mould the most recent development to an isolated rural style. Country names, a 'village' green, rustic fencing and even a gaily painted haycart were all part of their marketing tactics.

The number of house sales per site is generally low, averaging ten or less a month. The variation in the rate of sales depends partially on the inherent marketability of a site, on the overall state of the market and, especially, on the sector at which the development is being aimed. In the experience of one builder interviewed, approximately ten sales per month from a site could be expected for lower-range houses, six to eight per month for middle ones, and only four per month at the upper end of the market. In general, the larger the site the longer will be the development time; so one thousand starter dwellings on one site would for example take eight years on average to sell. It is common to find builders producing on a site for decades. Lovell Homes, in a well-publicized development near High Wycombe, took seventeen years to build 1070 houses (*The Housebuilder*, February 1981). Sales rates of these magnitudes usually make it imperative that the rate of housebuilding is geared closely to the rate of sales, to avoid tying up large amounts of working capital on site. They also mean that a significant proportion of a firm's land bank consists of plots on sites already being developed.

In financial terms, each site can be treated as being at a particular stage in its contribution to a firm's cash flow and profitability. There will be holding costs associated with the land's purchase price and with site development. Development costs can be substantial as the ground needs first to be cleared and levelled, and then serviced with roads, sewers, electricity, gas and water before housebuilding can start. One firm interviewed in 1979, for instance, spent £2m. on installing an access road to one of their schemes. Another suggested that total development costs ranged from £5000 to £45,000 per acre depending on ground conditions and topography. Initially, once development is started a site imposes large demands on working capital with little or no revenue from sales; after a while the site starts to generate a positive cash flow as sales revenue begins to outweigh additional costs, and eventually working capital is only tied up in houses as they are built. Each site, therefore, exhibits a clear time profile of costs and profits to the builder.

Revenue from a site is highly sensitive to quite small absolute changes in sales rates. So builders are constantly examining individual site selling rates and have to judge whether a downturn or upturn on one month's figures is a temporary aberration or a sign of longer-term market trends. On most sites prior to the start of building, firms have only a rough idea of how well the houses will sell, despite their market research. They usually try out a mix of house types on the first phase to see which ones sell best, and adjust the house mix of the rest of the scheme on the basis of that experience.

Uncertainty over sales on a new scheme influences the way it is built and the policy over selling prices. Generally two or three show houses are erected quickly and furnished to give the estate that almost lived-in look, then some firms wait for definite sales before building any more houses. The first houses built are at the entrance to a site to avoid the image of selling in a dubious wasteland. Builders often price the initial few sales relatively low to get some households to move in and then push up prices on later sales.

If site sales are poor rarely are prices cut except in the worst of slumps. In part, this is because of the anger generated in early purchasers who face an equivalent money loss and so the project gets a bad name. More important, the main effect of price cutting is a slight increase in the immediate flow of revenue, yet the loss of profit implied by such a cut could easily outweigh the additional costs of holding onto the completed houses until the market picks up. If building rates had been geared closely to expected sales the holding costs anyway should be low as they are related only to a few completed dwellings. Price cutting, nevertheless, does take place at the bottom of severe slumps when many firms are desperate for immediate revenue. In the mid-1970s slump firms in some cases lopped up to £1000 off selling prices. More common than reductions in actual price is allowing real prices to fall through holding actual prices steady when costs are rising. With either pricing strategy a firm's expected development profit from a site drops or, if the project is a big failure, large losses are made. One firm interviewed lost £3000 per house on an unsuccessful development, building the site out as quickly as it could in order to minimize the overall drain on profitability.

To avoid such disastrous situations firms use non-price inducements to encourage potential purchasers. If the problem is site specific, extras can be added to the houses, more landscaping undertaken or selective advertising campaigns used. If the overall market has turned sour, quota arrangements with building societies increase in importance. Some firms deposit short-term loans with societies short of funds if extra mortgages are allocated to their housing estates. Other ploys can be used which can be highly cost effective if picked up by the national media: some firms give cash rebates, free domestic appliances or petrol; others offer cheap mortgage interest rates for the first year; another introduced an unemployment insurance scheme whereby mortgage repayments would be paid if the purchaser was made redundant within two years of purchasing one of their houses. Such marketing methods have grown in importance with increasing market instability. For large housebuilders their cost is negligible and the return can be high. The most expensive inducements are 'we'll buy yours, if you buy ours' schemes, where firms buy and resell purchasers' existing houses during market downturns. Many large firms adopted this practice in the early 1980s, although some had to withdraw as sluggish sales of these second-hand houses began to absorb too much working capital.

As the number of house sales per site per month falls in a narrow range, the total number of houses built and sold by a firm is primarily a function of the number of sites under development. Output therefore is increased by spatial product differentiation. This may be achieved by building at more than one locality within a local housing market; especially if the sites are geared to different market sectors. But there is a point beyond which a firm cannot significantly increase its share of one market, so then output can only be increased by operating in more than

one area. Building on a number of sites has the added advantage of minimizing the impact on a firm's revenue of the possible failure to get good
sales rates on particular sites. As a general rule, the larger the firm the
more sites on which it is building and the greater is its geographical
spread. The need for product differentiation adds a spatial dimension to
land holdings. If, for example, a firm has a large number of plots in its
land bank but they are primarily located on a few large sites the firm still
needs to acquire land at new locations despite the apparent size of its
land holdings.

Whilst the spatial distinctiveness of each site enables firms to differentiate their product, the same characteristics also induce competition.
Builders with large sites at good locations have an overriding influence
on development in a local market. Competitors with poorer placed sites
in the locality catering for the same sector either have to build out their
projects rapidly before the competition comes on stream, hold onto the
site for many years whilst the better placed one is built out, or hope that a
major house-buying boom emerges. A similar position is created if one
firm manages to acquire a site in a location cheaply, either by having had
it in their land bank for years or through a fortuitous contemporary
purchase. They then set the price and pace of development by generating fears in potential competitors of undercutting their selling prices.

Fear of damaging competition intensifies the need for housebuilders to
get their calculations of future demand right and to spot gaps in local
housing markets carefully, especially where the gap is a narrow or
specialist one. Many firms prefer sites with a commuter type of demand
as this widens considerably the potential catchment area of purchasers,
making the market less precarious. One firm interviewed argued that the
best sort of site is one appealing both to commuters to a larger metropolitan area and to local workers. Each source of demand then counter-
acts any adverse effects that might arise with the other. If, for example,
local employment declines commuter demand may take up the slack,
and if rising transport costs put off commuters local demand still exists.

Spatial proximity need not necessarily lead to fierce competition
between builders. The ability to differentiate their product has made it
feasible and increasingly popular for builders to operate in consortia on
new large schemes. These consortia may take the form of a number of
builders formally combining to take on a large scheme, then jointly
applying for planning permission and providing the funding for the
basic infrastructure. Alternatively, the initial purchaser may sell off part
of a large site with existing permission to other builders and use the

profit from the sale to offset the development costs. Either way the probability of gaining planning permission is increased (so consortia are particularly popular for large tracts of 'white' undesignated land), and the initial investment in the development is spread between several builders.

With consortia the time required to build out a site is considerably reduced (possibly by years) as large sites are divided up into schemes built by distinct firms. As some sites have up to 10,000 plots this effect is often important. To avoid competition consortia consist of firms who have managed to differentiate their products from each other. Sales rates for individual firms are then similar to those achieved in their freestanding developments. The joint marketing of the site also generates additional demand at lower marketing cost.

The development process and company organization

Firms need good information about local housing and land markets. Getting that information necessitates close contact with local landowners, estate agents, surveyors, planners, builders and others involved in grapevines of information about a locality. Management can only gain that knowledge by frequent local visits and meetings. This puts a sharp limit on the geographical spread of a builder's operations. At most an enterprise can operate on a regional basis of, say, two hours' travelling time from its office, but many are limited to much smaller distances to avoid wasted time spent in travel. Firms operating in more than one locality have to have semi-autonomous regional divisions.

Given the problem of local market saturation, virtually any firm building more than 500 houses a year needs more than one regional operation. This regional network usually is overseen by a central management body. The common practice in housebuilding firms is to make each regional operation a separate profit centre. Some firms make their regional operations legally constituted subsidiary firms (e.g. Barratt and Tarmac), others make them profit centres only (e.g. Comben, Laing and Wimpey) within a unified corporate structure. (The difference in practice is minimal.)

Once a regional firm structure is set up, the need for close control over individual sites, the bringing of new sites on stream and the maintenance of a land bank put limits on the output a regional office can handle. Practice on operating size differs between firms depending on their organizational philosophies and the sectors of the market supplied. With higher-priced homes, for instance, more capital is committed per house and its turnover time is slower, so more control per unit of output is required, lowering the optimum size of a subsidiary. The largest subsidiaries produce up to 1000 houses a year but 500 is more common, with some producing only 250 houses a year when they deal in up-market developments. Once an organizational structure is geared up to a particular size, variations in output are fairly limited without generating substantial management inefficiency or high overhead costs per unit. This means there is a minimum size as well as a maximum one; one manager of a subsidiary interviewed said, for example, that 500 was their maximum annual output and 300 their effective minimum. There are consequently quite sharp discontinuities in overall firm size depending on the number of regions and types of market covered. This helps to explain the sharp jumps in output between the top twenty firms shown earlier in table 3.4.

The personnel involved in a regional housebuilding operation is fairly small, with usually less than thirty people in the office. One volume builder's regional subsidiary interviewed had only three people working there (involved in development only) and one of them was part-time! With the occasional exception, like the example just mentioned, the production process is also controlled from the regional office because building costs are an important determinant of development profitability. The reasons for regionalized firm structures and their size, however, relate primarily to development control not to production.

Expansion by large speculative housebuilders is an amoeba-like process of separating out new subsidiaries from the pre-existing organizational structure. Once a regional operation reaches a critical size, and expansion is still wanted, a new office is set up, based initially on part of the land bank and work-in-progress of the 'parent' operation. After a while the new subsidiary is made completely independent of its parent, and eventually the subdivision is repeated. In this way, for example, Barratt Developments grew from a local Newcastle housebuilder to a parent company with almost thirty housebuilding subsidiaries spread across the country. Once a firm is national in extent expansion results in an increased spatial density of its cellular subsidiary structure. The need for new subsidiaries to take over a pre-existing land bank arises from the diseconomies of operating with very low output levels. If a firm cannot split off a reasonably sized land bank in an area from its pre-existing stock it will actively look for acquisitions of local builders for their land in order to form the initial base of the new subsidiary. Only when all else fails will the firm start assembling land from scratch.

The classic way in which this type of expansion process takes place is through contiguous geographical expansion. But regional markets vary in their profitability, so more profitable gaps may be found further afield. Once again it all depends on a speculative calculation by the firm. Many large housebuilders, for example, set up regional operations in Scotland during the 1970s as a result of the North Sea oil boom and a government report that spoke of excessive profits being made by existing producers in Scotland (SHAC 1970). Similarly, one of the large housebuilders interviewed said it would not build in the South East because of the scale of the potential competition. Other firms took unionization into account before contemplating moving into an area; some firms said they would not build in London or Liverpool because of the strength of the building unions there.

The design of private housing[1]

House design is important at two stages in the process of housing provision: the first is the point of sale, the second the point of consumption when the user wants a useful and aesthetically pleasing living space and where there is a wider environmental physical impact. Design is of prime concern to the builder as it affects the ease of sale of completed dwellings. Obviously the purchasers' interests and tastes influence the ability to sell a house, but there is no necessary correspondence between consumer needs and tastes and speculative housebuilders' interests in design. This section explores this potential contradiction of owner-occupied housing provision.

For the speculative builder marketability and cost minimization pull in different directions with housing design. Design is an important component of the development process because, if used appropriately, it can enhance the revenue from a site considerably. Yet, at the same time, as an expensive physical activity builders try to limit its use and cost. Moreover, particular design solutions and structures influence construction costs and the intensity of land use. Individuality in dwelling styles and non-estate-looking site layouts, for instance, improve marketability but they raise construction costs and may lower site densities which increase land costs. These conflicting pressures have led to particular design solutions: a limited number of house types are used, for example, and their individuality is achieved by variation in their finish and external detailing.

Perhaps the most important aspects of design for builders are the layout of an estate and the mix of houses implied by it. Here the skills of predicting market gaps and market change come to the fore with design a central component of marketability rather than an aesthetic adjunct to it. Senior management kept considerable control over this aspect of design, using design professionals as a means to achieve desired layouts rather than giving them free reign over the conception of a scheme. Paradoxically, then, it is in speculative housebuilding that the design process is related most clearly to the needs of the immediate user. Instead of being given a rough brief of requirements and an approximate cost limit, as in so many other types of construction work, architects receive very strict instructions. More commonly than not, they are concerned only with aspects of a scheme rather than its conceptual whole, and often builders do not use professional designers at all.

The subordination of design to speculative profit has led to strong criticisms of the inappropriateness of the internal design of many private houses to the needs of their user households, and to criticisms of the barrenness and drabness of private estates. Planners tend to feel the same, imposing design limitations on planning permissions and issuing design guides of layouts and styles that will win planning approval, the most famous being one issued by Essex County Council in the mid-1970s. Speculative builders have frequently retorted that such comments and constraints are just sour grapes from disgruntled architects toppled from their Establishment role in the construction process and, more generally, the reaction of an intellectual and cultural élite who want to impose their tastes over the wishes of the population as a whole. They point out that they build only what will sell, and such freedom of exchange is claimed to provide the best expression of popular taste. The argument is flawed, however, as the statement 'if it sells they must have liked it' is tautological. On that criterion all purchasers of model T Fords had the same favourite colour: black.

Given the ideological opposition that generally has arisen between owner occupation and council housing, the most common design practice of speculative housebuilders has been to distinguish sharply their house styles from those of the council sector. In one area of the country, Liverpool, no neo-Georgian private housing is built because of the existence of a large amount of neo-Georgian municipal housing. Other general design rules are also followed. Individual family housing, in appearance if not in content, is virtually obligatory. Even single-person accommodation is now combined into this type of structure by placing four units under a traditional rectangular, pitched roofed, brick structure in back-to-back configuration. (This style was pioneered by Barratt under a typically incongruous name, the Mayfair, and quickly copied by other firms.) Traditional-looking external appearance is also important, although fashion over what period of history is the embodiment of tradition has varied: mock-Tudor in the inter-war period being replaced by neo-Georgian in the 1970s. Many of these features, moreover, depend on detailed styling rather than structural variation. A different styling of windows and decor, for example, can change an ordinary-looking house into a neo-Georgian one. Varying the colour of the roofing tiles, similarly, can help to break down the look of uniformity in a row of identical detached houses.

The changes in the owner-occupied market over the past decade have produced considerable changes in house layout and style. An unstable market requires a greater variety of house types on a site in order to reduce the risk of failure by spreading potential sales across different market sectors. The move up market has produced similar effects, heightening the importance of detached structures, no matter how small the gap between each house, and individuality as expressed for instance in non-row or stepped house configurations. Careful estate design can also increase housing density whilst projecting a scheme into a higher market sector. In this way residential land density has increased over time. Plot size is not a simple product of house price and land cost. Builders aim for market sectors and fix plot size accordingly. The increase in marketability with larger plot sizes frequently is small, and over time the sectoral norm for plot size is gradually whittled away.

The difference between estate layouts for private housing in the 1960s and layouts of the early 1980s is marked and reflects the changes in the housing market noted above. In contrast to present-day developments, the private housing estates of the 1960s usually had less varied mixes of houses in each scheme. Rows of uniform, terraced houses facing each other across standardized through roads or Radburn-style pedestrianized greens were common. Little attempt was made to use variations in layout, as is done today, to create a feeling of space and privacy at high densities. The 1960s were generally characterized by shortages of new housing so in such a suppliers' market builders would place emphasis on production economies and 'no frills', especially on first-time buyer estates, then the predominant form of new building.

What is most noticeable in current private housing schemes, apart from the variety of house types, is the predominance of what can be called the 'new vernacular' house style and estate layout. Through roads are rare, and cul-de-sacs abound, yet car access to each plot is still care-fully maintained. On schemes catering for more than one market sector it is clear that considerable ingenuity has gone into designing the layout in a way which segregates the development subtly into market sectors, and minimizes the need for owners of the more expensive houses to pass by the homes of their less affluent neighbours.

If comparison is made with recently built local authority housing a striking difference appears in the external appearance of new housing projects. Paradoxically, given the ideology of the soulless council estate, the finish on council projects is often to a higher standard. Better quality facing bricks are used, and far more expense is incurred on landscaping and shrubbery despite persistent attempts at quality- and cost-cutting over the past few years. The paradox is resolved by contrasting the nature of the two structures of provision. Public housing is a form of communal provision which by its nature tends to emphasize the joint-ness and communal nature of housing in its design forms.[2] Private housing, on the other hand, is an individualized form of provision. One result is that the purchaser is prepared to pay far more for features he directly and solely consumes than for other aspects of housing develop-ment. Communal characteristics like estate landscaping do not raise potential selling prices to any great extent because they are not con-sumed in such an individual way. Expenditure on these features, beyond a certain minimum level, will consequently reduce a housebuilder's profits.

With regard to the individual characteristics of owner-occupied houses, builders like to have only a small range of basic house types in order to limit construction costs. Each basic type is aimed at a particular market sector, so estates consist of various mixtures of them. Variations to each type are made by altering the trim and cladding material to take account of differences in fashion and regional taste (e.g. brick or recon-stituted stone external walls and style of window frame; even a change in the colour of the roofing tiles can considerably alter the appearance of a house). Builders tend to stick to basic designs for long periods. A survey of house design plans found that on average designs were used for about five years; one national firm said that one of its most common house types had a lineage back to 1955 (Leopold and Bishop 1981).

Repeated use of the same basic drawings with infrequent and piece-meal model changes is one aspect of attempts to standardize production and hold down costs. Another aspect is the shape of houses: rectangular structures with no breaks or protrusions from any of the walls is by far the cheapest house form, and it is adopted for virtually all house types. Bay windows and complicated floor shapes are rare, the latter because of complexities over load-bearing walls and roof trusses. Plain brickwork is also extensive as it is cheaper than large windows or fancy cladding. To avoid the 'little boxes' image the rectangular shape frequently is broken up by simple 'lean-to' structures added to one side. An entrance porch in this format has become a common feature in recent years, extending a few yards out from the front of the house with a pitched roof and meeting the main structure three-quarters of the way up the side wall. Such additions add variety but do not alter the simplicity of the load-bearing requirements of the main structure.

One area in which design form far outweighs cost considerations is in the propensity to build detached structures. Detached housing requires more land, has the highest servicing costs, and means that structural elements like foundations, walls and roofs cannot be shared. It has been calculated that the additional structural construction costs alone of building identical houses in detached rather than terraced form adds at 1980 prices £2000, or about 10 to 15 per cent, to the cost of each house (Leopold and Bishop 1981). Detached houses, however, are in a higher market sector, so the increase in selling price compared with a terraced house is much greater than the increase in building and land costs, especially when modern estate design techniques are used to keep up land density. Again, the magnitude of the detached premium depends on contemporary national and local conditions. One firm interviewed could sell terraced houses at £13,000 and detached at £22,000 on average in 1979; the gross margin on the former was £5000 (or 62.5 per cent) and on the latter £11,000 (or 100 per cent) on building costs including land. The premium on detached structures therefore can be considerable.

Quite detailed aspects of house design also influence the sector of the housing market in which a dwelling can be sold. Until recently most British housebuilders felt that detailed installation of kitchen equipment was a waste of time as it did not increase selling price (contrary to North American practice). A new sector of the market, however, has been opened up by producing small flats offering such all-in features. They enable an image to be created at minimum space standards: Barratt's Solo, for example, is essentially a small bedsit with kitchen, windowless toilet-with-shower and boxroom (the boxroom's image is transformed by calling it the 'dressing area'); all of it is contained within a 14 ft 6 in. × 17 ft 3 in. rectangular area (4.5m. × 5.4m.).

What is important are features that create an image for a dwelling and so fix the subsector in which it will sell. There need be no relation between the production costs of these image features and the additional price at which the houses are sold. Some, in fact, actually cost less than the lower market alternative. Three examples from the housebuilder survey illustrates the principle. One builder interviewed claimed that spending an additional £80 on higher quality, neo-Georgian style windows and doors added £1500 to the selling price of some houses the firm had recently built as it removed the 'starter' home image from them. Another firm tried to create a modern, spacious, unsegregated feeling to the interior of their middle-range houses by having an 'open' staircase and a doorless arch between the kitchen and dining room. These features actually saved £92 on the construction cost but were said to be important features enabling the houses to be sold at a premium. Finally, a third builder produced particularly spacious upper-market houses; the additional space cost an extra £500 to build but added £2000–£3000 to the selling price. (All three examples are at 1979 prices.)

Examples like the above should not be seen as formulae guaranteeing instant success to all builders; they were the particular means by which the builders in question projected their product into profitable gaps in their local markets. They do however illustrate a general principle of speculative housebuilding: profitability depends on getting the correct marketing mix. The inherent usefulness of those marketing features is secondary and possibly even adverse as they may allow other house features to be cut down.

It is well known that new private housing standards are usually much below those suggested by the Parker Morris Committee in 1961 as the minimum for the then typical standard of living of the working class. The detailed implications of this shortfall for users, however, are difficult to quantify as there is no acceptable definition of housing quality. Leopold, as an approximate but useful guide, compared plans for a sample of recent public and private three-bedroomed accommodation by using a design checklist devised by government officials in 1968 for the effect on the comfort and convenience of final users. Her results are reproduced in table 5.2 where it can be seen that new private housing is well down the success rating, trailing far below the public sector. This can lead to severe problems:

Applying this check list to the relatively smaller rooms of the private house plans makes it easier to pinpoint the side effects of constricted space. In many cases rooms simply cannot accommodate the basic furniture necessary for its stated function (e.g. double beds will not fit into the master bedrooms and allow sufficient circulation space on either side of it). Rooms that cannot meet the basic requirements obviously offer no flexibility for sensible alternative arrangements. This is, ultimately, a design limitation (imposed by cost constraints) that no amount of DIY applied by the owner occupier can overcome. And it applies not to starter homes (which often have only one bedroom) but to three bedroom homes averaging just under £30,000 (in November 1980).   (Leopold and Bishop 1981, p. 71)

So whilst design has become increasingly linked with the growing marketing orientation of speculative builders, associated with this trend has been a continuing fall in new housing standards. This fall is frequently blamed on rising land costs which, it is said, reduce the feasible

Table 5.2 Comparison of space standards and quality[1] in the late 1970s for three-bedroom houses in the public and private sectors

	Private[2]	Public[3]	Ratio of public/private
Areas			
Total enclosed area	67.58m²	90.45m²	1.34
Main bedroom	10.32m²	12.98m²	1.26
Third bedroom	4.65m²	6.50m²	1.40
Performance[4] (%) against checklist			
Overall	65	84	1.29
Partial breakdown			
Entrance	43	86	
Storage	33	83	
Kitchen	58	82	
Living areas	63	81	
Bedrooms	51	69	
Bathroom and WC	69	83	

Source: Leopold and Bishop (1981), p. 70

[1] Based on check list in *House Planning*, a guide to user needs with a check list, MHLG, HMSO, 1968; the higher the score, the more successful the design in fulfilling basic user requirements
[2] Private sector sample included eight representative house-type plans for four terraced, two semi-detached and two detached houses
[3] Public sector sample included eight representative house-type plans for seven terraced and one semi-detached house
[4] Performance based only on questions which could be answered 'yes' or 'no'; i.e. excludes questions which were not applicable or for which sufficient information was not available

space standards for houses that most could afford. Yet it has been seen here that the situation is far more complex than that, and related to the whole emphasis of builders' marketing orientation rather than to land cost alone. To blame simply the cost of land is falsely to attempt to externalize the reasons for problems arising directly from the nature of this structure of provision and the forces that drive it. Furthermore, new housing built by speculative builders is aimed only at a very small and possibly atypical section of the population. In 1979, for example, less than 0.5 per cent of total UK households would have bought a new owner-occupied house. Private housebuilding, therefore, cannot be said to provide 'what the public wants' but only what will sell at the best profit, which is only limitedly determined by some consumers' preferences and a truncated 'privatized' form of those preferences at that.

Land banking and development gain

Like other commodity producing enterprises speculative housebuilders need to hold stocks of their non-labour inputs if production is to proceed smoothly. With land the requirement is even more pressing because of the long gestation period between making an initial offer of purchase for a plot of land and its readiness as a fully serviced and developable site. Any builder wanting a steady flow of output therefore has to hold a land bank. Land banks, however, are not simply held for technical reasons associated with production; instead they are also a necessary part of acquiring and retaining large-scale development profits. Having considered the selling side of the development process an analysis of the concrete forms of this retention of development profit can be considered. To understand how land banks help to achieve it the nature of development profit must be considered in more detail.

In order to keep the analysis as clear as possible the nature of gains from land development will have to be defined more precisely than in the previous chapter. In that chapter the difference between house price and the cost of constructing the dwelling (including site servicing) was called gross development profit. Here it will be called development gain. This gain is divided between the landowner and builder: the former gets a price for their land and the latter a profit from housing development on that land. This terminology has the advantage of reserving a concept related to capital, that is profit, for the unambiguously capitalist agency involved, the speculative housebuilder. The class location of the landowner is uncertain. A landowner may be a capitalist, such as a farmer, insurance company or pension fund, but may not, as in the case of aristo-cratic landowners who are a non-capitalist type of landed property (Massey and Catalano 1978). The precise class location of landowners is of little immediate interest here. What is important is that the price they receive for selling their legal ownership of land is a revenue not a profit. This means landowners do not have to operate on profit-influenced criteria except when land ownership is being used as a form of capital. In other words, the actual social relation with respect to land is important in determining landowners' actions, not just the existence of landed prop-erty.

Development gain has been defined as equalling development profit and land price. The exact division between the two components is theor-etically indeterminate as it depends on the contemporary balance of power between builder and landowner. Other potential uses for the land, including the option of not selling it and keeping it in its current use, may influence the price the landowner will get for the land, but only as one factor among many determining the outcome of the struggle between the two appropriators of development gain.

To say that there is a struggle over an economic surplus between two categories of economic agents, of course, does not imply the necessary existence of some grand gladiatorial battle. The contest is more likely to take place within the confines of the mechanisms of the land market where both agents try to take advantage of the economic conditions they face. The speculative housebuilder builds for a general market where demand varies sharply (and is increasingly doing so, especially over the past ten years). Emphasis is placed, therefore, on maximizing profits through development gain and production methods are subordinated to this economic necessity. Landowners, on the other hand, do not have to invest capital but only to decide when is the best time to sell. Inflation has encouraged the holding of land as an investment asset but even so the land market is not immune to economic fluctuations. When the housing market is in a downturn land prices weaken and sometimes fall quite sharply. No builder has to buy a plot of land, and it is likely that more than one landowner wants to sell. So a degree of competition exists on both sides of the land market.

Most of the rest of this chapter will explore some of the more import-ant contemporary aspects of residential development that influence the conditions under which the struggle between landowner and builder takes place. But, because of the importance of land-use planning in Britain, the discussion can only be concluded in chapter 7 which deals with planning. In order to follow the empirical points being made it is important to note that the argument just given contrasts sharply with the usual approach to land values in which it is suggested that, at least theoretically, the value of land is always determinate (cf. the symposium in Hall 1966).

The usual way of considering development gain is to treat it as part of the increase in land values associated with housing development. When land is converted from its previous use to its new use as the site of an owner-occupied house, it is said to be possible to identify stages at which land value increases. With a greenfield site, for example, the first stage is the addition of a 'hope' value to the value of the land in agricultural use, at the time when it becomes probable that the land will be used for housing at some time in the future. This 'hope' value then gradually rises as the prospect of development increases. The second stage is the winning of outline planning permission, the third, site servicing and so on. When used for housing, land achieves its highest value when the profit-maximizing house types are erected on it. After a time the land value again changes as the dwellings become outmoded either through decay or changes in land-use patterns, and the site's land value then is its value in its best use minus the costs of demolishing the existing structures.

This approach has an obvious credibility in that the argument corresponds to differences in the prices at which land is actually sold; serviced sites with planning permission sell for much higher prices than equivalent 'green belt' sites. The approach, however, represents a generalization of these empirical reference points to all land whether or not it is actually transacted, and the generalization is made on the basis of a particular theory of land rent. Consideration of the 'land value' of a newly built house illustrates the point. The price of the house itself depends on the prevailing general level of house prices and the physical characteristics of the house relative to others; partially those physical characteristics concern the features of the built structure but they also relate to its spatial location. The value of the land plot on which the house stands is only notional as it represents a conceptual component of the price of the house itself. It is, moreover, a concept derived from a theory of rent which argues that land value represents all of the difference between house price and construction cost (including a competitively derived builder's profit). Land attracts to it all this potential surplus profit, it is argued, because of its general scarcity and locational specificity.

This theory of housing land rent is an application of the Ricardian theory of land rent discussed in chapter 4 which states that land value is the residual between production costs and selling prices. Land values consequently do not determine anything but are determined by demand and supply in the final product market. The theory is universal in that it can be applied to any land use and it is ahistorical in that it is indifferent to the recipients of that rent/increment in land value. Whether it goes to the initial landowner, the builder or to the state as a land tax does not affect the conclusions of the theory. Empirical justification for the theory is based on the generalization of a limited number of empirical reference points through the inductive process of applying the 'facts' to cases where they are not observed. Limited data about the price of land transactions therefore sustain the claim that increments in land value at each stage of the development process are universal. Yet the most important fact is ignored: namely, the nature of the ownership of the land at the point in question. Who actually owns the land determines whether a rent can be extracted for its use and the magnitude of that rent.

By making a break with the Ricardian theory of rent and its empirical generalizations it is possible to place rent in its social context. Land rent, either as a regular payment or capitalized as land price, is the charge imposed on land users by private landownership and the monopoly control it has over land. From the point of view of the user rent is a payment, for the landowner it is a revenue derived from having ownership of the land. In the structure of owner-occupied housing provision rent is the income received by the initial landowner from housing development. The initial landowner is the person who sells the land to the speculative builder; there could, of course, have been several 'initial' landowners who have sold the land to each other through a process of land dealing. But once the speculative builder becomes the landowner the rent relation ceases. Builders want land for its use value to them; it is necessary for housebuilding. They want to convert the land to housing and then sell the houses, they do not want land as a source of revenue in itself either by holding on to it or by selling it. They make more profit by being house sellers, not land dealers. When builders sell houses they receive revenue from which they have to deduct construction costs, overheads, interest payments and the purchase price of the land. Having done so they are left with their profit, partially from building but primarily from converting the land to a successful housing development; the latter therefore is a development profit. Any need to pay rent in the capitalized form of land price reduces their profit.

Land has a price therefore only when it is transacted for its use value as land. Other social agents might make hypothetical calculations based on that transaction price concerning plots they own or would like to own, but mere thought alone does not produce anything tangible. Houses do not have separate land prices; the land on which a house stands has been converted into a component of the house. This difference between land prices and house prices is not just a semantic one, as it fundamentally determines how key social agents operate in the housebuilding process. In particular, speculative housebuilders want to maximize the difference between land price and house price (subject to the turnover time of capital implied), as that maximizes development profit. They organize the whole of the housebuilding process to that end. Housebuilders consequently try actively to increase their development profit rather than act as passive receivers of a share of a Ricardian-style residual over which they have no control. The speculative nature of housebuilding, if nothing else, ensures their development profit is never reduced to zero through competition.

The opposition between land price and development profit suggests that it is wrong to think of speculative housebuilding as having two faces: an unacceptable one in land speculation and an acceptable one in house-building. Such radical liberalism is an attempt to divide the indivisible. This approach is shown in a frequently cited piece of mid-1970s financial journalism: 'Despite appearances, housebuilding is only partially the business of putting up houses. The houses are the socially acceptable side of making profits out of land appreciation.' (*Investors Chronicle*, August 1974).

Less dramatically it has been suggested that house development is a dual process of profiting from housebuilding and from land holding, where the needs of the two sources of profit come into conflict (EIU 1975, p. 40). This dualistic approach of separating land and building ignores the primacy of social relations. It wrongly tries to separate out land gains from land use, yet the unacceptable face cannot be removed whilst leaving the acceptable one unchanged. This theme will be taken up further in the chapter on land-use planning because the fallacy of being able to make this separation has fundamentally influenced the nature of planning policy.

The portfolio approach to land banking

Land banks help to increase development profit by keeping down builders' land costs relative to house prices. The obvious way to keep land costs down in an inflationary period is to hold onto land for several years after purchase before developing it. Some land banking achieves that function but the matter generally is more complicated.

A useful way of conceptualizing the nature of a housebuilder's land bank is to treat it as a portfolio of land just as a commercial bank or other financial institution has a portfolio of assets. In both cases, the portfolios consist of a spread of high-yielding but potentially risky assets (in the builder's case these will usually be sites of white land) and safer but less profitable assets that can ensure a steady cash flow and corporate stability. Portfolios also have a temporal profile consisting of assets with different dates of maturity and profit realization. In general, a land bank portfolio spreads risks and takes the pain out of speculation. Different building firms have different land-holding requirements depending on the nature of the capital they constitute (see chapter 3) and the sectors of the market in which they operate. A large portfolio of land has the added advantage of giving substantial market power. This market power is of crucial importance with respect to two sets of social agents: the planning system and landowners.

Most builders when asked by newspaper reporters or academic researchers how much land they hold reply that they have approximately a three-year land bank. This is the minimum holding they require to keep the building process going, given the time it takes to get planning permission and to bring land to the stage where building can start. Interest charges make holding onto larger stocks of land unattractive, they suggest. Given the bad publicity over land speculation by builders in the early 1970s and the continual political sensitivity of taxes on development gains and of land availability, this 'technical minimum' type of argument must be treated with caution as it is more suggestive of self-justification than necessarily of factual accuracy. Unlike public authorities who now have to publish registers of their land holdings (see chapter 8), speculative housebuilders' land holdings are a commercial secret.

Some evidence is available at a national aggregate level but the holdings by individual companies become known only in a piecemeal way through such sources as company reports and share prospectuses. This is unfortunate as it is impossible as a result to consider differences in land holdings by types of firm. Nationally the Private Enterprise Housing Enquiry (DoE) gives data on land held by housebuilders. In June 1980 the survey found that private housebuilders had 210,000 dwelling plots with planning permission in their land banks. For its sample (covering about two-thirds of private sector housebuilding) that represented about a 3½-year land bank at the 1980 housing starts level. The data for land held without planning permission are less accurate (and more probably under-reported) but crude estimates based on the acreage given in the survey show that private housebuilders held an additional 5½ years or more of white land. The grand total is a land bank of 9 years and this is likely to be a considerable underestimate.[3] A survey of land availability in Greater Manchester (HBF/DoE 1979) shows lower but still large holdings. In 1978 builders held a three-year land supply with planning permission and virtually another three years' supply without planning permission (there is no indication of whether land held on option was included in their assessment, which may account for the lower length of the land bank when compared to the National Survey).

Evidence for individual firms is scant and therefore has to be anecdotal. Larger firms tend to hold much more land than smaller ones, so they hold a disproportionately large amount of the total land and have even longer land banks than the industry average. Data for the largest firms were given in table 3.4, where reported land banks range from 2 to over 10 years' output. Some holdings are very large and some land is held for a long time. One builder interviewed in our housebuilders survey, producing around 400 houses a year, had a land bank that would have enabled them to build at that rate for at least another 38 years! W. Leech reported in 1978 that they had a land bank equivalent to roughly 15 years' current output, most held without planning permission. Firms like Wimpey and Laing frequently buy land knowing it cannot be developed for at least 15–20 years; Wimpey were reported as having 30,000 plots with planning permission in 1977 and another 15,000–20,000 on option (Simon & Coates 1978). More spectacularly, Barratt purchased a 23,000-acre estate in Scotland in 1979, although not all presumably for housing! Housebuilders can be seen to be substantial landowners.

The land actually owned, furthermore, does not represent all the development land over which a firm has effective control. It is common practice to hold options (or conditional contracts) on land for future purchase. With options the capital tied up in land is minimized and the return on that capital can be huge; EIU (1975) gives a good summary of the procedure:

> The conditional contract method of purchase has small variations from developer to developer. Naturally, the land purchased under conditional contract is land without planning permission at the time of

negotiation of the contract and initially an option price is paid calculated as a percentage (usually about 10 per cent) of the existing value of
the land. This option price is paid immediately. Future arrangements
entered into at this time seem to take one of two forms. The procedure
is that once outline/detailed planning permission has been obtained on
the land, the developer, should he take up his option, purchases the
land at a discount of about 10 per cent of the land value with planning
permission. . . . Alternatively, and even more advantageously for the
developer, some companies follow the practice of agreeing the eventual
purchase price of the land (i.e. when permission is eventually obtained)
at the actual time of purchasing the option, writing into the option
contract a stipulation that further incremental payments should be
made per annum at a percentage agreed upon at the same time, should
purchase not be made within say a 2–3 year period.

(EIU 1975, p. 45)

This EIU report gives a suggested example of purchasing a site with a
market value with planning permission of £100,000 per acre with an
initial option outlay of only £1000, and a subsequent outlay when planning permission is won of £90,000. This speculation on getting planning
permission for white land (i.e. land not designated in the local plan for
development) gives, when successful, guaranteed purchase of a site with
planning permission at a discount of £9000, or a gross return of 900 per
cent on the initial £1000!

Practice on the use of options varies between firms. The EIU report
quoted above found that on average 20 to 50 per cent of firms' land banks
were normally held as options, rising to as much as 80 per cent for some.
In our interview survey, some firms said options were outmoded because
of the introduction of development land tax in 1976 (although it is
difficult to see why). Others, however, actively pursued such arrangements; one firm had fifteen times as much capital invested in options as
they had in directly purchased land, which gave them a huge effective
land bank equivalent to twenty years or more of their contemporary rate
of output. Another firm used options where possible, finding them particularly useful when schemes involved land purchase from more than
one owner. The ability of any individual landowner to hold out for very
high prices is limited with optional agreements, as far less capital is
invested so the builder can easily pull out if the site cannot be assembled
at reasonable cost.

The overall size of large firms' land banks may hide significant spatial differences in their land holdings. In one area they may have large land holdings whilst being desperately short in others. It is common practice for speculative builders to sell (or swop) land in different localities to even up their portfolios. Much of this internal land dealing between builders is unpublicized and it makes it unclear to planning authorities precisely who is going to build what on any site. Sometimes transfers can take the form of one builder inviting another into a consortia on generous terms in exchange for land in another district, thus combining the advantages of consortia and land exchange.

A portfolio of land is continually run down as sites are released from it for housebuilding. The sites drawn out of the bank depend on the state of the housing market, the type of sites in the portfolio, and the ease of additional land purchase. Some sites might be built on almost immediately they are purchased whereas others might stay in a land bank for years. The overriding objective of a builder when deciding on land release and purchase is to maximize the overall development gain from the portfolio over time. This may be achieved by a rapid throughput of land or alternatively by holding onto a site for years whilst house prices and the locational desirability of the site rise. It all depends on the strategy of the firm and its speculative calculations. As a result it is quite feasible for two ostensibly similar sites to have totally different time profiles through the development process even when bought by the same builder. Land is rarely developed in the sequence in which it was acquired. A land bank should not be seen therefore as being like a physical store through which land moves in conveyor belt fashion with more sites being added at one end whilst others are taken off at the other.

Along with the desire to get the greatest profit is the need to maintain a steady cash flow for the enterprise. For cash flow reasons some firms divide up their land holdings into basic and marginal sites. Basic sites are ones attractive enough always to generate a steady rate of sales and hence revenue for the firm. Usually they are in the middle and upper ranges of the market although for some firms first-time buyers are their 'bread-and-butter'. Marginal sites are those where sales are less certain. In any year a firm will build on a number of basic sites to assure a steady minimum cash flow and build on a number of marginal sites selected on the basis of expected market trends over the next year. If the release has been well-timed, none the less, the profit made from marginal sites may be greater than from basic ones.

The 'ideal' site firms like to have in their portfolios varies. An ideal site for a volume builder, for example, was suggested by a number of respondents as being a well-placed 8-acre site for 100 dwellings at the lower/medium end of the market. Such a site provides a year's trading and a rapid turnover of capital as money is not tied up for long in site works and infrastructure. Others liked to have large sites once they had proved to be successful. These large sites are then a mainstay of company profits for five years or more and constitute prime basic sites. One interviewee also suggested that large sites had the advantage of taking the 'rough with the smooth' in housing market fluctuations so that infrastructure costs could be financed out of profits in boom years.

Builders, of course, cannot deal solely in 'ideal' sites; they have to take what land is available. Builders frequently complain about the poor quality of sites available for purchase and invariably attribute this to the planning system. In a number of localities complaints have been particularly vociferous in recent years about the groundwork involved in the sites available, either because of initial site conditions such as rock outcrops, poorly infilled land, or because of the high servicing costs associated with the site. One builder interviewed cited the case of being offered a 90-acre site by an insurance company at the then bargain price of only £20,000 per acre. Added to the purchase price, however, were many additional servicing costs: the county council specified that the developer had to provide spine roads, foul sewers, storm drainage and a school site; the Gas Board quoted £90,000 for gas installations, and water and electricity would have added more expense; so the project as a whole was unviable.

Builders' complaints about the non-ideal nature of the conditions in which they operate need to be put in context. They are a response aimed at changing the relative position of the individual enterprise in which the respondent wants only the non-ideal aspects of their situation to be changed. Yet the whole essence of speculation is to profit by spotting opportunities which have not already been discounted and incorporated by the market into price. Ideal sites, no delays, and a steady housing market would in fact reduce the opportunities for development profit as more potential profit from development gain would be discounted in land prices. If average development profits did actually rise in the short-run, furthermore, more producers would enter housebuilding and the increase in output would eventually lower house prices. So development profit would be squeezed from both sides. Implementation of action to resolve such complaints by builders would therefore result only in more complaints; this time about land prices being too high and house prices too low! Markets do not operate on the basis of the ideology of the individual even if some of their participants wished they would.

The market power of speculative builders and landowners

To be able to discuss adequately the relative market power of the two different groups it is necessary to know the characteristics of those groups and the pressures to which they have to react. The characteristics of speculative housebuilding firms were discussed in chapter 3. Unfortunately it is not possible to do the same for landowners as there are no studies or information concerning the nature of those who own potential development land. Knowledge about local land markets generally is poor. At a national level information about landowners is better, especially as a result of the work of Massey and Catalano (1978). But the nature of landowners as a whole is different from that of owners of development land alone. To know, for example, that large tracts of the Scottish Highlands are held by various branches of the nobility for the purpose of hunting and shooting does not help to clarify ownership in the outer metropolitan area of London. What the overall characteristics of landownership in Britain do show, however, is that most land is held by its owners either as a long-term investment or primarily for non-economic reasons (Massey and Catalano 1978). These characteristics of landownership reinforce the economic attractions of holding onto potential development land rather than selling it at the first opportunity. The absence of a need for frequent revenue forms a fundamental base for the power that landowners can exert in the residential land market.

Builders and landowners have a common interest in avoiding their conflict over development gain spilling over to overt antagonism, as neither would benefit from fundamental changes in the current structure of owner-occupied housing provision. The market form their struggle takes encourages the ideological presentation of common self-interest. In general, for example, landowners are perceived by builders as justifiably 'getting what they can' for their land in the contemporary market circumstances. In a number of circumstances, builders also can increase their development gain in an apparently harmonious fashion with landowners by offering them agreements that go beyond the formality of anonymous market exchange. For one type of landowner, owner occupier farmers, the sale of land on negotiated terms to a builder, for example, might help to avoid capital transfer tax for a farmer worried about this form of death duty or, alternatively, payment for the land sold can be arranged in ways which minimize the farmer's current tax incidence. The sale of one or two fields as housing land is a useful source of revenue but essentially incidental to the main business of farming. So builders actively woo local owner occupier farmers with potential housing development land on the edge of expanding urban areas. Similar mutually advantageous approaches can be applied to other types of landownership by builders.

Housebuilders and landowners, alternatively, may jointly bring land through the initial stages of the development process. Options on land without planning permission are good instances of this process. The builder can advise the owner of the best way to proceed or undertake the task himself; the proceeds of success then being shared between them. Another case of a joint approach is consortia between housebuilders and owners of large tracts of land. A number of large landowners, not surprisingly, are loath to sell outright or to accept options on sites without planning permission in situations when the uncertain prospect of development and the time horizon means that the land has to be sold at a substantial discount. Where the landowner is interested in land as a long-term asset it is particularly galling to see the major gain from housing development undertaken on land sold in that way accruing over the years to the builder as development profit. The result has been that some landowners have acted as developers themselves, selling off small serviced tracts to builders as the whole scheme goes along.

The difference between enhancing land prices and development profit comes to the fore here because few landowners have the expertise or ability to act as speculative developers. One respondent to the housebuilders survey told of a large landowner who invested substantial amounts of capital in infrastructure at one locality with the aim of selling off the serviced plot to builders. In doing so this landowner incurred large debts and so was no longer in the position of being able to wait for a long-term return from the land. In addition, the speculative builder's golden rule of minimizing working capital locked up in the ground was broken. The slump of the mid-1970s almost brought this owner to grief; he was saved only when a large national housebuilder came to the rescue. Since then the profits from housebuilding and land sales have been shared between them via a jointly owned development company.

Planning controls form part of the conditions constituting the relative market power of buyers and sellers in the land market. Owning land with little chance of getting planning permission for a change of use obviously excludes the owner from getting returns from development. Landowners often apply for outline planning permission as a test of their land's market value. This feature occurs more frequently on smaller sites (EIU 1975); on larger schemes the builder is in a stronger market position and so white land purchases by builders or the negotiation of option agreements are more common.

The relative market strengths of landowner and builder are not fixed but depend on the economic and political situation. The weakness of landed property during the 1930s and in the immediate post-war period was noted in chapter 2. Since the mid-1950s its relative position has changed dramatically. The growth of inflation in Britain, the associated poor profitability of industrial capital and the rapid growth of financial institutions looking for long-term investment outlets have all encouraged investment in housing development land. The rapid expansion of owner occupation, the high costs of housebuilding and the concomitant rise in house prices have, at the same time, ensured that this massive new investment in land was successful. Instead of being able to deal with landowners who were desperate to sell, as in the 1930s, builders now have to buy from suppliers in a position of greater market strength.

In view of these changes in landownership, the growth of large builders and their increasing ownership by long-term development capital over the past decade can be seen as a belated move into the productive side of housebuilding by similar types of capital to those having earlier entered the land market. Large builders are uniquely placed to make sure that development profit represents a high proportion of development gain, and the 1970s produced a set of circumstances that emphasized their advantages. Discussion of the development process has highlighted many of the reasons: only they have the size to benefit from the new emphasis on marketing and the design techniques associated with it; only they can diversify across a wide variety of localities and market sectors; and only they can operate counter-cyclical land purchasing policies in a land market where prices rise and fall sharply. The expansion of large producers has consequently helped to alter the development process in speculative housebuilding, yet these changes have also helped to create the crisis in production which the next chapter will consider.

6

The housing production process and the crisis of production

Introduction

The production process involves the purchase and assembly of the non-land inputs necessary for building and the subsequent construction of the houses. In many respects it is closely linked with the development process discussed in the previous chapter. Some features of housebuilding even straddle the distinction between the two. In this book, house design is considered as part of the development side of building, yet it could just have easily been considered as part of production. This ambiguity over the correct categorization of design does not arise with most other aspects of housebuilding. Both in the housebuilder's operations and in analytical terms there is a clear divide between development and production. Even so, an examination of production shows how influential are the needs of development. This interlinkage constitutes a major theme of the chapter.

Discussion of production is, of course, central to the arguments of this book for it is here that the reasons for the existence of escalating building costs and the resultant crisis of profitability in private housebuilding can be examined. The object, however, is not to see in production some ultimate cause from which all the problems of housing provision stem. Embarking on an analysis of a production process is not akin to the opening of the final gateway in the quest for the source of original sin. Housing production exists as part of a wider structure of housing provision and each aspect of that structure is influenced by the others in a mutually determining way.

It was argued in chapter 1 that speculative housebuilders are a combination of two types of capital, merchant capital and productive capital. This combination as merchant-producers gives them their economic strength. The control of production enhances the profitability of development by enabling it to be treated as a unified process of land acquisition and housing development, whereas the development process creates important and sometimes lengthy breaks in the land conversion pipeline from greenfield sites to completed housing estates which have important implications for housing production. Production, therefore, influences development and development influences production. The division between the development and production processes does not actually correspond to the merchant-producer aspects of the speculative builder's operations. The non-land inputs for house production also have to be purchased by the speculative builder. And here a particular set of exchange relations have also evolved. Examining the production process is consequently to look at another aspect of the merchant-producer dichotomy of the speculative builder. The structuring of production by one aspect of the merchanting role, land development, has implications for the other aspect, input purchase. It is the contradictions thrown up in the production process by the merchant aspects of the merchant-producer dichotomy that explain why there is a crisis of production in speculative housebuilding.

The physical nature of building places limits on the production methods that can be used. Some aspects of production, at least, must be undertaken on site: groundwork, road building and site servicing obviously are site specific. Yet above ground level, housebuilding is essentially the assembly of pre-manufactured components, such as bricks, concrete blocks, pre-cut wood and roof tiles. This assembly operation offers the possibility of standardization and substantial off-site prefabrication, both of which are essential preconditions for factory or industrialized building methods. The site specific aspect of some building tasks places limits on the possibilities for standardization. The tasks involved in groundwork, for example, are unique to each site and must be undertaken *in situ*. It is not possible, therefore, to produce houses like many other commodities in mass-production factories, each house coming off the production line complete and then simply being transported to the place of consumption. A house is only a useful dwelling once it has absorbed the socially useful characteristics of a plot of land, the most important of which is its location. Even a mobile house is not a home until it stops moving. The radical architect's dream of houses produced like cans of baked beans is consequently impossible. The housebuilding industry is not backward nor archaic because factory production is comparatively absent, even if factory-based techniques are universal in most other industries. On that comparison, it is simply different.

Whilst the production process does not create all useful characteristics of dwellings, only it can create the possibility of profiting from housebuilding. To do so production must be organized on capitalist lines and result in costs that are less than selling prices. So, like any capitalist firm, speculative housebuilders have a strong incentive to control and to reduce production costs. This is achieved by adopting the most appropriate managerial and organizational principles for any given technique and by using the most profitable technique. So discussion of the production process in speculative housebuilding has to centre on these two interrelated issues: the organization and control of the production process and the methods of production used, in terms of the reasons for, and consequences of, their existence.

Each aspect of the production process is considered with these criteria in mind. Discussion of them, however, cannot be couched in the dry technical terms of a management studies manual, for the different aspects of the production process can only be understood in terms of a three-way, antagonistic struggle between landowner, housebuilder and building worker. The last chapter showed how the development process was organized to place the housebuilder in a strong position against the landowner in the appropriation of development gain. Much of the way in which the production process is organized is a consequence of that struggle. So the pressures from it influence the relation between housebuilding firm and building worker. The firm wants to get the maximum labour out of the worker at the minimum cost but in ways that do not come into conflict with the requirements of gearing production to sales. The outcome is the use of particular production methods that rely on the handicraft production of simple, standardized dwelling structures. Those methods need a particular type of workforce of predominantly skilled and semi-skilled tradespeople who are often employed on a subcontracted and self-employed basis – the universality of which is unique to speculative housebuilding.

The workforce, not surprisingly, resists attempts to cheapen labour costs at their expense. But the historical evolution of trade unionism in the British building industry, which is fundamentally based on the individual site as the focus of trade union struggle, has virtually denied workers in private housebuilding the ability to use the trade union approach to workplace action. Without this, the only really effective weapon, other means have been used but with limited success. The result has been that the building workforce has borne the brunt of the need to vary output sharply, through casualized employment and, with the downturns of the 1970s and 1980s, increasingly long periods of unemployment. Over the long run, as will be shown later, this has led to a crisis in the reproduction of a building workforce of adequate size and competence, bringing in its wake new twists in the relation between employer and building worker.

This overview suggests a way that the argument of this chapter should proceed. The influence on production of the development process needs to be considered first, next the relation between employer and worker, to be followed by a discussion of the role played by the other major providers of inputs to housebuilding, the building materials supply industries. Then the recent introduction of timber-frame building methods will be considered. Having done that it will then be possible to provide a synthesis of the reasons for the crisis of production in the private housebuilding industry.

Accumulation, housebuilding and technical change

When organizing production speculative builders, in many respects, operate according to principles and pressures that are common to all capitalist enterprises. Yet, despite being subject to such general features of the capitalist mode of production, speculative housebuilding does have some characteristics which in combination make it a unique means of accumulation. Such distinctive features have to be considered when examining the determinants of the production process, so the latter cannot be regarded merely as an expression of some general law of capitalist development.

The specific relation to land generates a number of these features. One result of it is that the timing of the production of each unit of output is particularly important in determining the size of the profit that will be made. Housebuilders, in addition, have to resist the conversion of profits from building to land rent. The result is that production cost is only one element amongst many influencing the size of a housebuilder's profit. And, as the last chapter argued, it is unlikely to be the most influential; that role is usually taken by development gain. The general pressures inducing technical change within capitalist societies, stemming from the increased profits accruing in the short run to the innovating firm, are thus weakened in speculative housebuilding. Cost-cutting techniques are introduced only when they do not come into conflict with the other requirements of profitability.

This relationship between production costs and profitability affects technical change in the industry. Although many aspects of housebuilding involve technologies pre-dating the industrial revolution, technical change still does occur. But profitability requirements facilitate technical change only in specific forms, forms that minimize the capital tied up in production. What is ostensibly the least-cost building method need not necessarily produce the largest profit as such techniques generally depend on continuous and long production runs, which are anathema to the speculative builder. As technical change is generally an incremental process feeding off previously introduced improvements, the long-term consequences of this limitation on the form of technical advance may be substantial. At issue, therefore, is not simply whether or not the economics of speculative housebuilding have forestalled the introduction of a known better method of production. Instead, there is also the question of inhibiting the preconditions necessary for innovation. The preconditions are likely to be the interrelated ones of standardization, continuity of production and scale economies, and usually the last two cannot be achieved within speculative housebuilding. They have been met at least to a degree, however, in the industries supplying housebuilders with plant, equipment and building materials. So, not surprisingly, it is from these sources that most technical innovation has come.

The conclusion that speculative housebuilding has held back technical development through limiting the forms that it can take has to be one of logical deduction. As a statement about actual technical change it can only remain conjectural. There is no way of knowing what technical developments might have taken place if the preconditions for their existence had been met. The other predominant method of housebuilding in Britain, that of building to contract, also does not create their existence (Ball 1983), so comparison with it is of little use. Recourse to cross country comparisons, moreover, would at best bring out only differences of degree as similar structures of provision exist in other advanced capitalist countries. Non-capitalist countries, which at best have only been industrializing for fifty years in a hostile economic and political environment, do not have the economic resources to build to the standards that have become accepted as normal in the West. So, again, little can be derived from an empirical comparative approach. Empirical emphasis instead has to be placed on why it is so difficult to create those preconditions in the British speculative housebuilding industry itself.

A limitation on technical change is only one consequence of the lack of correspondence between minimizing production costs and maximizing profit in speculative housebuilding. Another effect concerns the relative efficiency of productive units within the industry. Most theories of competition between firms in an industry suggest that the most efficiently organized, technically innovative firms force others to adopt their methods or go into liquidation. Firms have an incentive to be more efficient than most for they will then earn higher profits; others will be forced to follow suit since the expansion of lower-cost producers will bring down the market price. The importance of development profit considerably weakens this effect. The market price of new houses is composed of development profit and land costs as well as production costs; firms with high production costs may consequently be shielded from the rigours of competition from more efficient producers through the profit they make in development. An astute developer need not be an efficient organizer of production, and competition cannot force him to be except within very broad limits. Whilst, therefore, it may be true that all builders have an incentive to adopt the profit-maximizing technique and to use inputs efficiently within that technique, there is no mechanism forcing them to do so. It is possible consequently for there to be wide variations in the productive efficiency of different housebuilding firms and, one would suspect, similar variations between the separate divisions of firms and possibly even between sites.

The primary importance of profit from development is likely therefore to be a major explanation for this sector of the construction industry of the wide variations in productivity between building sites that have been found in a series of studies since the war (Reiners and Broughton 1953, Forbes 1969 and Lemessany and Clapp 1978). These productivity studies covered only labour inputs into housebuilding and include case studies from both the public and private sectors. Yet a report that examined owner-occupied house prices (SHAC 1970) noticed a similar variation in final output prices that could not be fully justified by differences in land prices or in estimated production costs. This unexplained variation in house prices occurred even in a subsidiary survey they undertook of sixty-five almost identical houses built by the same builder, indicating that house prices are not forced down by competition to their input costs plus a 'normal' profit mark-up. These results illustrate the limited effect of price competition on productive efficiency.

On at least two counts – technical innovation and productive efficiency – it would seem that speculative housebuilding is not a conducive form of economic organization for the adoption of best-practice techniques. This conclusion has been reached despite the existence of profit making and competition, with whose existence productive optima are regarded so frequently as being virtually axiomatic especially by neo-classical economic theory. This conclusion about the production process can be expressed theoretically as an example of the dominance of the social relations involved in housebuilding over the physical ones, in which there is a struggle between builder and landowner over the conversion of development gain into land rent. Neoclassical economic theory tends to deny the possibility of land rent affecting methods of production and so consequently does not consider investigating this effect. Here it is seen as paramount.

Speculative housebuilding and the organization of production on site

The last chapter described how housebuilding is geared to the rate of sales with only a small number of houses produced at a time. Houses take a considerable time to build, eighteen months on average in 1979, so some building ahead of sales must inevitably occur. To some extent even that requirement can be minimized by installing the service facilities to the dwelling, and the foundations and ground floor concrete slab (known as building-to-slab), then leaving the erection of the superstructure and the internal fittings until they are required in the sales programme. Large sites are also built in discrete phases to reduce capital commitments. Within each site phase, access roads and other site services are installed first of all. This early construction of the road network has the advantage of easing the movement of materials and equipment round the site, particularly in bad weather.[1] There are consequently four identifiable and sequential stages in the production of private housing: first, general site clearance and groundwork preparation (sometimes phased on the largest schemes); second, phased completion of groundwork and site servicing; then the building of individual dwellings to slab; and, finally, the erection and completion of dwellings above slab.

Crucial to the understanding of the production process is the nature of the pressures creating this division into four stages and the relation of each stage to the rate of housebuilding. Where output is geared closely to expected sales, the time lapse between the completion of one stage and the start of another may be considerable. Other pressures, however, pull in different directions. Indivisibilities or economies of scale in production lead to pressures for larger volumes of output and for a dovetailing of each of the four stages into a continuous process. Large-scale, rapid production, however, might lead to stocks of unsold houses and hence large amounts of working capital tied up for long periods. So, unless sales are rapid, the minimization of working capital creates a third pressure. Working capital can be minimized if each stage quickly follows the next, but at a low total output volume. That pressure related to working capital however militates against being able to respond rapidly to sales; customers might have to wait for months whilst their houses were built − the antithesis of speculative building. In addition, scale economies in production could not be achieved. The way in which speculative housebuilders organize production is an attempt to come to terms with these contradictory pressures.

Small-batch production of easy to build, standardized house types is the compromise adopted by virtually all builders. The size of each batch varies between sites and firms, from less than five to over fifty, but generally it averages between ten and twenty-five dwellings at various stages of building. Within each batch there also is usually a mix of house types.

The vertical method of building

This small batch approach to housebuilding produces a distinctive method of building called the vertical method in which each house is produced essentially as a distinct entity. Production activity centres on the completion of one dwelling before proceeding to the next. So, in terms of the external structure, it goes vertically up from slab to roof before moving on to the next dwelling. Completion of one dwelling before starting the next is the extreme case that illustrates the principle, but the same method still operates with small-batch production if each task required for the batch is not undertaken as a unity. The sequential completion of houses is the main objective rather than continuous production within a well-organized division of labour.

Vertical building contrasts with horizontal building where the production process is organized so the tasks are worked through to maximize the continuity of trades. All the groundwork is done first, then the dwellings are gradually completed as a unity. Each trade moves down the line of dwellings as previous tasks are completed in a similar way to conveyor belt assembly, except that the workers rather than the product move down the line. An obvious contrast is the scale of building. Horizontal building requires the site to be constructed as a whole rather than in the small batches of the vertical approach. The clearest visual example of the classification as horizontal can be seen in the use of scaffolding which is linked horizontally between the dwellings being constructed, whereas with vertical building the scaffolding is erected in isolated units covering only a few dwellings at most.

These two methods of building are the physical consequences of different economic approaches to construction. With horizontal building an attempt is made to minimize production costs. Housebuilding to contract for local authorities is one case where the horizontal approach is used.[2] Vertical building results from the need to relate production to sales, so the site cannot be treated as a whole and nor can construction proceed as a continuous process. Only a limited degree of continuity can be achieved through building houses vertically in the careful timing of the sequential movement from one house to the next.

Another clear distinction between production on private housing sites and those built under the horizontal method is the location of site activity. Building the site at one go involves the large-scale movement around it of workers, plant and materials. To avoid congestion, mud-bound equipment and materials, and damage or destruction of already completed work, management has a strong incentive to organize the flow of work in a clearly defined spatial pattern. One instance would be to build towards the exits and entrances of the site. In private housing conversely the pressure instead is to get sales. Houses consequently are built at the entrance to the site first: few people would venture across a muddy building site to visit a show house and would only purchase once they could gain adequate and pleasant access. This need to show a completed front militates against efficient building practice.

Production levels and cost control

The broad requirement of gearing production rates to sales is achieved by adopting the vertical method but within it there is still considerable room for variations created by the conflicting pressures of sales, working capital and productive efficiency discussed earlier. Firms come to distinct conclusions about the trade-off between the three. There are different expectations of sales rates, ease of access to working capital and even different management philosophies – a marketing versus a production orientation, for example. These factors influence the size of each stage of production embarked upon and how closely stages follow on from each other, both of which have a considerable impact on productive efficiency. And, again, there is no reason to expect that the most productively efficient, least-cost sequence is the most profitable.

There are constraints on feasible output levels. To an extent the decision is taken out of a firm's control by outside agencies who insist that certain actions are undertaken as a unity. Planning permission, for example, might only have been granted once the builder agreed to install a spine road across the whole of the site at the earliest opportunity. The water authority might also want fresh water and sewerage pipes to be laid at one go. As a result indivisibilities and large initial amounts of working capital may be a precondition for starting development on a site, thereby encouraging the site's rapid completion. Further demands on working capital arise from the need to deposit performance bonds with the local authority to ensure that roads and other public facilities are of a satisfactory standard prior to their adoption at the end of the project. Usually arrangements with insurance companies minimize the cost of such bonds; the problem for the builder then becomes one of a borrowing constraint. Banks include the whole cost of such site infrastructure as part of the housebuilder's current capital commitments and reduce the firm's borrowing ceiling accordingly. The builder consequently may be unable to borrow further until the site is built out, again encouraging speedy completion.[3]

Other factors also play a part. With rapid increases in construction costs, for example, it can pay a housebuilder to install infrastructure and slabs well ahead of the expected programme of house completion. Subsequent rises in the cost of such work can at times make such actions highly profitable, even after the opportunity cost of the working capital required is accounted for. This example illustrates that even the timing of the stages of production can involve a degree of speculative calculation.

Small-batch production in many industries is the result of limited demand for specialist, bespoke products. This is not the case with speculative housebuilding. Small-batch production is adopted for its flexibility with respect to the rate of output rather than because of the bespoke nature of the product. Standardization of house types into a limited range of basic shells requiring a minimum of non-routinized building work is universal in speculative housebuilding. In fact, private housebuilding stands in stark contrast to the rest of the construction industry in the extent to which its product is standardized – a feature which has existed since the development of capitalism in the building industry at the end of the eighteenth century, as the uniform rows of housing in many nineteenth-century housing developments testify.

The use of standardized house designs means that construction costs should be well known beforehand. For each house type, the average labour and material requirements and acceptable variations around those averages are clearly specified from previous experience. Most large firms use computer-based cost control techniques, comparing forecast unit costs on each of their current sites with actual cost outcomes and taking action if costs get out of hand.

Standardization, of course, can only be achieved above ground. Below ground, site preparation, service and foundation work vary from site to site. Their cost, therefore, is not so easy to control. Particularly difficult sites might need expensive clearance, drainage or foundation work which initial site tests may not even fully reveal. Builders interviewed in the survey suggested that groundwork was the main area of cost un-certainty and variation. One, for example, said that typically site costs varied between £500 and £1000 per dwelling, a difference which may have a significant influence on profitability with lower-priced dwellings. The variability of groundwork costs has increased in recent years owing to more stringent requirements regarding foundation work necessary for NHBC certification. A series of structural failures following the build-ing boom of 1972–3 and the subsequent dry summer of 1976 led to this tightening of the regulations. The more solid the foundations the more likely is unforeseen additional groundwork.

Such potential cost problems concerning groundwork help to explain why private builders prefer well-drained, flat greenfield sites with a stable subsoil, as they require minimal groundwork. Such sites are scarce in a large number of localities so builders frequently have to make do with less than ideal sites. The potential problems associated with groundwork also mean that builders take a slightly different attitude to the labour process involved in this aspect of building. Site supervision and control is tight. Subcontracts are to specialist firms rather than on a general labour-only basis. The relatively expensive equipment (for a private housebuilding site) required for muck shifting and trench digging normally is hired complete with skilled operative.

Above ground, standardization comes to the fore. It has been possible for building firms to achieve standardization within small-batch production whilst minimizing the fixed capital tied up in productive equipment because of the method of production used. Handicraft assembly of 'traditional' but standardized and partly pre-assembled building components makes this possible. Traditional housebuilding consequently has survived in the speculative sector not because it was the already existing technique and speculative building is backward and conservative, but rather because traditional building methods are the most profitable; if they had not been there something like them would have been invented.

In recent years another technique, that of timber-framed building, has been widely introduced into private housebuilding and in many respects it fulfils the same economic criteria required in speculative building as traditional building. Discussion of that technique, however, is left until later as its introduction has been very much a result of the crisis of production associated with traditional methods. To explore in more detail what that crisis is, the building workforce must be considered in greater detail.

Building workers and private housebuilding

The most important issue related to the employment of building workers in the private housebuilding industry is the form their employment takes. The majority of workers are employed via labour-only subcontracting which is more commonly known as the Lump; generally workers also are self-employed. So instead of being employed directly by housebuilding firms as wage workers hired for their ability to work, most building workers formally are paid as subcontractors for services rendered. In practice, however, this distinction highlights the casual nature of building employment in speculative housebuilding rather than a unique form of relation between capital and labour. The reasons for the widespread existence in housebuilding of this form of employment centre on the struggle over control of the labour process between employer and worker.

Under a labour-only subcontract, the main contractor (in this case the housebuilding firm) provides the materials and most of the equipment required and the subcontractor is paid for carrying out the work. The labour-only subcontract differs from the predominant form of subcontract, known as supply-and-fix, in which the subcontractor provides the materials and equipment as well as the labour needed for the task.

Generally labour-only subcontractors provide only hand tools. The payment fixed in the subcontract may take the form of a lump sum for the completion of a specified task — in joinery, for instance, so many £s for the 'first fixing' of a given house; or a piece rate — so much, for instance, per 1000 bricks laid; or even an hourly rate, but this is quite unlike the usual basic hourly rate, in that it is the sole and comprehensive source of the payment due to the subcontractor.[4] Phelps-Brown (1968) found in a survey that the most common method of payment was a piecework rate and the payments made under this were usually made weekly. The labour-only subcontract may be let by a supply-and-fix subcontractor, so that it is then really a sub-sub-contract; and on the other hand it may not really be a subcontract at all, but be let directly by the client — as frequently is the case in private housebuilding.

The important principle to be recognized in labour-only subcontracting is that any payment made is directly related to the volume of work completed and the main contractor is absolved of any responsibility for statutory payments, these being the responsibility of the labour-only subcontractor. Within this broad definition labour-only subcontracting may cover a variety of arrangements embracing established firms operating as limited liability companies with a fixed place of business, or gangs under leaders or labour masters. These gangs may have a settled and continuing existence under a responsible leader who may also work with the gang at least part of the time but who also takes on the subcontract, pays the gang and bears the statutory responsibilities as an employer — stamping National Insurance cards, deducting PAYE, income tax, etc. At the other end of the scale, labour-only subcontracting consists of gangs, nomadic in character, constantly changing in composition, with all members functioning as self-employed persons.

The Lump is not confined to the building industry, it exists wherever casual employment is prevalent and trade unionism absent or weak. Even so, this form of employment is especially important in the construction industry and labour-only subcontracting is more common in private housebuilding than in any other sector. Within private housebuilding there are notable regional variations — the practice is most common in the South East of England and least in the North and Scotland. Although the Lump has grown considerably in importance in the other sectors of the building industry over the past twenty-five years, casual employment, piecework and lump-sum payments have been dominant features of employment in speculative housebuilding since its origins in the seventeenth and eighteenth centuries. These employment characteristics were a major reason for the distaste which the craft building unions of the nineteenth century had for speculative building. They exerted little influence over its conditions of employment; their battles instead were primarily with the Victorian building contractors and master builders, extending to private housebuilding only when such firms ventured there. The Lump has become a more formalized entity since then with the introduction of statutory employer obligations during the twentieth century. Many of the Lump's characteristics have only taken on major significance since the last war — income tax, for example, affected few workers until the post-war era. As employment legislation has increased so the significance of the Lump as a means of taking the employee/ultimate employer relation outside of that legislative framework increased; for instance, after the introduction of statutory holiday and redundancy payments and with the advent of selective employment tax (introduced in 1966 and subsequently abolished). Selective employment tax gave a major impetus to the growth of self-employment in the construction industry and with it the Lump (Phelps-Brown 1968). It should be emphasized, however, that the Lump as a means of circumventing employment and tax legislation is as much a way of identifying its existence as a reason for its existence. It will be argued later that such avoidance is merely ancillary to other fundamental economic reasons for its existence related to the nature of capital accumulation in speculative housebuilding (and elsewhere in the construction industry).

The use of labour-only subcontracting seems to have grown in private housebuilding over the past twenty years. Its absolute growth in construction as a whole during the boom years of the 1960s is well documented in official employment data. But its role in private housebuilding is less clear, as separate statistics are not available. Phelps-Brown (1968) found that a quarter of all payments for labour in private housebuilding were for labour-only subcontracting for Britain as a whole, rising to a third for larger firms building over 500 houses a year. Our interviews indicated a much higher proportion than this. To an extent the practice was difficult to quantify as some managers of housebuilding firms genuinely did not know how many workers were employed building their company's houses. This is not so surprising in the context of widespread subcontracting where trades come and go from sites having completed their contracted tasks. For the housebuilding firm employment is not the issue but the cost of the subcontracted work and its successful completion within the programme of site activity. The number of workers required to do the task, their formal qualifications, and even their individual competence is an issue internal to the subcontract gang rather than of interest to the firm which ultimately employs them.

Accurate data as a result are almost impossible to get. The same is true for distinguishing between the number of workers employed on supply-and-fix subcontracts and on labour-only ones. It is to be expected however that most subcontracts are on a labour-only basis for, with the exception of a few specialist trades, the housebuilder is usually able to purchase building materials more cheaply than the subcontractor through bulk purchase discounts. In such circumstances supply-and-fix is more expensive than labour-only.

With these provisos in mind, most firms interviewed indicated that subcontracting had increased substantially over the past decade. In a number of regions, it was suggested, many craft workers could only be employed on a labour-only basis. Only two firms interviewed employed more than half their workers direct and many others employed only a handful of workers direct on clearing up, carrying and remedial tasks and subcontracted the rest. Like the Phelps-Brown survey, labour-only subcontracting appeared to be more prevalent in the South than in the North West and East Midlands. Yet in the latter areas still half of the work and frequently far more was subcontracted. Firms working in Scotland also reported the widespread use of the practice there, whereas ten years previously it was virtually non-existent. The two firms who were major exceptions to the large-scale use of labour-only subcontracting were themselves unique in that they were also involved in large-scale public sector housebuilding and liked to switch workers between the two sectors to maintain a core of good, skilled workers. This concern for stability of employment is likely to have arisen because of the institutional features of public sector housebuilding where local authorities frequently try to ban the use of Lump labour, rather than because of any inherent preference on the part of the two firms themselves for directly employed labour.

Whilst it is clear that many workers in private housebuilding prefer working as labour-only subcontractors, obviously it is the employers who have the overriding influence on the form of employment of the workforce. With few qualifications, all of the managers of the firms interviewed stated strong preferences for using subcontracted labour. The main reasons given concern control over the production process in two interrelated ways: first, in terms of getting specific work tasks done at the lowest cost and, second, in maintaining a flexibility over the timing of production by minimizing the degree to which labour becomes a short-run fixed cost.

Traditional housebuilding requires workers with different skills to be sequentially involved in the erection of a dwelling with limited supervision and management. Because of the spreadout pattern of work on a site, supervision principally involves checking that tasks are undertaken as planned, and management functions are restricted to organizing the flow of work and materials. Workers consequently have considerable potential leeway over the detailed execution of their jobs and the pace at which they work, especially at times of labour shortage when the ultimate threat of dismissal can be used only sparingly by management. Building firms use payment-by-results methods to overcome this potential lack of control over the workforce. Workers then control and supervise themselves on terms specified by management as their wages depend on the amount and type of work they do.

By the judicious manipulation of piecework rates management consequently gains greater control of the labour process. This policy is likely to be all the more successful the greater the proportion of wage payment tied directly to piecework. Forms of employment contract like the Lump, therefore, do not simply influence the way in which wages are paid and how the other conditions of employment are determined, they fundamentally influence the struggle between management and labour within production over the pace and type of work done.[5] Direct employees' wages are governed by working rule agreements negotiated at industry level between employers and trade unions. This means a large portion of direct workers' wages cannot be tied to the pace of work; only about a third on average is a genuine piece rate. Under working rule agreements payments also have to be made by employers for nonproductive work time. The most frequently cited case by the managers interviewed was wet time, a payment to direct workers temporarily made idle by a shower of rain. Lump-sum payments to labour-only subcontractors, on the other hand, are solely based on piecework, either as individual workers or at the level of the Lump gang, whose individual members police the work rates of their fellow gangers. Within this payment by results framework, workers who are too slow, uncooperative or produce poor-quality output can have their subcontracts instantly quashed. So, paradoxically, by not employing workers direct, management gains a greater control over the labour performed.

The success of this form of control of the workforce was clearly brought out in our firm survey. Although all the managers interviewed said that Lump labour was more expensive than the total cost of employing workers direct for an equivalent time because of the higher rates that had to be paid, they all believed that the higher productivity of subcontracted labour significantly outweighed its higher cost.

Payment by results methods depend crucially upon the ease with which the results can be quantified. The standardized nature of speculative housebuilding makes this particularly easy. Bricklaying, for example, can be let at a rate per 1000 bricks laid; the more intricate skills of the bricklayer's craft over which there could be dispute concerning the rate for the job are not required. (Modern private housing does not generally even have a brick chimney.) Skilled workers themselves push for the simplest and most repetitive tasks — job satisfaction is sacrificed for the much higher wages received in repetitive work. The quality of work, however, cannot be quantified. But the simple nature of private housebuilding does not require high-quality craft work. Firms, moreover, can force subcontractors to make good faulty work at their own expense. Yet too much emphasis cannot be placed on quality without undermining the piecework system itself, and attempts to force subcontractors to make work good can lead to them walking off site. The Lump can justifiably be criticized for its adverse effects upon product quality. To cite the brickwork example again, a cursory visit to any recent private housing estate will invariably indicate even to the non-expert that emphasis has been placed on the speed of bricklaying rather than on its final aesthetic quality.

In addition to being able to control the pace of work, labour-only subcontracting also enables the building firm to control and to vary its price. Lump-sum contracts by their very nature are fixed-price contracts for completing specific tasks. Once having fixed the price, therefore, building firms know how much the task will cost. With direct labour, on the other hand, the ultimate cost depends on the management skills of the foremen and site managers as well as on the wages paid to the workers; its cost consequently is inherently uncertain. All directly employed workers, moreover, have to be paid the same rate for the same job. This is not the case with labour-only subcontracting. Each contract can be fixed at a different price if necessary. Two workers working side-by-side on the same job can be paid widely different rates, yet neither is likely to know so. If a job is urgent or a higher price has to be paid to induce the last few workers on to a site, this can be done by varying the lump-sum price. Price incentives can work therefore at the margin. With directly paid workers this is not the case. As all workers must be paid roughly the same rate, the need to employ an extra few workers forces firms to raise the general level of wages (or earnings if it is a bonus payment). All workers therefore gain from labour shortages with direct employment whereas only the marginal ones do with the Lump. The power of the workforce as a whole to take advantage of economic expediency is thus weakened when the Lump is prevalent to the advantage of the employer.

The other advantages of labour-only subcontracting for the building firm concern the need for flexibility of production and the associated minimization of working capital. Overheads for the firm are minimal with labour-only subcontracting, and working capital is expended only for definite pieces of work and after they have been completed. Advocates of the Lump argue that this flexibility feature is one of its main advantages (as it is said to be for subcontracting in general). It enables discontinuity of work at the level of the firm to be combined with continuity at the level of craft trade, because subcontract labour can move between firms when work with one comes to an end. This argument relies however on an implicit assumption of excess demand for labour. In periods of downturn in the construction industry this obviously is not the case, for when one subcontract is completed another is unlikely to be found for some time. The employment practice then becomes one where a discontinuity of work at the level of the firm is borne by an unemployed workforce rather than by firms as excess capacity.

Directly employed building workers are not permanent employees – labour turnover in construction is one of the highest of all industries – yet their cost to the employer is not so flexible as that of subcontracts. In the first place, direct workers are hired for definite time periods rather than for specific tasks. Consequently they can only be dismissed after due notice in accordance with the National Working Rule Agreement; usually at least one week's notice is required. Other working rule and statutory requirements also operate: holidays, temporary lay-offs, sickness and injury payments, travel allowances, dismissal procedures, redundancy pay, health and safety and employment protection legislation, for example. They all add to the relatively fixed element of wage costs and require office staff to administrate.

Whilst it is possible to identify quite considerable advantages for building firms of the Lump employment contract, that alone is not enough to explain its existence and expansion within private housebuilding and the construction industry as a whole. An advantage to an employer is generally a disadvantage to workers, so the historical conditions need to exist which enable such advantages for the employer to be realized. The central historical point about labour-only is that this Lump contract enabled building employers to retain the casual nature of employment during economic conditions which otherwise would have undermined it. The boom years in the construction industry of the 1950s and 1960s created considerable labour shortages, especially of skilled workers. In these conditions, without the Lump, building unions could have pushed up wage rates and improved working conditions, and possibly achieved their long-term goal of a decasualized industry of registered building workers. To an extent they were successful in raising wages but the sharp growth in the Lump considerably weakened their organizational strength, so the power of building trades unionism actually declined whilst employers clamoured for more workers. The bricklayers' union, for example, reported in 1965 that it was 'financially bleeding to death' (Austrin 1980).

Building sites are notoriously difficult places for trade unions to organize. The problems are particularly great when, as in Britain, the unions are themselves organized on craft lines and the site is the focus of local activity. On any site, workers will come and go, be demarked by trade and by employer, and employment practices on site can vary sharply between the firms operating there (as main contractors or subcontractors) and between the varied forms of employment contract that will exist. It is difficult to get unity between trades (especially between those formally skilled and unskilled), and individual workers are more likely to leave for better pay or conditions elsewhere than involve themselves in protracted disputes on one site.

Private housing sites exhibit these characteristics to a marked degree. The number of workers generally is low, few stay for long periods and the division into trades is great. Moreover, if a dispute starts the housebuilder can easily stop building because there are no penalty clauses for late completion hovering in the distance as there are within the contracting system. At site level, consequently, organization is virtually impossible; unionization in private housebuilding must stem from wider union practices. Local union branches need to be able to influence the employment of workers in their localities by insisting on union members only being employed under union-negotiated conditions. The weaknesses of the building unions have made this policy almost impossible to implement. In England, only in the Liverpool and London areas has this strategy been partially successful although in Scotland unionization has been far higher, at least until the mid-1970s.

The long-term weakness of the building trade unions in the construction industry as a whole is clearly brought out in figure 6.1 showing average earnings and weekly hours for construction, manufacturing and all industries and services. Despite being an industry characterized generally as one of acute labour shortage from the war until the early 1970s, earnings in construction are well below those in manufacturing except for a few

Employees' earnings in construction and manufacturing industries in relation to all industries and services. Average gross weekly earnings of full-time male manual workers aged 21 and over: October 1948–79: United Kingdom

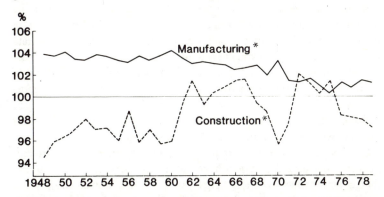

Employees' hours in construction and manufacturing industries in relation to all industries and services. Average total weekly hours of full-time male manual workers aged 21 and over: October 1948–79: United Kingdom

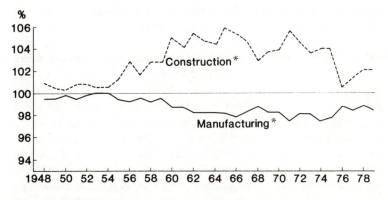

Figure 6.1 Construction employees' earnings and hours, 1948–79

Source: Department of Employment: October survey presented in *Housing and Construction Statistics 1969–79*, HMSO

* As % of all industries and services covered by the October survey

years after the momentous agitation of the early 1970s which culminated in the 1972 national building strike. For many years construction earnings have been below the national average for all industries and services, whilst fluctuating quite sharply within the short-run building cycle. To get those earnings, moreover, construction workers have had to work significantly longer hours, as the second chart in figure 6.1 shows.

In such circumstances the attractions of the Lump to many individual workers is clear. Employers faced with labour shortages are prepared to offer individual workers earnings well above the union rate and Lump contracts make this feasible. The gains to the workforce remain only at the level of the individual worker and are negotiated via the Lump with an individual employer, instead of union officials being able to negotiate high national wage rates which are then supplemented by additional local pay agreements won by shop stewards for all workers on sites in areas of acute shortage. Collective action via unionism could not win those gains because of the prior existence of the Lump. This fact encourages even more individual workers to go on the Lump, thereby further weakening union power, and so on in a vicious downward spiral.

It has been argued by many that labour-only subcontracting is a product primarily of excess demand with workers being attracted to it because of the higher earnings there (cf. Phelps-Brown 1968). This would imply that during downturns workers flood back into direct employment and trade unionism.[6] But it is difficult to see how trade unions, which have been weakened by the growth of labour-only subcontracting during a building boom, should suddenly be able to exert a strong influence on wages and conditions during a downturn. Individual workers might wish unions to negotiate for them then, but market conditions have turned against a union's ability to exert its power. Employers in such circumstances are also highly unlikely to succumb to the growth of trade unionism at the expense of the Lump. As argued earlier, labour-only subcontracting has definite advantages for them, so why should they accept the more expensive conditions of employment of trade union negotiated working rule agreements at precisely the time when the state of the market for labour-power has swung in their favour? More likely is a reduction in labour-only subcontracting rates of payment. So the Lump is a product of the weakness of trade unionism, not of the state of demand alone. The unions' weakness is determined by many factors amongst which must be included the jealousies of craft unionism.

In contrast to England, trade unions in Scotland have been far more successful until recently in excluding labour-only subcontracting, and the effect has been to increase overall earnings as the following report on the situation in the late 1960s makes clear:

> The labour market in England is therefore in a position to adjust to particular levels of demand through the presence of these marginal craftsmen. By comparison in the industrial belt of Scotland it is necessary on most contracts for men to be members of an appropriate trade union, and union membership is only available to 'time served' craftsmen. . . . In this situation, shortages of labour are difficult to overcome hence the higher earnings and more overtime worked in Scotland, increased overtime being an inducement offered to prospective employees rather than any requirement of a particular contract.
>
> (SHAC 1970, para. 103)

From the point of view of the individual worker on private housing sites direct employment has little to offer (at least in most of England and Wales) in terms of the influence of collective action via trade unionism. One result is that the fringe benefits of tax avoidance and evasion associated with self-employment increase in their attractiveness as direct employment has few alternative advantages. Workers consequently choose frequently to go self-employed rather than 'cards-in'. The pressures discussed above that lead to the impossibility of anything but an individualized, casual employment relation between worker and housebuilding firm thus receive their ultimate ideological vindication: workers actually end up choosing the form of employment created, preferring it to the alternative denied to them. Some 'freedoms' it would seem really are about nothing left to lose.

The reproduction of a building workforce

The argument concerning the economic position of workers in the previous section centred on the point that individual actions have systemic effects that weaken the position of all workers in the longer term, including the individuals in question themselves. The same argument can also be applied to firms in the industry via the effects on the long-term reproduction of a building workforce that can be hired, whatever the form of contract. Nowhere is this clearer than in private housebuilding. The short-term advantages for employers, outlined above, used to justify the widespread existence of labour-only subcontracting in the construction industry consequently are outweighed by the longer-term negative dynamic effects that are created. There are three interrelated effects that need to be considered. Two of them relate to the supply of building workers and the third to production techniques.

Many aspects of building work because of its handicraft nature involve skilled work. Those skills are learnt through formal apprenticeship schemes or via shorter training courses and the experience acquired over time on sites. Since the mechanization of most lifting and carrying work most workers are now either formally or informally skilled. An adequate pool of skilled workers therefore needs to exist for building to proceed smoothly. The casual nature of building work and the frequently poor conditions of work associated with it operate against the existence of such an adequate pool. During downturns in building activity many skilled workers made unemployed move to other industries and often never return to building even in the next upturn. The low historic level of construction wages relative to other industries plus the fear of further periods of unemployment makes a move back to construction an unlikely choice for most. Wages have to rise quite substantially in periods of shortage to attract workers back. This problem of building worker attrition has increased markedly during the 1970s with the overall decline in output and, especially in private housebuilding, with the increasing volatility of output levels.

The casualized forms of employment adopted by individual firms, like the Lump, strongly reduce the incentive to train workers. Few of the housebuilding firms interviewed in the survey, for example, had training schemes and of those that did most were very small relative to the total labour requirements of their housebuilding operations. Labour-only subcontractors are not directly involved in training either, so the traditional avenues of training are rapidly drying up (TRG 1981). To an extent they have been taken over by government sponsored schemes, especially via the Construction Industry Training Board to whom all private employers must pay a statutory training levy. Yet there is growing concern within the construction industry over the increasing inadequacy of labour supply except in severely depressed conditions.

The problem by its very nature is growing over time so that even mild upturns in activity may create bottlenecks and escalating costs. A good example is the mild upturn in construction activity during 1979, which led to a chronic shortage of some trades as employer questionnaire returns given in figure 6.2 show. Only 8 per cent of firms said brick-layers were readily available, for instance, at the height of the revival in June 1979, whereas fifteen months later when the industry was again accelerating into slump 81 per cent reported them readily available. It is

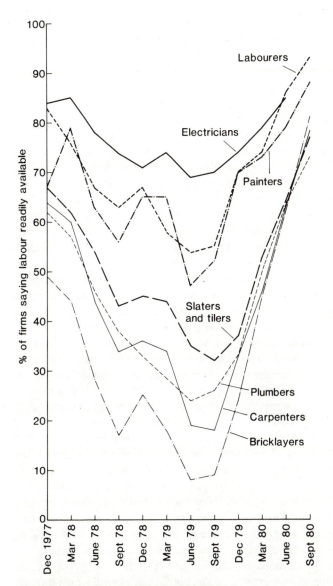

Figure 6.2 Building workforce availability by trade: quarterly employers' enquiry, December 1977–September 1980

Source: NFBTE State of Trade Enquiry

interesting to note that it is the least substitutable trades that became most short during the upturn (i.e. bricklayers, carpenters and plumbers). There is one exception to this rule, electricians, because they do not conform to normal employment practices. They are highly unionized, there are extensive training schemes for them, and frequently they are permanently employed by specialist subcontractors.

When labour-power becomes short during upturns firms start to use gangs of workers of poorer quality in order to achieve their target output levels. The timing of production to hit peaks in demand is especially pronounced in speculative housebuilding, so this sector tends to lead the way in the trend towards lower-quality work. The quality of work, moreover, is not the only problem: less-skilled workers also are less-productive workers. Novice bricklayers, for example, cannot lay bricks at the same speed as a skilled operative; costs therefore start to creep up as well. So one consequence of the loss of more skilled operatives is likely to be a fall in average labour productivity over time.

Many speculative housebuilding firms try to circumvent these problems by developing long-term relationships with a core of subcontractors, paying them well and on time, and also by trying to offer them as continuous a work programme as possible. Some firms even said that they would build slabs ahead of requirements so that labour-only subcontract gangs could see that there was more work available after the current batch of dwellings was complete. But these practices can only be an individual company response to a deteriorating overall situation.

One effect of a growing labour shortage has been a greater mechanization of certain site activities. One example is the increasing use of fork lift trucks to transport palletized packets of bricks around site and to lift them to the required height for bricklaying. A number of firms interviewed said that this practice had grown because it was now impossible to get labour-only bricklaying gangs to carry bricks as they used to. The cost has been transferred to the housebuilder and the operation mechanized as a result.[7] This example, however, is a rare case of labour-only subcontracting generating a change in site practices and building techniques. More generally, it produces an ossification of them into established procedures. Innovation usually requires a period of implementation when productivity and output actually fall whilst management and workforce learn the new process and iron out the inevitable faults. Such a stage is anathema to labour-only subcontracting as it results in a sharp reduction in earnings. A similar argument can be made for the development of new skills associated with new techniques. Even where innovation does occur with labour-only subcontracting there is a greater tendency for failures and subsequent remedial work as speed is of paramount importance to the operative. The form of control by management over the workforce made possible by the Lump therefore tends to perpetuate the pre-existing division of labour, and also the use of tried and tested materials and techniques.

Building materials

Innovations in building materials have always been a major impetus to changes in techniques in housebuilding. In the nineteenth century, for example, brick making ceased to be a site-based handicraft activity and became concentrated into large-scale brickworks. Later in that century there were revolutions in glass making and cement production. The inter-war period saw the advent of the concrete roofing tiles and plasterboard and the pattern of product development and innovation has continued since the last war.

Two of the most notable changes over the past twenty years have been related to plastics and woodworking. Moulded and extruded plastics, for instance, have revolutionized the installation of rainwater and plumbing systems above ground. Reactive innovations in clay pipes, however, have limited the penetration of plastics to underground piping. The housing drainage market is now dominated by one firm's (Hepworth Ceramic) 'Hep Sleve' system of clay pipes and fittings. With respect to woodworking, standardized window, door and staircase units are now universal. They are made in factories using computer controlled cutting machines and delivered to site ready for installation. Perhaps the most dramatic change has been trussed rafter roofs. They are lightweight, prefabricated timber truss units that have almost entirely replaced traditional roof truss construction over the past fifteen years. Made in factories, these units consist of thin timber members of uniform thickness fastened together by metal plates or plywood gussets. On site they can be erected by semi-skilled labour in a few hours.

One important feature of these developments associated with wood and plastic is that they have replaced far heavier materials. The load that the structure has to bear consequently has been considerably reduced, leading to further economies in foundation and superstructural work. Modern 'traditionally built' houses are now much lighter and less robust than their inter-war predecessors. Chimney breasts were made obsolete by central heating. The inter-war traditional heavy rafter roof also required solid structural support from all four outside walls and from internal masonry partitions. The trussed rafter roof is lighter and requires support from only two of the four walls whilst removing the need for internal partitions to help support the roof. Changes in house structures illustrates a general feature of materials innovations: the ones that are successfully introduced into speculative housebuilding are those that directly replace specific building components, initially lowering materials, and possibly labour, costs. They then may have a knock-on effect on other parts of the building process. But traditional building methods and the division of labour associated with them are not brought into question nor do they reduce the flexibility of the builder's working capital.

The larger housebuilders tend to be keenly aware of the benefits of materials innovations, scouring the trade press and producers' federations for knowledge of new materials and having centralized testing units to see whether the new products meet their requirements. With smaller producers, the process is far more haphazard. Traditional materials usually are used until the new ones have gained general market acceptance; conservatism and the fear of material failure outweighing the initial cost advantage of the new products. Failures in new materials, particularly when wrongly stored or erected, are quite common, so this product conservatism is not quite so irrational as it initially might seem. Even trussed rafters have had their long-term stability questioned after a series of expensive failures (Baldwin and Ransom 1978).

Many materials industries are highly monopolized. In 1981, for example, flat glass and plasterboard were each produced by only one firm (Pilkington, the world's largest flat glass producer, and BPB Industries); three firms had a 90 per cent share of the cement market – the market leader, Blue Circle, alone had a 60 per cent share and all the main producers have common price and marketing arrangements; two firms, Marley and Redland, had 95 per cent of concrete roofing tiles; four had 85 per cent of sanitary ware, where again Blue Circle is the market leader via its Armitage Shanks subsidiary; two firms, Ready Mixed Concrete, the largest world producer, and Amey Roadstone, a Consolidated Gold Fields subsidiary, had 45 per cent of ready mixed concrete, and in ceramic tiles the Norcros subsidiary, Johnson-Richards, had a 62 per cent market share (Savory Milln 1981). The London Brick Company (the largest brick manufacturer in the world) is famous for its dominance of brick making, with 45 per cent of total UK brick deliveries in 1979. It is the only producer of Flettons, the standard common brick.

The widespread monopolization of so many building materials has been argued to be a product of technical innovation, where continuous process plants generate substantial economies of scale, leaving space in the market for only a few large-scale producers (Bowley 1960). These large producers are then argued to be in a position to sustain technical innovation, so a succession of reports by the Monopolies and Mergers Commission and the Restrictive Practices Court have claimed that such monopolies are in the 'public interest'. Developments in the production facilities of the London Brick Company can be argued to illustrate these features: its currently planned factory at Whittlesey in Cambridgeshire will produce 5 million bricks a week, with an annual capacity equivalent to almost 15 per cent of total common brick production in Britain in 1979. Such new brick factories require less than a third of the labour force of earlier works. Massey and Meegan (1982) report in detail on the restructuring of the brick industry. Similar productivity gains have been made in concrete roof tile production in recent years (Savory Milln 1981).

Yet there is, of course, another side to monopoly suppliers: their ability to control market prices and the overall quantity that can be supplied. In the materials industries there is a need to distinguish between materials that are mainly imported (like timber) and those dominated by home producers (like bricks, cement, etc.). The price of the former (and their availability) is determined in world commodity markets, and the latter in the boardrooms of a handful of companies. It is with regard to these latter domestically produced materials that certain specific features can be drawn out. Such producers want to maintain a steady optimum output for the expensive fixed plant they have invested in. This means they will never set capacity to meet maximum demand. In booms, when stocks run out, long delays in materials availability may result. In such situations bottlenecks in the building industry appear and construction costs rise through the delays created by materials shortages. In downturns, producers again try to keep their plant running as closely as possible to its optimum capacity. This can be done by building up stocks; in 1980 for instance there were 1100 million bricks in stock, almost ten times the 1972 level. If the downturn in demand persists, plants are put on short time, say, on to a four-day week (and, in the long term, capacity is cut). Both stockholding and short-time working increase producers' overheads, and as monopolists they can pass on much of this increase in costs as price rises. They can also try to maintain overall profits in a similar way by raising the profit mark-up per unit of output.

These patterns of price response can be illustrated by looking at the price changes of certain materials from 1970 to 1980, which are given in table 6.1. It can be seen that there are wide variations in the rate of change of materials prices. By far the most volatile of those shown is timber; the world commodities boom of 1973 sent its price rocketing with a 75 per cent increase in one year; two years later when the world boom had long passed its price actually fell. The behaviour of prices of the other materials listed is much more sedate, following in the main increases in input costs as would be expected in monopolistic industries. If anything the rate of price change is negatively correlated with the volume of deliveries: the peak year of demand for cement, 1973, saw no increase in price at all! Yet as deliveries declined prices were pushed up rapidly, partially because of rising input costs (especially of energy in the wake of oil price rises) and presumably also because of the costs of excess capacity. Few industries, for example, could raise their prices by 40 per cent in one year at a time when demand was falling sharply, as occurred in the cement industry in 1975.

The private housebuilding industry is not the sole source of demand for building materials but it is still a significant customer. 55 per cent of London Brick's brick deliveries in 1979, for example, were to private housing (Savory Milln 1981). The needs of the private housebuilding

Table 6.1   Price increases for selected building materials, 1970–80

	Bricks		Cement		Building blocks	Timber (imported softwood)
	Price rise %	Deliveries (m.)	Price rise %	Deliveries (tonnes)	Price rise %	Price rise %
1970	9.8	6,356	16.7	17,014	10.4	5.0
1971	15.6	6,825	20.4	17,544	7.5	4.8
1972	13.5	7,023	5.1	17,883	12.3	6.8
1973	11.9	6,998	0.0	19,847	12.5	74.5
1974	16.7	5,011	14.5	17,388	15.3	32.9
1975	29.9	5,467	40.1	16,681	20.5	−8.3
1976	20.0	5,380	23.0	15,452	20.0	32.0
1977	18.3	4,717	18.7	14,389	19.2	20.5
1978	12.7	5,107	13.0	14,745	14.7	4.3
1979	15.0	4,909	19.4	14,938	22.0	16.0
1980	27.7	4,046	30.6	14,034	20.5	13.6

Source: HCS

industry, as dictated by the economic pressures faced by it, are clearly in contradiction with those of the materials producers. Speculative house-builders are geared up to rapid changes in output levels and increasingly have to be so. Materials producers need steady levels of demand. They do not get them from housebuilders but instead have to act as the materials buffer stock for the industry. The result is that in housebuild-ing booms materials shortages hold back construction and in housebuild-ing slumps the profitability crisis faced by builders is exacerbated. The latter occurs because material producers' unit costs rise when house-building is cut back and so those producers pass on the cost increases in price rises to the same builders already facing a slump. This means that builders' development profit margins are squeezed even further, so they cut back their output yet more. Once again, the organization of pro-duction in speculative housebuilding produces effects that go way beyond the confines of the individual enterprise to generate notable and adverse system-wide consequences.

The growth of the timber-frame building of private housing[8]

Over the past few years there has been a dramatic increase in the use of industrialized timber-frame building methods in private housebuilding. Some of the major producers like Wimpey, Barratt, Tarmac and Laing have almost entirely switched to this method, and overall it accounted for about 20 per cent of new output in 1980, and this is expected to rise to well over 50 per cent in the new few years.* This revolution in tech-nique would seem to belie the arguments presented earlier of a crisis of production; 'technology' has come to the rescue once again. But a more detailed investigation of this innovation reveals that it is very much an extension of the traditional organization of labour in speculative build-ing, with all its attendant problems, and its cost savings are small com-pared to earlier approaches.

Timber-framed building is the traditional form of building in Britain. It consists of a structural frame of timber onto which many different weatherproofing skins can be applied. During the seventeenth and eighteenth centuries brick structures replaced timber as the main form of building in Britain. In other countries where timber supplies remained plentiful, and low population densities limited the risk of fire, timber-frame remained the dominant housebuilding technique. In North America and Scandinavia timber is the most common form of house

* According to NHBC data its share in 1982 had risen to 34 per cent (44 per cent for volume builders alone).

construction. It is not surprising, therefore, that the recent innovations in timber-frame methods were developed in those countries, and they have been imported into Britain since the late 1960s, at first for use in public housing then later in the private sector. Whilst timber-frame has this long and continuous heritage, the new methods are very different from their predecessors. Traditional timber structures are heavy and require substantial cutting by skilled carpenters on site. Modern timber-frame structures are lightweight, are manufactured in kit-form in factories, and can be rapidly and easily assembled on site. With modern timber-frame, houses can be built from foundations to completion in ten weeks or less.

Computer design techniques and computer-controlled cutting machines have generated this revolution in timber-frame. Together they have enabled the evolution of wood-minimizing structures, using cheap softwood studs covered with thin plywood skins, which have high degrees of tolerance and structural strength. The weight of floors, roofs, etc. are spread through the structural members of the walls and partitions, so the superstructure as a whole is far lighter than with traditional housing. In contrast to earlier concrete industrialized systems such lightweight structures can be transported considerable distances, and they do not have to be assembled in large numbers to become economically viable. The design and manufacturing techniques of modern timber-frame also enable product variety within certain structural and standardized length constraints, so a large number of different house types can be assembled on site. All of these features enable economies of scale in factory production with low building rates of varied house types on individual sites.[9] Site assembly is sequential with the most popular method, platform-frame. First the ground floor joists are fixed, followed by wall panels and partitions up to the height of the first floor, onto which the first floor joists are fixed. Panels for the first floor are then erected and roof trusses placed directly above the vertical studs in these panels (see figure 6.3). The whole timber-frame structure can be assembled and made weather-proof in a few days, after which a brick (or other material) skin can be built around the frame, doors and windows added, the roof tiled and the second-fix internal work completed. Once finished the external appearance of the dwelling is virtually the same as a traditional house, distinguished only by the greater accuracy and sharpness of its dimensions.

To understand why the method is becoming increasingly popular in speculative housebuilding, its effect on the production process must be

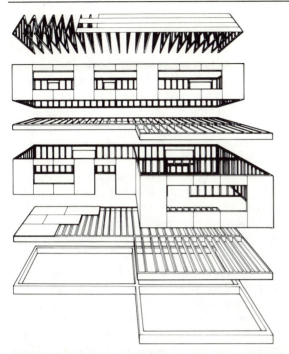

Figure 6.3 The sequence of erection of platform timber-frame housing
Source: The Housebuilder, February 1981

considered further. An important feature is the speed of site assembly, but associated with that is a change in the composition of labour tasks required. Traditional building is crucially dependent on two skilled trades to assemble the basic shell of a house: bricklayers to raise the load-bearing external and internal walls and carpenters to fix the joists, rafters, floorboards, stairs and door and window frames. As figure 6.2 showed these two trades are in shortest supply. Timber-frame is essentially an extension of the gradual off-site mechanization of the carpentry trades, one which has managed entirely to remove the functional role on site of those two trades. Their place is taken by a small number of semi-skilled factory workers working in highly automated factories, and on site by small gangs of timber-frame assemblers and nailers. There is conse-quently both a reduction in the amount of skilled labour required on site and a change in its place in the sequence of production. Other stages of production, for example, no longer have to wait completion of the brick and block load-bearing walls. They can now proceed simultaneously with the external bricklaying once the timber-frame shell is erected, or even in advance of the brickwork if necessary. This gives the house-builder much greater leeway over the organization of trades in situations of local or temporary labour shortage. Timber-frame therefore helps to generate a greater reliability of production rates by limiting the problem of a temporary non-availability of trades. This not only helps production to be closely geared to sales but also increases the potential bargaining strength of employers in periods of labour shortage, shown earlier to be a growing feature of private housebuilding.

At the same time, like other building material based innovations, timber-framed methods enable subsequent knock-on rationalization of the labour process beyond the initial erection of the building's load-bearing shell. The technique has encouraged the complete elimination of wet trades from inside a dwelling, for instance. The nailing of standard-ized sheets of plasterboard, that need no cutting to fit awkward shapes, replaces the plastering trades. Plumbing and electrical fitting is simpli-fied and rationalized by being forced to conform to the rigours of the timber-frame structure (holes cannot arbitrarily be knocked in it). Externally the facing material can be varied depending on the relative costs and availability of labour and materials. So, whilst it does not remove the basic handicraft nature of housebuilding, timber-frame enables a considerable reorganization of its constituent components and via standardization imposes greater work disciplines on individual trades.

By conforming to, rather than transforming, the pre-existing relation-ship between building firms, the workforce and materials suppliers it has been possible to introduce timber-frame to private housebuilding. The great advantage to builders is that it enables them to concentrate on getting the maximum development profit because of the method's speed and productive reliability. Yet timber-frame's technical compatibility with these pre-existing relations also means that it helps to exacerbate the problems created by them. Earlier it was argued that private house-builders rely on having pools of unemployed labour-power and materials supply to give them productive flexibility; pools that are organized so that firms can pass on potential short-run profit reductions as price reductions to input suppliers yet avoid profit rises being soaked up in rising input costs. Housebuilding firms' success in evolving that organ-izational framework (particularly with labour-power) created, it was argued, the general longer-term problem of the reproduction of these input pools. Timber-frame puts an even greater emphasis on short-run flexibility and on organization of inputs to accommodate it: reducing the period of production without increasing the overall level of output implies greater idle time or reduced capacity working for workforce and materials producers alike. So, whilst the technique may enable some lowering of the use of skilled labour on site for individual firms, it exacerbates the problems of production (and its continued reproduction) for the structure of provision as a whole.

This issue can be seen in microcosm by looking at the ways in which timber-framed buildings are erected. There are various possible relationships between housebuilder, designer, timber-frame producer and erection gang. The builder, for example, may provide the drawings to the manufacturers or, more commonly to get scale economies, will order one or more variations of a standardized package. The supplier delivers the units to site where they may be erected by the builder's own workforce or on a supply-and-fix basis. A skilled gang is required to assemble a timber-framed building (albeit with a different type of skill from more traditional tasks) and the work must be done under knowledgeable supervision. If housebuilders use their own labour force and management, trained to carry out the task adequately, they lose the flexibility of a casualized workforce or incur high and recurrent training and learning costs if new gangs are periodically brought together. An important component of the workforce becomes a short-run fixed cost whose efficient employment depends on steady output rates. Yet owner-occupied housebuilding is not conducive to steady production rates. If the builder opts for the supply-and-fix alternative they enter the realm of a small handful of manufacturers faced with high short-run fixed costs (particularly if they have a supply-and-fix workforce) and a monopolistic pricing strategy. Building firms face consequently their traditional production dilemmas. Opting for self-employed gangs or a learning-by-doing approach with the workforce currently available does not resolve the problem either, as cost and quality control uncertainties become acute.

As an industrialized system timber-frame needs industrial standards of accuracy of assembly. Already there has been some disquiet over its erection standard. The National Housebuilders Council found in an extensive timber-frame site survey in 1980 that on 15 per cent of sites there were no nailing instructions; on 13 per cent inadequate erection instructions; on 15 per cent damaged or incorrectly placed internal wall vapour barriers, and on 14 per cent similar problems with outer breather paper (*Construction News*, 12 February 1981). The Council plan to step up their guideline requirements, and the failures that will inevitably ensue from such erection faults will lead to progressively more restraints.[10] A casual workforce cannot erect such systems adequately, yet decasualized the workforce becomes an expensive overhead cost with the sharply fluctuating output of this speculative industry.

The importance of these probably adverse future trends is emphasized when it is realized that timber-frame building offers little or no production cost advantage over traditional methods. As with all cost comparisons in building it is difficult to get accurate data, yet for all of the respondents in the interview survey of housebuilders, timber-frame's production cost advantage was minimal at best. What persuaded the larger housebuilders to switch to the technique was overwhelmingly the reduction in working capital resulting from quicker production times. Less gross development profit therefore is swallowed up in interest costs. Its speed, moreover, makes it possible to gear production even closer to sales, and the high degree of thermal insulation that can be achieved with the system has subsequently been used as an important sales feature. The switch to timber-frame by housebuilders consequently is one of careful economic calculation based on prevailing interest rates and the relative costs of different types of labour-power and materials. An earlier growth in use of the technique, for instance, in 1972–3 was brought to a sharp halt by spiralling timber costs (see table 6.1).

During the latter half of the 1970s the price of imported timber fell relative to bricks and cement and this helped to spark off the second major move by private builders into timber-frame production. The reasons for these relative price variations were discussed earlier and they highlight the importance of a set of historically unique circumstances that easily can be reversed during a subsequent economic upturn (or by a world shortage of timber). Such unique circumstances have been compounded by a crisis in timber merchanting. Most of the major timber-frame producers have evolved out of firms specializing in traditional timber merchanting and joinery manufacture. The collapse in overall building work throughout the 1970s, combined with the volatility of imported timber prices, the high costs of stockholding and a trend towards direct customer sales by North American and Scandinavian growers, hit these firms badly. Their structural plight forced the UK's two leading timber merchants, Montague L. Meyer and the International Timber Corporation, into a protective merger in 1982. Timber-framed systems were for merchanting and joinery firms both a means of diversifying and of creating alternative uses for wood. As with many other building innovations, public housing was initially used as a testing ground and a means of building up necessary plant, expertise and marketing networks in the protected cost-plus environment of public sector package deals. Between 1973 and 1976 the public sector output of timber-framed houses more than quadrupled from under 3000 to over 13,000 units completed. During this period a handful of producers built up strong market shares. Nine producers produced over 80 per cent of the record 1976 output. Regionally some market shares were even greater. Yet after 1976 the public sector market collapsed, dropping to under 5000 units by 1978 (Cullen 1982). Producers were back where they started in 1973 except for all the additional deadweight of the expenditure used to build up their timber-frame businesses. They then began in earnest to try to attract the private housing sector. Whilst meeting with some considerable success in terms of market share the continued decline in housing starts meant that in 1980 only about the same number of units were sold in total as in 1976.[11]

The point of this tale is that speculative housebuilders are switching to timber-frame in circumstances where supply conditions are strongly to their advantage: wood prices are relatively cheap and timber-frame producers have kept prices keen to gain market share. Those conditions will not always last, yet even so the production-cost benefits at best are still only marginal. If the supply situation changes firms may well try to switch back to traditional methods, but the earlier move to timber-frame will have worsened the conditions for reproducing the ability to use the traditional techniques and hence their cost. Whatever happens, therefore, construction costs are likely to carry on rising.

Long-term changes in housebuilding costs − a synthesis

This chapter has argued that the production process of speculative housebuilding is not subject to the same pressures for technical change, or even for uniform efficiency, as frequently is the case in other industries. In addition, in the longer term, the nature of the industry has had adverse consequences on its own existing technical state via effects on the inputs to the production process. It is for these reasons that it has been argued here that there is a crisis of production in speculative housebuilding. The increasing instability of the housing market in which speculative housebuilding operates, moreover, is exacerbating these adverse tendencies. So the crisis in production is growing over time and it is liable to get still worse.

It should be stressed that this conclusion has been reached through an examination of the nature of the whole structure of housing provision within which speculative housebuilders operate. The crisis has not been caused by some especially inept 'rogue' or 'cowboy' producers. To an extent the crisis could even be claimed to have generated a greater efficiency within the industry in terms of profit-making. It has definitely contributed to the change in the nature of the firms that operate within private housebuilding, mapped out in chapter 3. However, a greater awareness of ways in which profits can be made does not necessarily lead to lower production costs over time. The crisis of production has been created if anything because individual firms have become increasingly adept at organizing their production processes to maximize profits. This apparent paradox occurs because the crisis is a structural one that stems from the whole nature of current owner-occupied housing provision.

One implication of this conclusion is that it is difficult to see how any minor change in the way housebuilders build houses can overcome the problem, nor even whether a greater stabilization of owner-occupied housing demand would do the trick. But raising these issues is to drift into the politics of policy prescription; housebuilders however have not waited for policy prescriptions to be found and implemented in legislation. They have tried within their own sphere of influence to overcome or at least counteract this crisis in production. They have done that in a number of ways. One is to introduce a completely new industrialized building system, that of timber-framed building. Another is to attempt to increase the overall size of development gain, and also their share of it at the expense of the landowner. The planning system is now a major site of their struggle to increase development profit. Finally, like all construction firms, they have taken advantage of the depressed years of the early 1980s to squeeze their input purchase prices. Workers have had their piece rates cut in this attempt to bolster profitability. Speculative builders' long-term relation to their workforces, however, has meant that firms could not take advantage of the slump to push through new work practices on a workforce weakened and demoralized by the threat of unemployment, as has been common elsewhere. Speculative builders already had the ideal employment practices to enable such changes, so the slump offered no added advantage. But there is a sting in the tail of their success: speculative builders cannot force workers to work for them, and the terms they offer make workers increasingly reluctant to do so. At the root of the current housing crisis is an impasse in class relations.

7

Land-use planning and speculative housebuilding

The significance of planning

There is a great academic divide between studies of housing and studies of planning. Although housing is by far the largest urban land use (Best 1981), planning literature tends to be concerned only with the spatial distribution of housing and the effect of planning policies on the land market, whereas housing studies tend to ignore spatial questions by focusing on state legislation and subsidies related to tenures and households. This division is unfortunate as it has led to one-sided and, therefore, wrong views on some important issues associated with owner occupation.

From the planning side there now exists a widely held misconception that a 'containment' of urban development by the planning system has been the prime cause of the changes in the spatial pattern of new development since the 1940s and simultaneously has been the principal reason for a sharp rise in residential land prices that is said to have forced up house prices (cf. Hall *et al.* 1973). On the other hand, the neglect of the planning system by most housing studies is a mistake because planning does have important positive influences on housing provision as well as the claimed negative effect of raising land prices. One instance of planning's significance for housing is its effect on the size and location of state expenditure on the built environment. Although this type of expenditure may not necessarily be earmarked directly for housing schemes, without the provision of roads, public transport, sewage, water and other facilities housebuilding is impossible. In so far as owner-occupied housing schemes benefit from state infrastructure expenditure they are in effect being heavily subsidized by the state. The massive decentralization of urban populations with which the growth of owner occupation has been associated would quite simply not have happened without large-scale state infrastructure expenditure, so there has been a close correspondence between the suburban bias given by planning to that expenditure and the continued growth of owner occupation.

The importance of state expenditure on the built environment is one aspect of planning's role in intervening in the determination of the overall level and spatial distribution of social activity. One important consequence is its impact on the size and distribution of potential land development gains. With minimal taxation this gain is available to the private interests that develop and own new building structures. In relation to owner occupation, therefore, planning intervenes in the struggle between landowner and speculative builder over the creation and appropriation of development gain. In the light of the analysis of previous chapters it can be seen that this intervention is at a crucial point in the structure of owner provision, so its theoretical consequences and its empirical effects need to be clearly understood if the current state of owner occupation is to be explained. This chapter suggests that, because of the use of an inappropriate theory of rent, the intervention tends to be misunderstood by planning theoreticians and practitioners alike.

Land-use planning obviously is not just about housing, it concerns other land uses as well. But the nature of the present British planning system has been strongly influenced by successive political conflicts over working-class housing provision. Land-use planning emerged as a significant political issue in the early years of this century when it was proposed as one solution to the environmental squalor and high cost of urban working-class life. Later, broader strategic questions of transportation, employment location and rural protection were added to the remit of planning, but their addition was structured by the political frameworks and compromises created by the previous centrality of the housing development issue. Moreover, the recent attacks on the role of planning have again centred on housing through the effect of development control on owner-occupied housebuilding (see chapter 8). Consideration of private housing development, therefore, brings to the fore the limitations and failings of successive political compromises over planning, especially in their attempts to accommodate land market interests at the expense of those of final land users.

Examination of the significance of the British planning system on the structure of owner-occupied housing provision necessitates making a theoretical break with the general institutionally based conception of planning. It is widely recognized that land-use planning is associated with conflicts of interest in which some lose and some gain: a non-Paretian trade-off of options from the consumer-choice perspective of neoclassical economics, or conflicts between competing interest groups from a sociological or political theorist's perspective. But the nature of those conflicts and the way in which land-use planning is related to them takes one general form that misses the real significance of land-use planning. The state in the form of the planning system is seen as an arbitrator of land-use conflicts. As an arbitrator it may be regarded as benign, as it usually is in mainstream planning discourse, operating in the broad interests of 'society' above the parochial self-interests of the immediate participants in land-use conflict (cf. Eversley 1973). Alternatively, it may be treated as a biased one: a state bureaucracy operating against the interests of 'liberty' in the eyes of the Right (cf. Hayek 1960), or a charade played out to support the interests of big business and land-owners according to much Left radical literature (cf. Kirk 1980).

Whatever the colouring given to the arbitration framework its fundamental basis is theoretically flawed because an attempt is made to analyse land-use planning in static distributional terms. Yet planning is part of a process of altering the overall physical environment. The extent to which it succeeds in doing this changes the nature of some conflicts, removes a few and creates others. (The role of land-use planning in generating a built environment which has facilitated the decentralization of urban populations is a case in point.) Obviously the extent to which the material conditions are altered varies over time and much of that variation is outside of the control of the planning system itself. Even so the impact of planning is more than distributional. An analysis of planning and of the consequences of different types of planning, therefore, has to go beyond a comparison of a check-list of gainers and losers.

Not only does planning alter the physical basis of conflict, it also creates a political framework through which conflicts are fought out and, in addition, ideological discourses within which the process of planning is conducted. Any community group faced with the awesome prospect of presenting evidence at a planning enquiry is acutely aware of the jargon and the limits to action and debate imposed by the rubric of the planning system (cf. McAuslan 1975).

The issue of politics raises the second major difficulty with the arbitration approach: the state is separated out from the class nature of society in its role as arbitrator. Its actions then become potential instruments for intervening in conflicts whose existence and form are pre-given. In other words, the static distributional approach to land-use planning necessitates an instrumentalist theory of the state. The state is an invariant machine that operates in the interests of which group, or groupings, manages to control the levers of power. Such an instrumentalist view will be avoided here; instead what will be emphasized is the planning system as a site for struggles over land use, the outcome of which may lead to their transformation. Part of that transformation may also involve fundamental changes in the apparatus of planning itself. Conflicts over land use, for example, led to the 1947 Town and Country Planning Act which created the modern land-use planning system. Subsequent political struggles have produced significant changes in the role of planning and currently threaten its effective powers over land development, as chapter 8 will show. The significance of land-use planning, consequently, is in the precise contemporary role it has in influencing the dynamic of struggles over land use and its impact on their outcomes.

The main topics of this chapter concern the historical development of the British planning system, planning's role in the struggle between landowner and speculative builder, and the politics of planning. Many of the conclusions reached differ from those generally held because of the difference in theoretical perspective outlined above. In particular the question of land rent is given a central place in understanding the limitations and possibilities of different forms of land-use planning, whereas this item has tended to be neglected or segregated in much of the planning literature. First, however, it is necessary to clarify what land-use planning in Britain is all about.

The modern planning system

The hoary old question 'what is planning?' appears at the beginning of virtually any study of the land-use planning system. This is not surprising as no precise definition of it can be given. It is not a theory but an administrative practice: an intervention by one part of the state apparatus into land-use activities. What planning is and what it does, therefore, depend on the nature of that intervention. Defining the contemporary nature of land-use planning consequently depends on understanding its results, in terms of its impact on land uses, as well as describing its formal administrative procedures.

The ambiguity about the content of planning is reflected in the widespread consensus, both politically and theoretically, over the need for some form of land-use planning. The consensus is based on the recognition that an unrestrained private land market can produce severe social and economic disruption because of the spatially fixed characteristics of land. Adjacent uses may be incompatible; land developed for one activity fixes it in that use for a long time period so it may influence the whole pattern of development of a wide locality for years to come; finally large-scale state expenditure on land-related infrastructure (roads, etc.) has to be organized into useful spatial forms. The significance of social interaction in the determination of land uses creates, in the terminology of neoclassical economics, externality and public good (also known as collective consumption) problems (Foster 1973 and Harrison 1977). Planning, therefore, has a unanimously agreed role in improving the functional interaction of land-use activities: nationally, regionally and, of greatest concern, at the level of the town or large city.

Beyond a broad consensus over the need for some sort of planning, wide differences emerge because of the inevitable contradictory role that planning has. Theoreticians on the Right, for example, are concerned with the threat to private property that land-use planning implies. Planning restricts the freedom to use private property as its owner sees fit, and ends up threatening the very existence of the market mechanism and private property itself. For Hayek, one of the theoretical mentors of the Right, for instance,

> The issue is . . . not whether one ought or ought not to be for town planning but whether the measures to be used are to supplement and assist the market or to suspend it and put central direction in its place.
>
> (Hayek 1960, p. 350)

Planning, however, is a non-market intervention that has to question the criteria of the market if it is to change the outcome in terms of land use. This means that certain aspects of the market must be suspended and replaced by central direction. The whole operation of the market itself is put into question as no amount of adjustment to the market can actually determine the final land use until the point is reached where the planning authority fixes the use directly. The interactions between land uses are too great and the consequences of land-use change too global for any partial adjustment of a market parameter to succeed. The result is that land-use planning must always have an uncertain relation to the private land market because their criteria of operation are so different. Champions of the market mechanism, therefore, can give only token support to the notion of planning: a conclusion towards which Hayek himself is forced. 'All endeavours to suspend the market mechanism in land and to replace it by central direction must lead to some such system of control that gives authority complete power over all development.' (Hayek 1960, p. 354).

The ambivalent relation between planning and the land market is extremely important as it makes it impossible to see the development of land-use planning solely in functional terms. Instead function, economic and ideological interest in land-use change, and political struggle have intertwined to create the present-day planning system. The outcome in Britain has been a characteristic attempt to play ostrich with compromises that resolutely ignore the contradictions between planning and the land market.

Land-use planning in Britain is concerned almost solely with changes in land use. It therefore is primarily development planning. The planning system is based on a series of legislative measures, the most important of which is the 1947 Town and Country Planning Act. Plans themselves are formulated on a two-tier basis, following legislative changes at the end of the 1960s. Structure plans are drawn up at county level in England and Wales (and in regional districts for Scotland). These plans are broad statements of land-use policy for the county in question to which detailed local plans have to conform. Local plans are drawn up by the lower of the two major tiers of local government, the district councils. Professional planners, employed by the local authority, formulate the plans on the basis of policy directives from their local council, plus guidance from central government and conclusions derived from something called the planning process. The latter consists of surveys of local economic and social problems in relation to land use, followed by their evaluation by planners in terms of goals formulated through the political process and by planners themselves. This leads finally to a translation of these goal-related evaluations into statements about desirable and possible land-use change: the 'plan'.

The plan consists of a number of elements: a written statement and schematic key diagram in a county-wide structure plan, a description of primary uses in the proposals map on a topographic base in a detailed local plan, and finally illustrative material in supporting policy documents. None of them is a hard-and-fast statement of actual possible development at any single location. They are at best imperfect guides requiring careful interpretation when making decisions in development control (Davies 1980). The planning process, moreover, takes a long time; so for many areas plans are out-of-date or non-existent. A number of counties, for example, still do not have approved structure plans, and a large number of the associated local plans still need drawing up. In 1980, for instance, the statutory development plan for Reading remained the 1957 plan! (Davies 1980).

Land-use plans in Britain, therefore, are schematic and indicative; with planning acting as a guide to development rather than instigating it. Planners try to predict likely development, either in terms of feasible public expenditure or likely private development schemes, and steer it in directions that fulfil planning objectives. In this process a clearly defined land-use plan plays only a part. Planning has to come to terms with the social activities whose spatial location it is trying to influence with only limited land-related means at its disposal. So hope, compromise and inducement on an incremental basis are at the heart of the planning process.

There are two other formal links in the plan-making process. They concern central government and public participation. Central government has a number of roles: as overseer vetting draft structure plans, adjudicator between opposing local authority plans, and co-ordinator generally trying to ensure that local planning policies fit in with the policies of the government of the day through circulars, consultation and the planning appeals procedure. Public participation takes place through public inquiries over development schemes, and also through local authorities 'consulting' the populations in their areas when drawing up the initial plans themselves.

The way in which the policies formulated by land-use planning are implemented takes two interlinked forms, one public and the other private. Co-ordination between state agencies usually ensures that public expenditure on the built environment corresponds with planning policies. Yet, with the partial exception of the new towns programme, the planning authorities are in no position actually to instigate public expenditure programmes: their volume and location generally is decided by the political, economic and administrative procedures and pressures affecting the spending body. Divisions within the state apparatus consequently have an important influence on what is feasible in planning terms. Public expenditure also plays an important part in encouraging and facilitating private development. It provides the services and infrastructure that make any area a viable proposition for private development. Given current tax laws in Britain, moreover, the profits or betterment from development made possible by such public sector investment mainly accrue to those private development agencies. This land gain obviously is a considerable inducement to private developers to go along with the planners' brief.

The other main element of plan implementation is a statutory one: all new development in Britain requires planning permission from the local planning authority. This situation has existed since the 1947 Town and Country Planning Act which nationalized land development rights. The need for planning permission is the principal instrument for implementing planning control over private development. For those who are refused planning permission the Department of Environment is the quasi-judicial court of appeal, via a network of planning inspectors. The Secretary of State for the Environment can make the final decision on appeals and his/her conclusions are unquestionable within law.

Statutes and case law precedents define the criteria on which planning permission may be refused. Mainly they refer to physical matters such as compatibility with other land uses in the locality, road congestion and the appropriateness of the land use. But associated with them is a catch-all phrase, derived from the 1947 Act, that refers to 'any other material conditions'. This phrase has been important as it enables development to occur where the other more rigid planning criteria might forbid it. It is a complement to the deliberate vagueness of the plans themselves.

The vagueness in planning criteria is of particular importance for suburban owner-occupied development. During the 1970s a series of studies showed that, of land granted permission for residential development in areas of high demand, less than half had been designated for residential purposes in land-use plans (cf. JURUE 1977 and EIU 1975). Considerable bureaucratic discretion therefore is built into the planning system, a discretion that has enabled apparent rigid state control over land use and predominantly private sector development to co-exist. It is a form of co-existence, however, that has tended to benefit developers. It has also heightened the importance of the ideologies of the planning professionals who are the principal dispensers of that discretion. As Foley (1960) and Glass (1959) showed only a few years after the introduction of the modern planning system a variety of often conflicting beliefs guide planning practitioners; yet overridingly there is a belief in the technical superiority of land-use planning whatever the social context. As will be shown later, this brand of apolitical technicism that pervades the planning profession has produced specific urban forms and a staggering neglect of the distributional consequences of land development. With wide scope given to official discretion, and minimal public accountability associated with it, the characteristic repressive paternalism of British state bureaucratic structures is firmly implanted within the current planning system.

A clear example of the consequences of seeing planning as a technical exercise is the way in which the influence of financial gains on land-use development has been marginalized within mainstream planning discourse. The problem of land is predominantly seen as one of its availability for the planned land-use changes. The question of economic gain from development has been effectively pushed out of sight. Land taxation is a technical, fiscal matter of little interest to planning as long as it does not interfere with land release. Land nationalization is virtually unmentionable: an idealistic dream with dangerous left-wing connotations. The land values question is part of the history of planning rather than a central element in its current operation.

No compensation for loss of land value can be claimed on refusal of planning permission. Land is frozen in its current use by refusal and the owner simply has to accept the consequences of such a refusal on the potential selling price of his land. Before 1959, land compulsorily purchased for public use was compensated at its existing use value and, since then, at the estimated 'market value' of the land. Increases in land price have also been subject to a chequered history of 'betterment' taxation: at first, in the 1947 Act, 100 per cent of the estimated betterment was taxed away; then, after 1953, for many years there was no tax, and subsequently over the past decade land sales have been subject to either a capital gains or a development land tax. Two attempts by Labour governments in the 1960s and 1970s to alienate a much higher proportion of land gains to the state were watered down prior to legislation and then made administratively inoperable after enactment. The effective tax rate is at present extremely low, although variable in its impact. But this chequered history indicates that the land values (or compensation and betterment) problem will not go away.

A loss of land gain through taxation or nationalization obviously is a threat to the economic interests of both speculative housebuilders and landowners. They therefore are hardly likely to regard land gain as solely a technical or distributional matter. So the failure of the planning system to remove the effects of land gain has limited the effectiveness of planning. One irony of planners' attempts to wish away the land gain issue, consequently, has been that mounting pressures for higher land gains are helping to produce a crisis of planning, with planners having diminished control of land-use change. But technicist views of planning and attempts to fudge the implications of state intervention in the land market are not new; they are a central component of the history of planning. Yet external forces, especially ones associated with periodic crises in housing provision, keep forcing a reformulation of land-use planning practice. Much, therefore, can be learnt about the British planning system from its history.

Urban crisis and the development of land-use planning

Land-use planning developed through a series of sharp ruptures with pre-existing social policies. The process was not one of steady progress. An initial advance towards planning in the years prior to the First World War, for example, received substantial setbacks in the following decades. The legislation of the 1940s reversed that situation and the 1980s are producing yet more changes. The varying fortunes of planning to a considerable extent are a political product of contradictions between the efficient spatial organization of urban working-class life and the rights of private property embodied in private landownership. Those contradictions are extremely complex as notions of efficiency mean different things to different people, reflecting opposing class interests. Conflicting political demands over working-class living standards have been diverted in part into a series of ideological positions over types of land-use planning supported by shifting alliances of divergent interests.

The political alliance in support of the 1947 Act was a peak that no planning measure before or since has been able to achieve. New problems and forms of class antagonism were emerging and it became increasingly essential for the state to respond to contemporary changes in the spatial pattern of social activity. Landed property had consistently put up barriers to any real form of effective state intervention for decades, so by the 1940s it became imperative that something should be done about its power. That confrontation was one of the most important results of the 1947 Town and Country Planning Act. Yet, by nationalizing development rights, the 1947 Act only resolved one aspect of the conflict between landowners and other social groups. By introducing a betterment tax the 1947 legislation tried to fudge the question of private landownership and development gain. And that fudge still exists. The following overview of the development of planning will try to highlight the importance of these social antagonisms.

The origins of the planning system can be found in the nature of nineteenth-century British cities and the ideologies of reform that grew up within them. The social consequences of rapid urbanization and extremely low incomes for much of the population produced plenty of empirical material for critiques of unplanned development. Many vital social functions were poorly provided or non-existent. Housing, sewerage facilities, clean water, traffic congestion, conflicting and noxious land uses all constituted severe problems for the town dweller. Those associated with direct personal consumption particularly affected the poor but others, like disease, were less respectful of class boundaries.

Towns for most of the nineteenth century did not need large-scale state action (apart from a police force) to make them viable centres for capital accumulation. The contemporary characteristics of capitalism did not require state expenditure on the reproduction of the workforce nor on transportation infrastructure. Railways in Britain were financed by private capital, and with industry tied very much to rail and water transportation and to steam power, production had to be located near to transportation facilities with workers' housing packed tightly around it. The expanding commercial and financial sectors also had to be tightly and centrally located for similar reasons. After the 1835 Municipal Reform Act the local bourgeoisie controlled most large towns with little possibility of formal political opposition from other social groups. There was strong resistance by them to central control over a wide range of issues that were regarded as local matters. So for most of the nineteenth century and frequently beyond, local bourgeoisies had effective control over state intervention in urban life (Fraser 1979). Public expenditure therefore tended to be limited to the founding of a constabulary, to conspicuous public works and symbols of oppression (Leeds, for example, built a jail first, then a town hall), and to improvements in water and sewage systems. New grand thoroughfares additionally could be used to break up potentially threatening concentrations of the poor in central city districts. There was substantial land-use change within Victorian cities; yet it took place in an *ad hoc* way and principally by private interests, sanctioned by Parliament where necessary. Little effective resistance was aroused by wholesale urban redevelopment schemes, the most notable of which were new railway stations that caused substantial disruption to pre-existing land uses (Kellett 1969).

Throughout the nineteenth century growing populations, rising standards of living and improved transportation led to the expansion of towns. This expansion was not gradual but tended to occur in sharp waves of development, especially with housing. Urban rents followed a similar staggered path but they did not fall back to their old levels at the end of each building boom. So over time they rose, and by the 1900s rent on buildings had reached 10–11 per cent of total national domestic income (Offer 1980). Landowners, housebuilders, financiers and the owners of existing town property and the land on which it stood all benefited from this urban growth. They also benefited from the low level of rates charged by municipalities until the 1890s, as this stopped rent revenue from being syphoned away in property taxation.

State involvement in suburban development was limited primarily to belated regulations setting down minimum by-law standards of building. Similarly powers to condemn the worst existing slums were gradually strengthened but still remained weak. Both types of activity principally were health measures rather than concerned with the standard of living of the mass of the working class. Back-to-back housing for instance was forbidden because of the lack of through ventilation it created. Spatial building forms were changed by the gradual spread of building by-laws designed to stop the building of unhealthy cheap housing erected as inner courts inside the residential blocks delimited by existing thoroughfares. Building by-laws specified the need for street frontages, back yards, and minimum street widths. The profit-maximizing built forms adopted by speculative builders in response to these regulations produced the monotonous grid street patterns of terraced housing that came to be so severely criticized by housing reformers at the end of the nineteenth century.

Urban policy in the nineteenth century, therefore, was piecemeal and reactive. It centred on the opposing concerns of the bourgeoisie with the physical threat of the working class and the impact of expenditure on the rates levied by local authorities. Little attempt was made to direct or to encourage new development for fear of the expense, so comprehensive land-use planning was out of the question.

The closing years of the nineteenth century changed all this, because of shifts in the relations between capital and labour. Capital accumulation by this time had entered a new phase. The 1870s had marked the end of the world ascendancy of British industrial capital. Financial capital was still strong and growing but its activities principally were directed to the Empire and the export of capital. British manufacturing industry had to undergo substantial restructuring in order to sustain its momentum in the face of growing world competition and substantial changes in technique. But it had to do this in an economic climate that was increasingly unattractive. Some industries, including some of the staple export ones, were comparatively successful but many opportunities were missed.[1] The late nineteenth century was a period when expansion in the capitalist world economy centred around coal, chemicals, steel and electricity, with production methods increasingly based on mass production, large factories and the scientific management techniques of intensive labour utilization associated with Taylorism and, later, with the production line pioneered by Ford. Newly emerging capitalist economies in Europe and North America dominated in the changes. Even so some significant changes took place in Britain. What is of interest here is the new requirements of capital, generated by these changes, in terms of the social reproduction of a labour force and the spatial location of production.

Whilst changes in production methods greatly increased the scale of production in many industries and created whole new branches of production, the transportation requirements of those industries generally still tied them to their traditional urban locations. Yet the changes within the workplace progressively altered the type of labour power needed for production. The big new factories required large workforces, and much of that workforce had to be skilled. The new production processes also usually needed a stable, well-disciplined workforce that was literate and physically capable of sustained work over long time periods. The existence of this type of workforce could only be achieved with better education, food, health and housing for the mass of the population from which those workers were drawn. At the same time, the changing nature of many branches of production, especially in their scale, aided the formation of workers' organizations. The outcome was that pressures began to mount for improved living standards for the working class, an important element of which was housing.

If changes in the national economy were pushing for improvements in housing conditions, the existing structure of rental provision could not deliver them (see chapter 2). Concern over housing conditions mounted in the decades prior to 1914, despite a large building boom in the 1890s and considerable improvements in public transport which enabled far more workers to live in the inner suburbs. The problem was that rising urban rents made it impossible to finance more housing consumption without large increases in wages, whereas attempts to expand housebuilding led to sharp rises in building costs, so large increases in supply occurred only in periods of economic boom when demand was high. New housing production, therefore, could not provide sufficient supply to satisfy growing demand. So the years of economic stagnation of the Edwardian period saw a growth in housing distress as well as falling real wages and mounting economic militancy by workers (Pollard 1973). A growing contradiction between an increased social need for new housing and an inability of the contemporary structure of provision to satisfy it, therefore, began to create an urban crisis of mounting dimensions.

The growth and finance of municipal expenditure in the Edwardian era exacerbated working-class housing problems, reflecting the breadth of the urban crisis. Even though after the 1890 Housing Act local authorities began to build housing for general working-class needs, problems of municipal finance held back new housing schemes. Ironically general pressures for increased municipal expenditure also restricted new housebuilding for the private rental sector. The reform of local government in the 1890s changed its powers and functions. One result was a rapid increase in municipal expenditure financed principally out of rate income. Local government expenditure in the UK virtually doubled between 1890 and 1900 in real terms and continued to rise rapidly until the First World War (1890 £51m., 1900 £99m., 1905 £128m., and 1913 £134m., at 1900 prices). 70 per cent or more of this expenditure was financed from rate income. Rates were a deduction from rents received by property owners. So growing municipal expenditure on the new requirements demanded of an urban area, by pushing up property taxes, dampened even further the prospect of new housing provision by reducing its profitability to prospective landlords. So the already strained link between increases in rents and more housebuilding was weakened further by the expansion of municipal expenditure and the means by which it was financed.[2]

The urban development process was coming increasingly into contradiction with the requirements of accumulation by industrial capital. It was forcing up the cost of reproducing the labour force and even questioning the possibility of adequate minimum physical living standards. Through pressures on housing rents, in particular, it influenced the struggle between capital and labour over working-class living standards. In these circumstances, general political agitation against housing landlords and urban living conditions was bound to increase. The forms agitation took depended on the political grouping from which it emanated, yet taken together these demands made some sort of state action imperative. The problem for governments in the years prior to 1914 was that any sort of action (including inaction) came into conflict with at least one major political interest. The most obvious solution, for example, was large-scale exchequer subsidies for working-class housing to rent but, as bourgeois commentators were quick to point out, they in the long term would destroy the economic mechanisms of private housing provision. The state, therefore, would be faced with an open-ended obligation to finance working-class housing against the economic interests of private landlordism and bourgeois concern for 'sound' public finance. State subsidies would also have the politically undesirable effect of providing a practical demonstration of the inability of the private market to provide one of the basics of life. Some of the new ideologies of town planning fashionable at the time seemed to offer a way out of the dilemma by suggesting the possibility of politically painless reform. What is of interest is how broad support was gained for them, and how the social conflicts which they tried to ignore led to their failure as effective reforms.

Once the conditions developed to make land-use planning an attractive political option, the groups previously agitating for the introduction of planning came into their own as they set the ideological language and terms in which debate and legislation would be carried out. Reform groups were predominantly middle-class in character and their dominant ideology represented an extension of the earlier nineteenth-century mixture of philanthropic social concern and fear of the power of an urban mob that could destroy the society from which they derived their advantageous social position (cf. Stedman-Jones 1971). Physical reforms were seen as a means of alleviating the plight of the poor and also weakening the potential for disruption and revolution by the masses. This orientation was combined with an anti-urban ideology (Glass 1959) based on imitation of the contemporary landed aristocracy and nostalgia for a mythical by-gone age of rural peace and harmony. At the end of the nineteenth century reform movements started agitation for substantial state involvement over planning and housing issues. Yet it was not until the end of the 1930s that it was accepted that the state must play a leading role. Prior to that it was believed that private initiatives could succeed; so demonstration of that possibility was a principal objective from the 5 per cent philanthropy housing movement of the second half of the nineteenth century (Tarn 1973) through to the garden suburbs and cities of the first thirty years of this century.

A variety of planning schemes were proposed. Some architects began to draw idealized cities but there were two main types of practical proposal: freestanding new towns, of which Ebenezer Howard's Garden Cities were the most significant, or the municipal overseeing of suburban development in the form of town extension schemes. Associated with these two alternatives could be included municipal ownership of development land and innovations in the design of working-class housing, based on lower density street layouts, cottage style single family dwellings and efficient but economical internal design (Swenarton 1981). Architects in the 1900s put a considerable amount of effort into new designs of working-class housing estates. The partnership of Parker and Unwin was the most influential. The style that evolved, with its references to idealized pre-industrial English villages, enabled the planning movement to present a clear visual image of its environmental aims with which to capture popular approval.

Town extension planning with no land acquisition quickly won out in a unified planning reform lobby by 1907–8. Its lack of offence to private interests and minimal increase in state activity appealed to the dominant figures in the planning lobby, whereas the inclusion of house design on garden city lines as an afterthought and half-hearted encouragement of garden city experiments enabled others with more radical proposals to save face and offer their support (Sutcliffe 1981). Stripped of the garden city nomenclature this is, of course, the type of suburban planning that has existed in Britain since the 1950s.

The significance of this form of land-use planning in limiting direct state involvement in housing provision was emphasized strongly by the earlier reformers. Horsfall, a capitalist philanthropist and principal publicist of suburban planning, wrote in 1900 that

> What is needed by the authorities is not the power to buy up large tracts of land or the power to build workmen's dwellings, it is the power and the intelligence needed for the right use of the power, to make and to enforce the strict observance of plans for large areas round every town which is growing.
>
> (Horsfall 1900, quoted in Sutcliffe 1981, p. 70)

Like many other contemporary social reformers, Horsfall cited the more advanced state of planning in the large cities of the arch-imperialist rival, Germany, to support his case.

Another philanthropic industrialist, this time in charge of Unionist (Conservative) Birmingham's Housing Committee, John Nettlefold, took up Horsfall's arguments with enthusiasm. He used them to support his reversal of Birmingham's commitment to suburban council housing, and by 1905 became a well-known national publicist of extension planning, using the term he coined, 'town planning' (Sutcliffe, ibid.). Hampstead Garden Suburb, started in 1906, confirmed the respectability and anti-public-housing significance of this form of planning in Establishment circles via the publicity and aims propounded by its founder, Dame Henrietta Barnett. The adroit use of the architects of Letchworth garden city, Parker and Unwin, to draw up the development plan and the fashionable country house and imperial architect, Lutyens, to design the central communal buildings, visually highlighted the advantages garden suburb ideology could have for the ruling class of an Empire.

The attitude of planning reformers to speculative housebuilders was ambiguous. The housing reform tradition of the nineteenth century had shown that no private interest (even with a lower 5 per cent return) could build enough working-class housing, and that voluntary organizations could not fill the gap. Given their aversion to state provision, these reformers could only hide their lack of an effective housing strategy by repeating earlier reformers' pious hopes of housing co-operatives and self-build. This theme was also echoed by the pre-war Liberal government with little effect on housing provision. So, in practice, the housing field was left to the speculative builder.

The ideological and political significance of 'town planning' was quickly grasped by various political groupings. Surveys, like Booth's and Rowntree's, of the plight of the poor had provoked Liberal consciences but, although continuously in power after 1906, their ideological principles meant they could do little about them, apart from champion the 'town planning' idea, which they did in the 1909 Housing and Town Planning Act. The Liberals, and particularly Lloyd George, were keen to attack private landed interests but not the rights of private property nor private housing landlords. So this measure was ineffectual, like the land tax of the 1909 Budget that was meant to complement it. Lloyd George's later Land Enquiry Committee of 1913–14, drafted principally by Seebohm Rowntree, continued the trend by refusing to recognize the contradictions thrown up by private landlordism.

Once the Unionist (Conservative) Party had been converted to the strategy of Social Imperialism under Joseph Chamberlain's influence it too was prepared to intervene directly into urban working-class living conditions. The Social Imperialists' response to the relative decline of the British economy was to propose a major redirection of British economic and foreign policy (Gamble 1981). They envisaged the future of Britain as based on Anglo-Saxon racial supremacy in an enlarged

Empire that would form a largely autarkic trade bloc impenetrable to foreign competition through a tariff system of Imperial Preference. Urban conditions in Britain constituted a major barrier to this strategy as they seemed to indicate the home stock was showing signs of racial degeneration contrary to the supremacist scheme of things. Although the Boer War fitted closely into the Social Imperialists' strategy it also shocked them, amongst others, because of the poor physical state (later shown to be exaggerated) of potential urban working-class army recruits. The maintenance of the large armed forces necessary to create and defend the Empire seemed at risk. Urban reform consequently became a central concern of this group, one which had the added advantage of winning working-class support to their cause. Chamberlain had earlier, in the 1870s, been an energetic reforming Mayor of Birmingham (Briggs 1968).

Intervention to protect capitalism, not surplant it, was the Social Imperialists' aim. The result was they only wanted action where they felt the market failed to operate. Nettlefold's support, mentioned earlier, of town planning in the suburbs and housing rehabilitation in the inner city would fit the strategy. So does the reaction of the Moderates (alias Conservatives) to council housing after their takeover of the London County Council in 1907. They ran down suburban council housing developments but increased overall housing expenditure by adopting inner city rebuilding projects, because slum clearance was obviously beyond the scope of private endeavour whereas it was hoped suburban development was not (Swenarton 1981, p. 30). Again, since the 1950s the linkage of council housing with urban renewal has been a firmly entrenched land development strategy of successive governments.

Working-class organizations in the 1900s could not be carried along with the policy of town planning once they became aware of its anti-state expenditure bias. Horsfall gained early support for his town extension plan proposals at a meeting held during the 1904 Trades Union Congress. But the principal working-class housing organization, the Workmen's National Housing Council, denounced the planning movement when the latter's support for the Liberal government's inaction on housing matters became obvious.

Working-class agitation over housing prior to 1914 is difficult to quantify. By the end of the nineteenth century the working class had both the power and the means to influence urban policies and politics, especially since the enfranchisement of a substantial proportion of the male working class had changed the nature of local politics. Most of the party political predecessors of the Labour Party tended to discount the importance of housing issues (including the Fabians as well as the Social Democratic Federation and the Independent Labour Party), but local working-class housing organizations and trades councils did agitate over immediate living conditions as did the Workmen's National Housing Council (WNHC). Wohl (1977) suggests that workers themselves took little interest in housing matters, citing as evidence the exasperation of working-class organizations trying to agitate for improvements as, for instance, the WNHC which in 1901 wrote 'whoever else takes the housing question seriously, the mass of those most affected by it − the working people − have not done so' (quoted in Wohl, ibid., p. 320). The later success of the Labour Party over housing issues and the mass agitation of 1915 over rents, however, do show that when the circumstances were right consciousness could be raised. Throughout the 1900s the WNHC, and later under its prodding the newly formed Labour Party, campaigned for direct exchequer grants for council housing.

The period from 1900 to 1914, in summary, was one of class stalemate over an intense urban crisis. In the creation of the stalemate the notion of town planning played a significant, albeit diversionary, part. It proved impossible (theoretically as well as practically) for revolutionary organizations to raise working-class political consciousness through agitation over urban living conditions. This failure was important as it closed off one way of broadening the scope of the socialist revolutionary critique of capitalist society, as Engels had been aware seventy years earlier (Engels 1969). If a way of politicizing the urban crisis had been found by revolutionary groups, the success of revolutionary socialism as a hegemonic strategy might well have been substantially enhanced, if later spontaneous working-class reaction in the First World War against their housing problems is any indication. On the other hand, state action on urban conditions also failed to become a component of a counter revolutionary strategy. Such action, as later periods were to show, would have reinforced the credibility of the ideology of harmonious potential social reforms and weaned key groups away from support for revolutionary struggle. As has been seen, significant elements of the ruling élite, including wide sections of both main political parties, were prepared to accept the inevitability of major state intervention into working-class urban life. The ideology of town planning, however, enabled them to avoid confronting the political consequences of such intervention.

Town planning as an ideology failed to deal with the contradictions between private rented housing provision and the contemporary housing needs of society. And it even failed to confront its own problematic relationship to private landownership. These failures subsequently led to its temporary demise as a reformist social programme. Sustained working-class agitation towards the end of the First World War and a fear of social revolution forced on a previously reluctant government the need for a large-scale programme of centrally funded council housing. The 1919 Housing Act with its open-ended Exchequer subsidies was the result. The 'homes for heroes' campaign was conceived by the Cabinet as part of a counter revolutionary strategy (Swenarton 1981). The housing programme was to be a concrete reference point for claims that working-class life would be improved in the post-war years. So the programme was finally introduced that many town planning ideologists earlier had tried to avoid, adorned ironically with a design format taken from the planning movement – suburban estates on garden city lines.

The collapse of working-class power in 1920–1 led to the curtailment of the housing programme less than eighteen months after its adoption, but mass council housing was here to stay. The ambivalent relation between state housing and the planning movement has also remained. The subsequent emergence in the 1920s of owner occupation as the new dominant structure of housing provision revived interest in the potential role of planning in town extensions. But planning's relation to either state power or private landownership had yet to be satisfactorily sorted out by its proponents. The problems faced by the other, comparatively minor, town planning experiment of the Garden City highlights the difficulties to be faced. Whatever is thought of the British town planning system, none the less, its origins as a form of collectivism, let alone socialism, are pure myth.

The origins of planning and the Garden City Movement

The existence of a growing crisis of capital accumulation within the late nineteenth-century urban environment enables Howard's Garden City idea, propounded at the turn of the century (Howard 1898), to be placed in its historical context. Whilst his proposal shares many similarities with other planning proposals it did also constitute a radical break with them because the Garden City idea did aim to alter the structure of the social relations involved in working-class housing provision. Like many reforms, none the less, the logic and consequences of the Garden City scheme are unclear in the form in which they were initially presented. The proposal is a classic planning ideology of the type described earlier: a technical solution prefaced by a critique of contemporary capitalism, yet devoid itself of any notion of class antagonism. It did, however, have the merit of pointing to the contradiction between adequate housing and other urban facilities and high, privately appropriated land rents.

Howard's Garden Cities of Tomorrow were to be self-contained communities of about 30,000 people enjoying high standards of amenity provided in the city or the surrounding countryside, with workers fully employed in capitalist (or, Howard hoped, co-operative) enterprises. Both firms and people were to be decanted from the overcrowded cities. Land uses were to be spatially separated by function in the city itself; each city was to be developed in groups of six around a comprehensively planned central city; and the city grouplets in turn were to be linked throughout the country by a rapid transportation network. Segregation of function and urban hierarchies, hallmarks of modern planning, were therefore built into Howard's proposed urban system.

The distinctiveness of Howard's plan related to landownership. Land would be purchased at agricultural value and owned in perpetuity by the city. (The land was to be held in trust by four gentlemen of responsible position!) By owning the land, development of the city could be paid for out of future rents. Money would be borrowed to finance building of the infrastructural network of the city and then paid off with municipal rental income. Rents could be kept low but even so Howard envisaged that the whole of the city's income would be financed by those rents, with a surplus remaining sufficient for constructing future public works and for providing the population with old-age pensions and sickness insurance. Rather than intensify the conflict between capital and labour, land rent would be used to ameliorate it. Capital could have a contented, healthy and productive labour force and infrastructural facilities without high taxation or the upward pressure on wages of high housing rents. Workers could have good, cheap housing and a better living environment in surroundings uncongenial to agitation for more profound social change. Land-use planning and municipal landownership, therefore, were joint and indivisible aspects of the Garden City scheme.

The Garden City has had an important influence on planning thought since its conception. The post-1946 policy of building New Towns was derived almost entirely from the Garden City idea, especially in its putting together of landownership, planning and infrastructural investments rather than in terms of the original garden city design forms associated with Parker and Unwin. Similarly, much of the emphasis of proponents of effective land-use planning (which came to be called 'positive planning') has centred on public agencies acquiring land, or being able to direct its exchange at current-use value, with all the better-ment derived from changes in land use going to the state. The two attempts, however, by the Garden City Movement to create new towns through private subscription, at Letchworth in 1902 and Welwyn Garden City in 1919, raise other issues important to the understanding of contemporary planning. The two towns did eventually emerge but as a planning policy they were a failure. Howard's mixture of Henry George style land economics and private capital investment proved to be unworkable. The major problems were economic. They concerned, on the one hand, the scale of the initial expenditure required and the sub-sequent time profile of the revenues from the enhanced rents and, on the other, the role of private capital as financier and builder.

To highlight the financing difficulties, the Garden City scheme has to be compared with contemporary suburban development. Many of the problems of nineteenth-century suburban expansion were associated with its haphazard nature, its profit-maximizing but cramped street layouts and, with the partial exception of projects for the wealthy, the lack of public infrastructure facilities and buildings. Many nineteenth-century speculative developers relied on the pre-existence within city structures of such facilities, or they gradually grew up after the development was complete, or more likely they never appeared. Suburban expansion, through a spreading out of an urban area, consequently was not only a demand related factor, associated with transportation, but also was the cheapest way to build. Howard wanted to go against the econ-omics of such speculative building by creating self-standing towns, pre-dominantly for the working class, with high-quality layout and public amenity. This was not profitable for private capital. If it had been it would have been done before. Large sums of money have to be invested in a new town: to buy the land, build up the infrastructure, erect buildings, etc. Yet little profit from the investment would be seen for ten years or more and even then the scheme might fail. Not surprisingly, Howard had difficulty attracting private investors to his Garden City companies. Although professing to believe in utopian socialist ideals, Howard had to delegate administrative control to a well-connected member of the legal profession, who helped to persuade Lever and Cad-bury (of Port Sunlight and Bourneville fame) to finance much of the development of Letchworth (Fishman 1977). These three gentlemen held similar views about the role of land-use planning to those professed by the supporters of town extension planning discussed earlier. So

Howard quickly had to abandon his utopian perspective, which he seems to have done with little regret, paving the way for the unified political front presented by the town planning lobbyists.

Despite such patronage, both Letchworth and Welwyn were chronically undercapitalized, compromises had to be made to those financial exigencies, and both towns took decades to complete. Welwyn, in fact, only staved off bankruptcy in the 1930s because of large state aid (Schaffer 1972, ch. 1). Only the state has the financial means to undertake such projects because it does not have to operate directly on capitalist criteria related to profitability. Slow recognition in the inter-war years by the proponents of land-use planning of the scale of infrastructure investments helped to force them to accept the necessity of central state direction of the planning process.

The financial difficulties of the pioneer Garden Cities encouraged the use of speculative housebuilders. The development of the towns had consequently to accommodate their needs. Builders obviously build to make profit, so compromises in styles and layout proved necessary to induce them to build. More significantly the rate of building in a garden city came to depend upon its profitability. One consequence was that the development of a town had to take a long time as no builder would want to flood the market with houses. This extended further the payback time from the improved rents to be derived from the initial investment. Moreover, builders will want a share of development gain as their profit, so if a garden city authority in such circumstances wants building to be done they may have to give most of the development gain to the speculative developer. In this way, the principle of municipal appropriation of increase in land values becomes threatened, and with it the viability of the garden city project itself.[3]

So whilst they were not actually state agencies, these two experimental Garden Cities highlight the dilemma of state intervention in building development. State involvement raises the profits to be made from development through a better ordering of land uses and physical infrastructure provision, yet if private development interests are involved most of the gains from public investment will accrue to those interests. Development gain is at the centre of state involvement in development. Who gets it and how they get it will determine the distribution of the economic gains from development and also the feasibility of any development. The result is that by having to accommodate itself to private development, land-use planning is transformed in its nature.

One final historical irony of Howard's initial scheme emphasizes the importance of this public planning private appropriation dichotomy. First Garden City Ltd was the company set up in 1903 to build Letchworth. Funded by share subscription this company paid no dividend for over forty years and so hardly constituted blue chip stock. Yet the rises in land values of the post-1945 era encouraged a small group of investors to buy up the shares of the company with the aim of selling off its assets (i.e. Letchworth) at a high profit. So Howard's first scheme to remove individual private land property from urban development ended up in the hands of land speculators! Amidst much embarrassment and political concern a Bill was rushed through Parliament in 1962 to municipalize Letchworth. The land speculators achieved their aim, however, as they were given £3m. in compensation (Schaffer 1972).

The fact that the Garden Cities Movement can be used to illustrate many of the problems associated with land values and present-day planning is not surprising. The economic principles of the Garden City idea became the model for progressive planning practice. Its influence can be seen in the New Towns programme, as mentioned earlier, and also in a series of widely dispersed attempts to bring suburban development land under public ownership by acquiring it at current-use value or publicly appropriating all the subsequent development gain. The public appropriation of land betterment was proposed by the 1942 Uthwatt Committee as part of their practical solution to the problem of compensation and betterment. It appears again in the 1967 Land Commission Act, and yet again in the 1975 Community Land Act. But, apart from Letchworth and Welwyn, such planning practices had to remain part of the ideological armoury of the proponents of planning until 1947, as no earlier government was prepared to introduce effective planning reforms. Howard, however, did help to found one of the most influential propaganda lobbies for planning, the Garden City Association, which ended up being called the Town and Country Planning Association; so it is not surprising that the garden city ideas had such an ideological influence on subsequent events.

The containment of urban planning

The urban crisis which threatened to be such a major impediment to industrial capital at the end of the nineteenth century and in the first years of this century gradually receded. With it a major impetus to state involvement in land-use was lost, and the planning movement had to wait in the political sidelines for most of the inter-war period content with ineffectual reforms like the 1932 Town and Country Planning Act.

There were a number of reasons for the reversal. Liberal governments in the years prior to the First World War had failed to dislodge the power of private landownership. Yet by the 1920s economic change seemed to indicate a permanent decline in the power of urban landed interests as the application of new technologies finally broke down the land-use patterns of earlier times. The revolution in power for industry that came with electricity freed industry from its previous locational constraints.[4] Electricity also led to substantial improvements in public transportation, with growing networks of tramlines and underground railways from the 1900s onwards. By the 1920s and 1930s motor transport also began to have a considerable impact on the transportation of goods and people. The movement of people was mainly affected initially through the introduction of the motor bus but later increasingly through personal car ownership. So the suburbanization of industry and dwellings ended the late nineteenth-century relationships between urban development, urban rent, industry and labour analysed earlier. Political conflict over housing issues was also considerably lessened. The inter-war housing boom improved the housing conditions of the politically strongest groups. Housing conditions in inner city and traditional industrial areas might still be poor but the mass unemployment of the 1920s and 1930s sapped the strength of the working class to force much change.

The failure to come to terms with private landownership also made ineffectual the limited planning legislation introduced between 1909 and 1932. Those Acts were a sham because of their land compensation procedures. Under them mere formal acceptance and publication of a town plan could result in large public expenditure on unforeseen amounts of compensation to landowners denied the right to develop or sell their land to its most profitable use. Local authorities consequently were reluctant to introduce planning schemes for fear of the subsequent compensation claims, virtually none of which could be recouped from a betterment tax on landowners gaining from the plan. So, not surprisingly, by 1942 only 3 per cent of the land acreage of Britain was covered by operative planning schemes, and according to Cullingworth (1979b) in the draft plans that covered about half the country in 1937 enough housing land was zoned to accommodate 350 million people!

Inter-war speculative development continued to follow the minimal infrastructure and poor public facility principles of suburban sprawl noted earlier, with little control by the state over its form or location. Yet the argument for better planning, at least temporarily, had been reduced to one of an answer to land-use chaos rather than the more politically compelling one of a way out of social crisis. Events had contained the planner not the private developer.

The reasons for the 1947 Planning Act

The very features that had led political pressure away from agitation for the state involvement in urban development, however, ultimately helped to create the preconditions for the post-1945 planning legislation. Suburbanization, the motor car and industrial decline had by the end of the 1930s forced the land values question back into the political limelight. It is important that these pressures are clearly understood as they fundamentally affect the interpretation given to the role of the post-1947 planning system. Emphasis in many explanations of the 1947 legislation is put on the industrial change and economic crisis of the inter-war years and its effect on post-1945 society. After the Second World War it is said that the state had to have a regional policy because for military, economic and political reasons the drift of industry to the south had to be stopped. It has been argued as a result (e.g. Hall 1975) that land-use planning was introduced primarily as part of regional policy following the recommendations of the Barlow Commission in 1940. Direct regional policy would move industry to the previously depressed regions and land-use planning would halt the spread of the large cities in the south which made them such attractive places for industry to locate. This regional policy effect was undoubtedly important but sole consideration of it leads to the conclusion that the prime function of land-use planning is that of urban containment: in other words, to the view that the land-use planners' ideology of containing suburban sprawl coalesced in the immediate post-war period with contemporary political expediency associated with industrial location. Bureaucratic power was then given to that planning ideology with adverse consequences for urban development ever since.

The most well-known and authoritative view within this framework is a monumental study published in the early 1970s called *The Containment of Urban England* (Hall *et al.* 1973). This study concluded that the planning system has produced a particular pattern of urban growth: 'Its outstanding features are urban containment, the development of suburban communities isolated from employment and other opportunities, and rises in land and property values – an unexpected effect of the system' (ibid., p. 378). The price set for urban containment therefore is very high. The ideal of a good urban environment is posed as being in direct conflict with cheaper housing and better employment opportunities. This resultant conflict can also be extended to the actions of social agents: private house builders are in conflict with local planners over land allocation (cf. Hall *et al.* 1973 and Underwood 1981), and 'citizens' are in conflict over town planning issues (Simmie 1974), for example. The planning system is therefore conceived as a barrier that a social group must overcome or use against others in order to achieve its ends, with some groups paying the costs and others getting the benefits. The role of the planner also has been 'contained' to that of the negative function of allowing or forbidding development.

So explanation of a historical event, the 1947 Act, has become intertwined with specific conclusions about the current nature of the planning system and theories of how to study planning. Trade-offs and conflicts undoubtedly exist and their effects are differentially spread through distinct social classes. But what is remarkable is the way in which they have been posed in this approach: the planning system is regarded as all powerful; little consideration is given to other social and economic pressures. Examining the other strand of pressures leading to the 1947 Act, associated with the 'positive' role of the state in urban development in its organization of land-use activities and its vast expenditure on the built environment, leads to different conclusions. Planning is certainly a site for struggle between particular social groups but that struggle is structured by an economic and political dynamic that determines the possibility of and the necessity for some form of state intervention into the land development process.

The closing years of the 1930s was a period when two decades could be looked back on which contained dramatic social changes and clear examples of the inability of the separate parts of the capitalist system to function as a socially coherent whole without substantial state intervention in economic and social life. These characteristics were reflected in land use. During the 1920s and 1930s about 1 million acres of land in Britain was converted from agricultural to urban use; that represents a 50 per cent increase in the previous total acreage of towns in Britain (Best 1981). The peak period was in the 1930s during the speculative housebuilding boom. Agricultural interests came into conflict with development interests as some of Britain's best agricultural land disappeared under the spreading suburbs. Considerable disquiet was also raised over the loss of countryside as an amenity and at the visual drabness of most new suburbia. The state, moreover, was forced into the development process in order to make these new suburbs feasible living places, via its statutory servicing and road functions. Yet the expense of providing these facilities was made far higher by the unplanned, dispersed nature of development. The motor car, in turn, was forcing money to be provided out of a reluctant state exchequer. Unlike the earlier introduction of public mass transportation, such as railways and tramlines, it was paradoxically impossible to envisage private capital providing roads without intolerable restrictions on access and traffic flow. Private, individualized transportation was predicated on the state providing the means by which to implement it.

Large investments began to pour into new road schemes and a national network of trunk roads was started. Yet suburbanization continually created greater demands on the road system, and the actions of speculative builders actually reduced the movement of traffic on existing roads through ribbon development. By building along an existing road the builder avoided the new road costs of estates. But such ribbon development meant that existing roads could not be widened or improved and it generated additional congestion. Moreover as soon as a new road in open country was built there was an enormous incentive for builders to erect houses along it. 'We shall ultimately be driven again to by-pass some of the by-passes' warned Unwin in 1931 (quoted in Sheail 1979). Yet governments were not prepared to stop ribbon development as that meant limiting the rights of private property in land. Limitation could only be achieved through confiscation or by providing large sums of money as compensation for the loss of development rights. The procrastination surrounding the passage of the Restriction of Ribbon Development Act 1935 and its complete ineffectiveness brought out this contradiction clearly to contemporary observers (Sheail 1979). Planning could not proceed until something was done about the land question.

The mass destruction of the Second World War took those contradictions to new heights. It was clear that the state would have to play the leading role in reconstruction; argument was limited to the degree to which private development was to be encouraged. Moreover, by the end of the war it was also clear that a new phase in capitalism had been reached. The state in future would have to intervene more directly into economic and social life. Expenditure on the built environment would be a major element of that involvement. In part this intervention was necessary to create the economic conditions for expanded accumulation but, at the same time, the heightened political consciousness of the working class had to be accommodated. Planning was claimed throughout the depths of the mass destruction of the Second World War to be a necessary component in a victorious, new post-war age when the horrors of mass unemployment, slum housing, decrepit nineteenth-century urban structures and inter-war suburban sprawl would be swept away (Backwell and Dickens 1979). Such pronouncements, of course, could just have been state propaganda aimed at sustaining the morale of a war-weary population. But their propaganda success did help to generate a post-war political consciousness pushing for the fulfilment of those ideals.

A whole series of strands consequently came together in the 1940s pressing for the introduction of effective land-use planning. An urban crisis existed at least as acute as that of thirty years earlier. The physical fabric of towns was antiquated, worn out or destroyed. There was a chronic housing shortage after 1944 which private agencies could not begin to solve. And there was a broad constellation of political forces pressing for change. Given the contemporary strength of working-class power there could be no return to pre-war society. Housing subsidies and this time planning reforms could be used again to demonstrate the feasibility of the reformist path to social change. Yet still the problem of land compensation and betterment remained. Land speculators had already been active in the war years buying up bombed sites at low prices and waiting to sell them at the war's end.[5] Temporary legislation limited their profiteering but the land problem had become acute. The economic power of landowners, property owners and building developers had to be spiked if reforms were to succeed and planning was to regain any political credibility. Building controls, council housing programmes on garden city lines and the 1947 Planning Act were the post-1945 Labour government's attempted solution to the problem. Each component had only a short life.

The intellectual basis of the 1947 planning legislation

The two major reports that formed the intellectual justifications for the 1947 Planning Act, those of the minority on the Barlow Commission (1940) and Uthwatt (1942), both suggested forms of land nationalization as 'best' solutions to the problem of planning. Barlow suggested nationalization of development rights, leaving the general question of compensation and betterment to a subsequent expert committee. This latter proposal led to the setting up of the Uthwatt Committee on Compensation and Betterment. It felt that the

immediate transfer to public ownership of all land would present the logical solution – but we have no doubt that land nationalization is

> not practicable as an *immediate* measure and we reject it on *that ground alone.*   (Uthwatt Report, para. 47, my emphasis)

By treating land nationalization as the 'logical solution' the Committee wanted to exclude private landed property from the development process. This would lead, as they saw it, to the removal of the conflict between the development process and the activities that needed new buildings. With public landownership the state could orchestrate development in the interests of land users alone, as the problem of compensation and betterment and the barrier it represented to effective land-use planning had been solved once and for all. The new landowner, the state, would not want the maximum land price possible before any individual development could be undertaken. Betterment would accrue to the state but only that amount commensurate with socially optimal land uses as designated by the plan. At the same time compensation would not be required for loss of potential development rights in land fixed in its current use by the plan. The criteria of the private land market would be abolished and with them the relation between land ownership and use turned on its head, because no longer would the interests of landowners dominate those of land users. As owner, the state would orchestrate land uses into the best possible spatial pattern and then get any betterment left after that land-use configuration had been achieved. Land gain would finally become a true residual product of the spatial distribution of economic activity.

The immediate impracticality of this land nationalization solution referred to by Uthwatt was not so much a reference to the war that was raging at the time as a recognition of the political power of the private landed interest and of the cost of the compensation proposals for land acquisition. Presumably it was recognized that land nationalization could easily be whipped up, as it had been in the past, to be regarded as a threat to the very existence of private property and the capitalist system. The Committee presumably was also aware that this was not true, consisting as it did of three eminent members of the legal profession and a vice-president and past president of the Chartered Surveyors' Institution. Their more practical solution was an attempt to avoid such political furore whilst keeping the main development features of the full nationalization programme. It suggested that all rights to development should be vested in the state and all land nationalized as and when it was actually built upon. Owners of land nationalized were to be paid the current-use of the land (that is its value without development) plus compensation for severance, disturbance, etc. When, after nationalization, development was undertaken by a private developer the land would be leased to the developer with covenants attached to ensure they built on terms agreed by the planning authority. A rent would be charged for the lease but the Committee was vague about the actual criteria to be used in assessing the rent. The Committee also saw no objection to a private developer paying an inducement to a landowner to release land for one particular development, as long as it conformed to the plan and the state formally acquired ownership of the land and assigned the lease to the developer. Owner occupiers who erected houses on land they owned would be totally excluded from the scheme.

The Uthwatt proposals are one stage in a clear heritage of the ideology of modern land-use planning in Britain. Many of their ideas have roots in the Garden City Movement; for instance, private landownership as a barrier to planning and public ownership of development land (though nationalization had replaced municipalization). The 1947 Act, in turn, implemented legislation akin to some of Uthwatt's proposals, although it made them far more difficult to operate through trying to avoid land nationalization altogether by introducing a betterment levy instead. One clear strand remaining from the earlier heritage in Uthwatt and the 1947 legislation was the technical superiority of planning. This view is based on a belief in a fundamental harmony of interests in society over land-use once the problem of compensation and betterment has been solved. There was little concern over the political process of planning, given this perspective, as it assumes that there are generally obvious bureaucratic solutions to planning problems which can be devised by a technocracy subject to overall political control and rudimentary democratic safeguards. The notion of such a harmony in class societies is misplaced but that criticism leads to a debate over land use. What is just as crucial is whether or not the problem of landownership and land-use planning had in fact been solved by Uthwatt's practical proposals. The history of planning in Britain since 1947 shows it has not, yet the heritage of Uthwatt has lingered on, as can be seen in the two attempts to nationalize development land by Labour governments and in the principles of betterment taxation.

The common threads outlined above exist because of a shared ideology about the nature of planning and land. In it the land values question can be solved by not compensating landowners when development is forbidden and by seeing betterment as accruing to the landowner alone which can then be removed or at least reduced by nationalization or some degree of betterment taxation. But nationalization only removes the private landowners from the development process, it does not exclude other private agents like speculative housebuilders, who also profit from development gain. The alternative policy of land betterment taxation obviously faces similar difficulties. Land-use planning does not simply have to come to terms with the economic interests in development of landowners but also with those of other private agencies profiting from development. The post-1947 planning system has had to accommodate both landowners and developers.

In the Uthwatt-style formulation of the post-1947 planning system, once the question of compensation and betterment has been resolved it is claimed that the undesirable financial implications of land development cease to exist. Landowners can no longer impede the implementation of the plan and no socially inequitable financial gains accrue on implementation of the plan. In place of control by the land market there is now overall control by the plan. In the new hierarchy the plan is drawn first and then subsequent public and private development happily conforms to it. This obviously is wrong, not simply because of its political naivety but also because of the neglect of land developers' acute interest in development profit, shown for housebuilding earlier in chapter 5, which depends on timing, location and above all on getting some of the development gain which the planning reformers had hoped to syphon away.

Land nationalization in isolation is not quite the logical solution Uthwatt claimed it was. This is because the claim is based implicitly on an inappropriate theory of land rent, yet that Ricardian theory of land rent still dominates thinking on land issues. The problem can be seen most clearly in Uthwatt's reasoning associated with the attempt to keep down the compensation payment to landowners for the loss of their development rights. The Committee argued that the overall amount of betterment would not be altered by the introduction of planning, it would just shift between land sites; and also that the land market considerably inflated the true value of land that might have been developed in the absence of planning. In doing so they tried to reduce the problem of private landownership to a technical one of a multiplicity of owners (or of the need for a unification of landownership, in their terminology). Their argument has a number of flaws, yet it does illustrate clearly why land does not have one technically determined, economically residual value and why struggles over the determination of its actual value fundamentally influence the nature of land use and the cost of providing that use. So it is worth going over Uthwatt's reasoning again.

Uthwatt on compensation and betterment

Compensation is a payment to a landowner for loss of certain rights in land or for its compulsory acquisition. Betterment is an increase in land values, either in general or more specifically as the result of a public investment such as a flood prevention scheme. It has been argued (and enacted in various laws since the seventeenth century) that increases in land values should be taxed to help pay for the investment in the public investment case or, more generally, that betterment is an unearned increment which may be taxed away without affecting the level or distribution of economic activity in the economy as a whole. Betterment in the context of the debate over land value taxation in Britain since Uthwatt has been related to the wider definition of betterment: all increases in land values are said to be the result of general economic activity (the 'community' in the phraseology of the 1947 Act) not of an individual landowner's actions. This latter position of course, stems back to the Ricardian theory of rent; radical Ricardians, like Henry George, would have suggested taxing away all of the landowner's rental income not just that attributable to betterment.

According to Ricardo, and to subsequent neoclassical economic theory, permanent surplus profits will accrue to production on lands at particular locations, and being specific to those land sites they can be appropriated by landowners as rent. Betterment, therefore, is the capital value of rent increases arising either from a change of land use or from increases in profitability in the current use. Because it is a capitalized flow of rental payments, future increases in betterment may be anticipated in the current market price of the land. Compensation is, in effect, payment for the loss of potential future betterment. With uncertainty over the timing, location and profitability of future development the anticipation effect may give land a 'hope value' based on the expected probability of development. The notion of hope value was important to Uthwatt's explanation of the land market.

It can be seen that central to the logic of compensation and betterment is the notion of a single value, or price, for a plot of land, based on the current and expected surplus profits derived from the economic system as a whole that will accrue to a site by virtue of the economic activities that take place there. Competition between economic activities for the scarce resource land will ensure that all land-specific surplus profits accrue as rent. The rate of profit on capital using those lands will tend to equality across lands because the land-related surplus profits are appropriated as rent: so the economic system has also been made theoretically determinate by attributing a single value to each plot of land. As most land is not marketed at any point in time, the land values of many sites may be difficult to assess in practice even if apparently not in theory. A valuation profession has grown up to smooth the transition from theory to practice.

In trying to explain why claims for compensation made planning impossible in the inter-war period the Uthwatt Committee stumbled across a phenomenon that brought the whole notion of a unique value for a plot of land into question. Instead of following through its logical consequences, however, they simply used it as a justification for limiting the amount of compensation to be paid upon nationalization of development rights. Compensation, it should be remembered, would be paid mainly for land not currently on the market, the loss of value through being unable to develop had therefore to be assessed for most sites rather than based on their selling price. Uthwatt introduced the notions of 'floating value' and 'shifting value' to derive their global compensation sum; the former concept is the one with the most interesting consequences.

Floating value can best be explained in terms of a simple example: suppose a specific area is currently in agricultural use and could be converted to housing land for 10,000 dwellings. A once and for all development of 1000 dwellings is going to be built in the near future on one of the ten farms into which the area is presently divided. Each farm would be equally suitable and no one knows which one will finally be bought for the development, although all owners are keen to sell. The result will be that the value of the land in housing development floats over all ten farms, and each of the ten land areas acquires a hope value based on the probability of the float fixing on them. So the selling price of each farm will reflect the hope that the development will settle there. Purchasers with the most optimistic view of the development taking place at a particular site will be prepared to pay the highest price for the land, outbidding others with more pessimistic views. It is to be expected therefore that optimistic speculators will bid up the price of individual sites above their expected value if the actual location of the development was random. Like a sweepstake, therefore, if each farm is sold individually the sum of their enhanced selling prices is greater than the likely price of purchase for housing development. Adding more realism and uncertainty does not alter the picture; the sum of individual lands' selling prices will still tend to be greater than their total value with development. In the land market some land sellers' hopes are dashed when the float does not fix on them. The market does not compensate the losers; unsuccessful land speculators therefore have to sell at a loss or more likely will hold on to their land until another development scheme comes along.

When compensation is being assessed for the loss of development rights arising from the introduction of effective town planning, Uthwatt argued that the inflating effect of floating value should be ignored. Otherwise landowners would unreasonably be compensated for the speculative hope value of their land as well as for the actual loss of development rights. Uthwatt's argument can be extended further. Landowners when putting in compensation claims for the loss of development rights have a great incentive to exaggerate the probability that the float would have fixed on their land. Landowners' claims therefore will tend to be even greater than the inflated hope value of the market. So the sum of the landowners' claims is greater or equal to the sum of individual selling prices at a given time which, in turn, is greater than total land value after development.[6] There is an asymmetry in land values depending on the nature of the purchaser and the structure of landownership.

The argument about floating value helped to minimize a tricky political problem. The political consensus over the need for more effective land-use planning emerging during the war years was founded on the need to buy out the opposition of landed interests and ideologically protect the sanctity of private property. To expropriate the development value of land without any compensation was a dangerous precedent to leave to posterity and any future left-wing government.[7] But the compensation to be paid to landowners could be astronomical, running into thousands of millions of pounds, which financially was out of the question. Uthwatt's argument put an upper, but still high, limit on the compensation required. Their theory of land values enabled them to argue that the global compensation sum should ignore the inflating effect of floating value and be based only on value in development, and the latter would be only a half to a third that containing floating value, they estimated (Uthwatt Report, para. 80). In other words, the global compensation sum was to be fixed by reference to the state as purchaser, the lowest of the three potential sums. The whole question of compensation is thrown into disarray, however, by the asymmetry in land value.

Any notion of 'just' compensation based on one clearly defined value collapses. It might be argued morally that gamblers' hopes should not be compensated, yet they would have been if they had been lucky enough to sell the land with the right to development and its hope value intact. If, alternatively, all land value is regarded as the product of an 'unearned' increment, why compensate at all? To quote Uthwatt on planning: 'this involves the subordination to the public good of the personal interests and wishes of landowners'. This argument surely is just as good a case for no compensation as for the proposals they suggested. It had been used, after all, for over a century by land reformers before Uthwatt. Removal of floating value prior to compensation is a half-hearted application of the principle, satisfying no one. Furthermore, many earlier restrictions on land use had not been compensated for, nor had the biggest affront to real property ever imposed in Britain when housing rents were frozen in 1915. Compensation is a political not a judicial, logical or moral matter.

Uthwatt views on compensation, none the less, did broadly prevail in the 1947 planning legislation. A £300m. compensation fund for development value losses arising from the 1947 Act was set up (equivalent to more than £2500m. at 1980 prices), but not without some criticisms: one MP called the 1947 Town and Country Planning Bill 'the Town and Country Bounty Bill', whilst another suggested that if the government had £300m. to distribute on compassionate grounds he could find far more worthy recipients (both quoted in Leung 1979). The incoming Conservative government limited the compensation handout in 1953, an economic expedient made politically possible by the simultaneous abolition of the 1947 Act's betterment tax.

Uthwatt's views on betterment involving nationalization of development land, on the other hand, were not enacted; instead a 100 per cent betterment tax, called a development charge, was introduced. (Some have suggested it was introduced to sabotage the likely continued existence of the 1947 planning legislation; cf. Ravetz 1980.) The idea of the development charge was that all land should be traded at its current-use value. Yet no mechanisms were introduced to ensure that this would be the case or that land would be traded at all. The merits and demerits of the short-lived 1947 betterment legislation have been well chewed over (cf. Parker 1966, Cullingworth 1979b and Leung 1979). But the proposals of Uthwatt have not, and in many respects they are more interesting. The first part of Uthwatt's analysis of betterment was concerned with one possible objection to their compensation scheme, that the global sum of development values might be reduced by the introduction of comprehensive planning. Uthwatt countered this objection with the notion of shifting values: planning only shifted development from one site to another (to one preferable in planning terms), it did not alter the level of development nor consequently its value. To quote:

> Development values as a whole, however, are dependent on the economic factors that determine the *quantum* of development of various types required throughout the country, and as planning does not reduce this *quantum* it does not destroy land values but merely redistributes them over a different area. Planning control may reduce the values of a particular piece of land, but over the country as a whole there is no loss. (Uthwatt Report, para. 38)

Critics of the Uthwatt proposals were quick to point out the flaw in this argument: development is spatially specific, location at one point could easily result in less surplus profit than elsewhere and, hence, produce a lower land value. Office development on the Isle of Skye as opposed to central London is a clear example. The example of one negative shift could be extended to show in theory that the global sum may be diminished by planning, and this is what the critics wanted to demonstrate (Leung 1979). But planning could also raise total development value and, in fact, is more likely to do so. The widespread pressure for the introduction of effective planning, after all, resulted from the negative global consequences of individualized, unplanned expansion; consequences that took the form of congestion, the loss of the best agricultural land, non-conforming uses, under-utilized infrastructural investment and so on. A whole series of pecuniary and non-pecuniary negative externalities, to use the neoclassical economics terminology, could be removed by planning. When the introduction of state land-use planning is combined with the expansion in state expenditure on the built environment that has occurred since the 1940s, it is clear that the additional betterment has been enormous. For, with private landowner-ship, the 'unearned increment' derived from such state activities ends up in the hands of the landowner or developer.

Not only Uthwatt was wrong on this score. Its mistake, moreover, *under*estimated the case for the transfer of betterment to the state. More recent commentators, by ignoring this upward effect on land values of state intervention, have considerably *over*estimated the costs of planning. It is simply not true to say, as the *Containment* study (Hall *et al.* 1973) mentioned earlier said, that a rise in land values necessarily is a cost of planning. The rise in land values could just be part of the distribution of the financial benefits of planning. It, perhaps, is a distribution that would not be favoured by many, but that is why the question of land gains remains so central to the question of land-use planning. The critique should be of the nature of landownership not of planning as such.

Uthwatt's proposals for collecting betterment were very different from the 100 per cent betterment levy of the 1947 Act. Having partially recognized the asymmetry of land values, the Uthwatt Committee was keen in their proposals to introduce 'neither dual ownership of the land nor divided control' (ibid., para. 49). The 1947 Act managed to do both: a plot of land is owned partially by its private owner and partly by the state as its development rights were nationalized in 1947, and it is controlled partially by its private owner (who decides when to sell, when to change its use, etc.) and partially by the state (which has to issue a planning permission for that change of use or make a compulsory purchase order). The problem for Uthwatt was that such ownership distinctions would create a division of interest: privately appropriated betterment against planning in the public interest. For Uthwatt nationalization of development rights meant nationalization of the right to develop not, as in the 1947 proposals, the right to forbid development. For undeveloped land the only feasible way round this dual ownership problem, Uthwatt felt, was nationalization of all land required for development. The state would thus acquire the 'owner's interest' and so could amalgamate it with development rights into a unified ownership. So any land would have solely private ownership prior to development and solely public ownership on, and subsequent to, development. This unification of ownership 'made certain that an occasion should not arise for the acquisition over again of the development rights in the same land' (ibid., para. 103).

Uthwatt's proposal was to purchase land at its current-use value prior to development, then use it for public purposes or lease it to a private developer. The 1947 legislation, on the other hand, ensured that *de facto* the development rights in land would have to be reacquired time and time again, as dual ownership meant that the land value associated with the right to develop remained with the private owner: hence the need for the 1947 Act's development charge. Subsequent repeal of the tax in 1953 and later, in 1959, of public purchase at current-use value left all the land gains in the hands of the private owner. Dual ownership thus again led to an asymmetry in land value: zero for the public owner and all for the private one.

For land already built on Uthwatt did not propose nationalization. Even so land use still had to conform to the town plan, and compulsory purchase would be used to ensure conformity (including, of course, in redevelopment) where necessary. To remove the private incidence of betterment annual increases in site values were to be taxed at the rate of 75 per cent. The remaining 25 per cent would encourage owners to invest in their sites in the most efficient way. Owner occupiers as well as ground landlords were to bear this tax, in a way similar to schedule A tax prior to its abolition in 1963. Schedule A, however, was a tax on income, whereas Uthwatt's tax was on increases in land value which is the present value of future rents. The tax was consequently a tax on capital with widespread implications, although it is unclear whether the Committee was aware of this. The merits and demerits of these developed land proposals are not of interest here but the suburban ones are, and Uthwatt's proposals do seem to get to the nub of the relationship between betterment and land-use planning. But despite its recognition of the possibility of more than one value for a plot of land arising from an asymmetry in ownership or interests, Uthwatt's betterment proposals do rely on the notion of actual development producing one enhanced land value. Betterment can clearly be recognized and defined as a result. Yet the compensation sum for lost betterment had varied depending on whose position was taken; this alone would suggest that the magnitude of betterment is not so clear cut as it would seem. The next section, in fact, will suggest that it is theoretically indeterminate.

Housing development and the planning system

To treat betterment as a single value assumes that all the surplus profits accruing from development are fixed in the land; the accumulation and investment strategies of activities taking place on land are not therefore influenced by the existence of increases in land values. The land market simply allocates land to its best (i.e. most profitable) activity. This is a central tenet of Ricardian rent theory. Yet earlier chapters by examining the activities of landowners and housebuilders showed that their activities are strongly influenced by the possibility of large 'surplus' profits accruing to them through development. The notion of betterment, in other words, tends to obscure the effects of social relations that are so crucial to the understanding of the land development process and planning's role within it.

The transfer of ownership from landowner to speculative builder and then to owner occupier is a movement of land between social agents with different objectives. There is a struggle between builder and landowner over the conversion of development gain to land rent (or land price). The notion of a single amount of betterment obscures this struggle and the important effects produced by it. The planning system is one arena for the struggle over the formation, realization and distribution of development gain between builder and landowner. Both the drawing up of a plan and its implementation via development control and state expenditure influence the relation between landowner and builder. The planning system is not, therefore, a public constraint on private interests but a place where private interests compete.

Planning is a political intervention in the land market. Levels and locations of state expenditure are politically determined. Other interest groups that are represented politically, furthermore, can intervene in the land allocation process. Amenity groups, for example, can ensure that no additional building takes place in certain areas. These political interventions have consequences for both landowner and builder. Severe restriction on land release in areas of high housing demand will, for instance, probably tip the market balance in favour of the landowner. Land-use planning can therefore enhance the monopoly position of a few landowners. Examples like this have helped fuel the view that the planning system has been the principal cause of rising land prices.

Planning, however, can also aid the builder. Planning permission is required for building but, at best, the planning system can only influence the long-term pattern of land development. In the medium to short run it will be housebuilders that actually determine which land is built upon. Land-use planning therefore does not remove the phenomenon of floating value. Planning authorities generally try to determine land release for a 4–5-year advance period and most of the economic effect of hope value (given probable discount rates) occurs during this time period. No landowner consequently is faced with the certain prospect of development on land they own with planning permission and so they do not have an absolute monopoly.

Contemporary land-use planning in Britain further aids the position of the builder by the form it takes. Changing spatial patterns of suburban development, encouraged by mass car ownership and large-scale road building, have generated nodal suburban growth rather than a continuation of the concentric expansion of urban areas of early periods. As part of this nodal expansion pattern, large-scale block planning 'release' of land at points designated for suburban growth has occurred, whereas piecemeal, small-scale expansion has been discouraged. The ability to buy large tracts of land with planning permission at a limited set of locations has favoured large builders and weakened the power of local landowners and of smaller builders who rely on smaller sites. Only when most of the land at a designated node is in the hands of one landowner (or a few who collude) will landowners be able to dictate the terms of development; otherwise market power tends to be on the side of the builder.

The discretion given to the planning authorities to grant planning permission and the high proportion of land given permission without being designated on the initial plan have also influenced the development process and the distribution of gains from it. The vagueness of planning criteria, the high chance of success, and the potential size of the development gain encourage many builders and landowners to try for planning permission on land traded at agricultural prices because it is not allocated for development in the plan (known as 'white' land). This process favours the builders rather than the initial owners because they tend to be the ones with the resources and expertise necessary for success. The divisions between the levels of the state apparatus involved in the planning process add to the apparent arbitrariness of this process, as each level is subject to different political pressures. Central government, for example, has periodically forced local planning authorities to release far more land for residential development than the latter had previously designated (cf. Mackay and Cox 1979). Such windfalls are unlikely to have been previously discounted in land prices and, again, this tends to benefit the builder.

Land banking is important for the builder in their relation to the planning system. In the first place a land bank stops the need to get outline planning permission and detailed planning permission for a scheme from becoming an absolute barrier to immediate building. The need to get planning permission is frequently seen, especially by representatives of housebuilders, as imposing a delay on development. Planning permission however is just part of the process of transferring property rights in land to the housebuilder. As the 1947 Town and Country Planning Act nationalized development rights, planning permission is the transfer of ownership of that right from the state to the builder (for which no rent is charged). At the same time the builder has had to acquire the ownership rights to the land in its current use (for which a rent is charged). Both processes take time; delay exists only when either process takes an unusually long time. In a study of land supply in South East England, for example, it was found that transfer of ownership between private bodies accounted for more total time delay in getting land to the detailed planning permission stage than any other single factor; in fact, it represented 25 per cent of all time delays (EIU 1975).

The existence of hope value gives speculative builders another reason for holding large land banks. They can then buy land which does not have the prospect of being built upon for a few years and hence has a lower hope value. In this way the threat of monopolistic land price rises is weakened. Much of what would be hope value in the land market therefore floats around inside a housebuilder's portfolio of land, enhancing the development profit not the initial landowner's revenue. This feature is reinforced by the fact that much actual development land eventually given planning permission is white land not previously listed as available in planning terms. This increases to a considerable degree the continued existence of hope value, as the value of this land prior to being given planning permission is based virtually all on hope. The magnitude of this type of land is enormous, and it casts considerable doubt on the efficacy of planning control of speculative housebuilding. In their study of land availability in the South East for a two-year period in the early 1970s the EIU found that over half of the land granted planning permission was of this type:

> In all four county areas studied less than half the planning permissions granted in the period since October 1972 were on land identified at that date as available in planning terms. This phenomenon is the result of the peak demand conditions in the period referred to, and to the inherent difficulty of anticipating precisely which parcels of land are available to come forward for development so that they can be designated in planning terms. *Such a procedure is in any case hardly 'planning' in the normally understood sense.*
>
> (EIU 1975, p. 6, my emphasis)

The EIU report also explains how builders go about getting planning permission for white land:

> The larger developers seem fairly confident of their ability to judge correctly the likelihood of obtaining permission on a site and indeed a number of them are willing to put substantial resources behind getting land zoned for development, such as offers of improvements to infrastructure and amenities to the local authority to a greater extent than they as developers could reasonably be expected to provide, or indeed which would be commercially viable were the land purchased with

planning permission obtained, and margins obviously that much tighter.   (EIU 1975, p. 44)

The characteristics of floating value, the vagueness of criteria used for granting planning permission to a potential residential site, and the far larger scale of the permissions granted than would be implied by formal plans alone, make a nonsense of simple claims that local planning authorities have failed to designate enough land for development to meet demand.

Planning control does place limits on what is possible in a development. There can be stipulations on house types, tree preservation, estate layout, landscaping, estate design and the use of facing materials. Yet planners can only impose what is feasible. In practice this means laying down stipulations which still enable the builder to make a profit; thus, in general, planners can only follow the market rather than lead it. Builders frequently use precedents set by more lenient planning authorities to get adverse decisions elsewhere overturned on appeal. When Barratt Developments, for example, introduced their very small 'Mayfair', a one-bedroomed starter house, in the late 1970s, some planning authorities tried to stop the introduction of this house type in their areas by using legislation banning back-to-back housing. By citing favourable precedents from other localities Barratt was able to overcome these objections.

The need to obtain planning permission similarly influences the location of the land which can be developed by private housebuilders. So, instead of suggesting that planning has created a shortage of residential land, it could be argued that planning has caused land price rises by 'anchoring' floating value. Instead of a large number of sites on the fringe of an urban area having a hope value associated with the probability of their being developed in the near future, it is suggested that some sites now have a certainty of being developed as they have been designated as such in the local plan. By removing the mechanism of floating value where any one of a large number of sites may be developed, builders are now forced to deal with a far more limited number of landowners who have land with planning permission. As monopolists these owners can hold out and force up prices. To argue, however, that the post-1947 planning system has entirely removed the phenomenon of floating value is to give the planning system a power it does not have. Development plans designate land that may be developed for housing purposes, generally for a five-year period in the future, with the plan periodically being updated and revised. Yet planning control is only negative, it cannot say whether that land will be developed or when it will be developed. This uncertainty means that floating value is still generated.

Even in the absence of effective planning constraints, floating value would only have been attributable to land that had a probability of being developed within a finite time period (the length of time depending upon the implicit discount rates used). With the advent of effective planning control floating value will now be attributable to land that is sold with planning permission, to land that is designated as being available for housing development in the plan, or to land that has some probability of getting permission even though it is not designated in the plan. So in any locality, unless there is an absolute, unalterable ban on development, which is rare, substantial areas of land have some potential for development in the foreseeable future. Any individual landowner consequently cannot exercise an additional monopoly power simply because of having land with planning permission. They have to control a lot of the potential development land before they succeed in doing that. And if they control much of the potential development land they will anyhow be in a strong market position whether or not the planning system exists.

Despite the arguments above, empirical evidence would appear to support the view that the post-Second World War planning system has forced up land prices. Hall *et al.* (1973) are adamant that insufficient land had been designated over the years, causing a 'distorted' land market and forcing up house prices. Their conclusion has gained widespread acceptance but is based only on the fact that land prices and house prices have both risen in the post-1947 era. Correlation does not necessarily imply cause, yet Hall *et al.* provide little explanation for their assertion beyond a crude correlation of land and house prices. Their data, moreover, are based upon the decade of the 1960s; later behaviour of the housing and land markets has even brought into question the posited correlation between rising land prices and house prices (see chapter 4). What Hall *et al.* omit to consider is that housing land prices are deductions from the potential profits of housing development. Land prices consequently tend to rise as development profitability rises and fall when it falls. Paradoxically, a figure in Hall *et al.* (1973) provides data for 1962 to 1969 that enable calculation of the relation between changes in development profitability and land prices to be made for the period they consider (their figure 6.3, volume 2). It would seem that only in one year, 1969, did construction costs rise faster than house prices, so throughout the

1960s the profitability of housing development rose. It is hardly surprising that land prices rose as well because landowners acquired by that means some of the increase in development gain. The exceptional year of 1969 itself could have been caused by landowners holding off sales in the hope that the 40 per cent Betterment Levy introduced in 1967 would be repealed after the general election early in 1970. (Their hopes turned out to be right. The Betterment Levy was repealed by the incoming Conservative government in July 1970, on the grounds that it had reduced the supply of land!)

This discussion, by highlighting the existence of a conflict between housebuilder and landowner over the conversion of development gain into land values, suggests that it is not possible to consider the effect of land-use planning on private development and the land market without examining the nature of this distributional conflict and the conditions which structure it. Yet this is precisely what most analyses of planning and the land market have failed to do, including the influential *Containment* study mentioned above (Hall *et al.* 1973). It is not possible, for example, to deduce that rising residential land prices exist because of planning restrictions without examining the housing market. Nor can complaints by builders about land shortages induced by planning constraint be taken as proof of such shortages. Because of their structural position, one of the limits to builders' profitability will be perceived as a shortage of land, but for them land will always be short as long as their share of development gain is limited by the demands of the landowner (which will always be the case unless land prices are zero). The question whether planning affects land prices is unanswerable and must be replaced by examination of how land-use planning intervenes in structures of building provision, be they associated with factories, offices, houses or any other form of built structure. Those structures and their relationship to the rest of the social system, moreover, change over time, so the role played by planning also will change.

Planning does have effects on land-use development but not in the way envisaged by the architects of the 1947 legislation. There is no development hierarchy with the plan sitting at the top and private developers busily realizing its objectives after a little prodding from the planning profession. Viewed in terms of the declared aims of the planning system, land-use planning has been an abject failure, as many demoralized planners seem aware. But those aims are not very relevant when examining the impact of the planning system. What is far more important is how it intervenes in specific structures of building provision.

Planning and political struggle

The conclusion reached at the end of the previous section was based on statements about the logic of land-use planning's position in the struggle between builder and landowner over the appropriation of development gain. That does not explain how the political process enables that struggle to play such an important role, particularly as the interests of land users and land developers do not necessarily coincide.

The importance of this point was seen in the evolution of planning described earlier in the chapter. The explanation started off with planning as a possible reform in one sphere of the conflict between capital and labour in Britain in the first decade of this century and ended up describing its position three-quarters of a century later in another conflict between housebuilders and landowners over the appropriation of development gain. The transition to this latter conflict emphasizes the importance of the way development occurs in determining what planning is about. The effect of state land-use planning on relations between capital and the working class has been influenced by the continued existence of private development and landownership, so it is not possible to analyse land-use planning simply in terms of its intervention into the struggle between capital and labour, or of any other political groupings of land users alone.[8]

What is interesting about the politics of planning is the extent to which it has facilitated the political articulation of development interests whilst those of users have remained muted and displaced. Political leaders, civil servants and local government officials have adopted the mantle of knowing what land users want. Their links with the interests of the groups they claim to represent are usually extremely tenuous, and their ideological formulation as being in the interests of the 'nation', 'city' or 'town' invariably helps the interests of capital above all else. Even in a city like Sheffield, which has been politically dominated by Labour for years, such political outcomes are produced. A good example is its council building programmes of the 1960s which involved substantial inner city redevelopment. A major impetus to these schemes was the shortage of skilled labour in the local steel and engineering industries. In their implementation the layout and design of the new structures was the product of 'experts', usually a combination of local officials and private design practices and contractors, making no reference to the working-class households affected by the redevelopment schemes or housed in the new projects. The huge deck-access Park Hill project of the mid-1960s is a classic instance of which Dickens and Goodwin (1981) provide more detail.

This effect of misrepresenting the interests of the populations nominally being served is common to most aspects of state activity, but it is particularly pronounced in land-use planning because of planning's nature as an administrative practice. As planning is done by local government, whilst being overseen and structured by central government, a distinction can be drawn between politics and the planning process at national and local levels. Nationally, planning is an aspect of state intervention in the built environment as a whole, the most important item of which is public expenditure. Since 1945 there have been huge amounts of state expenditure on the built environment (some data are given in the next chapter). The planning system has helped to create this vast activity, directed it to specific localities, and co-ordinated it with private development. Nationally, therefore, political pressures on state expenditure and planning are closely intertwined.

The changes in state policy towards housing are good examples of such interlinkages, particularly the increasing emphasis on owner occupation. This shift has altered the politics and role of planning in the housing sphere. After the re-emergence of owner-occupied housing provision in the early 1950s the spatial distribution of new housing development, and hence planning's relation to it, diverged on tenure lines. New council housing was related to comprehensive redevelopment and overspill schemes. But intense political protest in the suburban fringes restricted the extent of new council suburban development (Cullingworth 1960). Divisions between metropolitan and suburban local government were successfully used to stop council suburban expansion, and the respective planning departments played important roles. The failure to open up the suburbs forced council housing to remain in the city centre and raised densities. New town developments, which included large proportions of council housing, were the only exception to this trend. Their location similarly was the product of intense political activity rather than 'objective' planning principles. (The designation of Swindon as an expanded town in the mid-1960s is a good example (Levin 1976)). With the containment of council housing, first spatially then later absolutely, the planning of housing development has become concerned principally with the spatial expansion of owner occupation, especially in designating improvement areas and in suburban land availability. Shifts in emphasis between tenures and their contemporary structures of provision consequently lead to commensurate changes in the planning system. Yet, at a local level, these shifts in planning practice rarely feature as significant political issues.

For the majority of the population land-use planning, if it means anything to them at all, is a distant and seemingly eternal entity; a part of the local council which is only of direct interest when its proposals impinge on localities of immediate relevance to them. Most direct relations between the populace and planning authority consequently tend to be confrontational, consisting of local groups mobilizing when their interests are threatened by land-use development proposals. Participation exercises in the formulation of land-use plans usually elicit minimal popular response, except from a handful of established, well-heeled, local pressure groups (cf. Damer and Hague 1971). This reaction is hardly surprising as land-use planning has little or no direct control over resources. It facilitates, forbids and co-ordinates yet rarely can it initiate anything. So the planning department remains faded into the`backcloth of the local administrative apparatus. In much of its activity the planning department is able to proceed as its senior officials see fit, within the broad guidelines of council policy. Frequently the latter are extremely vague on most planning issues and planners are aware of which items might be contentious and hence 'political'. This means that in most of its activities local planning is routinized and de-politicized. The ideologies and jargon of planning and its professional functionaries have further reinforced this quality of planning in the post-1947 era.

The formulation of a structure plan or its rejection by the Secretary of State for the Environment hardly creates a ripple in the world of local politics compared to, say, a bus fare or rate rise. Controversy only occurs when this routinized procedure comes up against the interests of groups with the means to articulate their protests politically. The politics of planning consequently is reduced to the politics of a limited set of pressure groups: developers versus conservationists is the classic instance, the suburban fringe the classic location.

Planning cannot however escape being a part of the local political process. Yet, in a number of respects, this inevitable politicization of planning reinforces its eternal, abstract qualities and so its apparent lack of direct political relevance. The aims of plans become general statements of long-term political goals directed at all sections of the local electorate – vacuous, well-meaning statements, imbibed with a consensus perspective: full employment, more investment, better housing, additional schools, improved transport and an upgraded environment are sentiments splattered around the forewords, introductions and strategy statements of most structure plans.

In part such sentiments are a product of the current political function of planning and the technical way in which problems are confronted. The planning system must take land-use problems in the immediate empirical form in which they are presented, without questioning those forms or attempting to explain the cause of the problems. Yet, despite having no adequate explanation of their existence, planning must assume those problems can be solved and try to implement land-use proposals that help towards their solution. Implicitly or explicitly it must claim therefore to know their solution, and the claimed solution has to have a land-use component and involve little or no social conflict. This procedure, implicit within planning practice, is obscure however, as only the land-use component of the solution needs to be stated. In this way the potential incoherence or illogicality of the solution gets lost or hidden beneath jargon and good intentions, thereby diffusing any potential political opposition to the proposals.

This approach can be seen in the response of many planning authorities to the problem of rising unemployment. Most large cities have experienced sharp job losses in manufacturing industry over the past twenty years. The policy statements of most structure plans note this fact for their area, conclude that more private investment will create more jobs, and draw up transportation and industrial land availability plans accordingly – oblivious to the questionability of whether or not such strategies will create more jobs or to their role in reinforcing class exploitation in British society.

Such arguments should not be construed as implying that planning has no effect on land-use development; on the contrary its role is substantial. But, because of the misspecification of the problems with which it is dealing, the effects are different from those envisaged in the plan itself. One particular irony has been that by failing to recognize the existence of social conflict the planning system has diverted criticism away from the dominant agencies involved in those conflicts on to itself. Problems which it claims to be able to solve become 'a failure of planning' when they still exist.

The absence of political struggle over the fundamental issues with which planning is dealing encourages the system to move in the direction of accommodating the interests of the providers of the built environment, especially private developers, as they are the means by which the development goals of the plan can be implemented. But in trying to encourage development by private agencies, planning continually gets caught in the contradictions raised earlier of attempting to plan a private market. Planning, in at least some of its aspects, must go against market forces yet it has to ensure that the beneficiaries of those market forces still function. Builders, developers and landowners have to be coaxed to follow the plan by the lure of profit. Yet, by definition, some of the most profitable development schemes are forbidden by effective land-use planning.

This contradiction leads to the ultimate irony that one of the main beneficiaries from planning, developers, are also one of its major political opponents. They both want to go along with planners' suggestions at some locations and go against them at others, because both strategies lead to the maximum development profit. Landowners that lose out from a plan by being unable to sell their land for a new use will also feel aggrieved. Others who wish to retain existing land uses, like local residents and preservation societies, will oppose land-use change. The politics of planning consequently has all the appearances of a pressure group confrontation. Yet it is more than that, as those confrontations are structured by wider forces. Suburban development, for example, is influenced by events and changes in the owner-occupied market.

Just as it is impossible to understand the owner-occupied housing market without looking at residential land development and the role of planning within it, so it is impossible to understand changes in the politics of planning without understanding changes in the development activities which it is trying to influence. With this conceptional point it is now possible to consider why the current problems of owner-occupied housing provision are leading to the collapse of planning control.

8

The demise of planning
control

The political crisis of land-use planning

Planning issues are rarely important news items and, when they are
reported, coverage is restricted to the traditional themes of conservation,
new roads, commercial development and expressions of horror at post-
war comprehensive redevelopment schemes. This lack of coverage is an
indication of how little planning issues feature in popular political
debate. All the same there has recently been a large-scale political
onslaught on the functions of the planning system as they emerged over
the twenty years or so after the 1947 Act. Most importantly the ability of
the planning system to influence land uses has been severely weakened.
The change has not been towards greater popular involvement and conservation of pre-existing community structures as hoped for by many in
the early 1970s, but instead to the advantage of the developer. This
theme could be explored for all types of land use – shops, factories and
offices as well as for housing. Here emphasis will be limited to housing
and it will be argued that the current crisis of owner-occupied housing
provision has played an important part in the general political onslaught.[1]

The political struggle over planning has principally taken the form of a
complex technical debate over planning delay and land allocation. They
are hardly newsworthy items but the results of that debate are likely to
lead to further urban decay and suburban sprawl. The process will be a

piecemeal one extending over many years as developments take place or fail to take place at particular locations. By the time the spatial transformation is widely recognized it will be history not news; another failure of British social policy, a crisis of land-use change.

The debate over development control has been protracted, having lasted for almost ten years since the speculative land boom of 1972–3. Development control is the negative function of planning which gives it power over land use through being able to forbid development by refusing planning permission. The planning system has been charged with creating unreasonable delay in processing planning applications. The claim has also been made of a further tightening of restrictions on suburban housing development, producing an overall land shortage. So the planning system is under attack over its efficiency and its policies, and obviously if the mechanism can be shown to be inefficient the credibility of its policies becomes tarnished.

Polemical and academic literature over the past seven years since the first, interim, Dobry Report (1974) has produced a written debate running into millions of words. Much comment has been a direct part of the political process, with government sponsored reports (e.g. EIU 1975, JURUE 1977 and Nicholls et al. 1980), House of Commons Committee Enquiries (8th and 11th reports of the Expenditure Committee 1976–7 and 1977–8), and DoE circulars of advice to local authorities. On the whole, the planning system has been said to work quite well (Davies 1980 and Underwood 1981 provide comprehensive surveys of the debate). Not all authorities are as efficient as others, but relatively simple procedures have been introduced to encourage them to improve and all the political fuss has anyhow concentrated the minds of planning authorities in that direction. But the debate should not be seen simply in technical terms. Much of it centres on the social conflicts over new development in which planning intervenes. It is a debate about whose interests should have greatest weight, for instance a developer or local objectors. The debate ultimately is a political one, to which there has been a substantial response by the post-1979 Conservative government through legislation, advisory circulars, statutory controls and exhortation. The end-product has been a surreptitious, but profound, weakening of planning controls over private housing development. Little controversy has arisen over this change because the formal apparatus of land-use planning is intact. It is primarily its effectiveness that is weakened.

In many ways the changes in the planning system are more fundamental than the failures of the better-known attempts to divert land gains to the state via the Community Land Act of 1975 and the Development Land Tax Act of 1976 (Massey and Catalano 1978). Both pieces of legislation were a direct consequence of the land and office development boom of the early 1970s. The Community Land Act was repealed in 1980, whilst development land tax has had even further loopholes added to what was already an easily avoidable tax. These attempts to appropriate land gains for the state altered the situation established in the earlier 1950s against the developer. In many ways the failure of the two pieces of legislation was inevitable, given the way they were implemented (Massey and Catalano 1978). But the attacks on development control go to the very heart of the post-1947 system by limiting even the negative power that planning has had over land use. Given that the attack on planning has been going on for almost a decade, with a piecemeal but continuously sympathetic response by successive governments, this demise of planning control cannot be simply blamed on the right-wing market ideology of the post-1979 Conservative administration. There is a structural malaise in the planning system and in its relationship to land development agencies.

On two counts over the past few years, taxation and planning constraint, the potential gains from development available to be shared between landowner and developer have increased. Previous chapters showed how this increase in gain came just in time for many hard-pressed speculative housebuilders. But to say that the planning system has been transformed at the behest of speculative housebuilders is insufficient. They constitute only one component in a much wider political struggle. What has happened instead is that the conjuncture of economic forces and political alliances which led to the creation of the post-war planning system has collapsed, enabling the position of speculative housebuilders to be strengthened out of all proportion to their direct political influence. Planning as a result is in profound crisis. Any credibility previously entertained for attempts at the physical designation of land uses by a relatively insignificant department in local authority bureaucracies is evaporating. As a recent pamphlet by the Council for the Protection of Rural England noted 'Planning itself is in the melting pot' (CPRE 1981).

The undermining of the notion of land-use planning as the prime determinant of land uses was inevitable. Like the Keynesian myth of being able to fine-tune a capitalist economy to ensure full employment and prosperity, it was based on false premises. Planning's lack of a firm theoretical base, as outlined in the previous chapter, meant that once the conditions disappeared which gave it the appearance of being able to control land uses, it would be subject to a political buffeting it could not resist. Its power had centred around the ability of the state to orchestrate land-use development through large-scale public expenditure on the built environment. Once that expenditure started to dry up in the early 1970s development control was not as easy as it had once seemed. Uthwatt's strictures on the need for the state to control the right to develop rather than just the right to forbid development came home to roost with a vengeance.

The suggestion that the role of land-use planning in residential development has been substantially weakened may surprise many. Formally the planning system still exists with the same broad features. If anything its formal role has increased since the advent of structure planning and associated attempts at greater public participation. They were two reforms of the 1960s whose major impact was not felt for almost a decade. Recent planning legislation has done little to alter this basic framework (see Barrett and Underwood 1981 for details). As explained in chapter 7, however, the planning process involves a considerable amount of ministerial and local discretion. Substantial changes have occurred in this hazy yet crucial area of planning, with the result that it is becoming increasingly difficult for the planning system to direct development in any other direction than would have occurred in its absence. This is the fundamental crisis of planning.[2]

Changes in the planning system cannot be divorced from the long-term crisis of British capitalism and the spatial restructuring associated with it. Political reaction to that crisis has helped generate large-scale cuts in state expenditure and industrial change has altered the level, type and spatial distribution of economic activity. Changes in the relationship between speculative housebuilders and the planning system are the product of these general social movements and of internal contradictions in the structure of owner-occupied housing provision. To explain how in combination they have led to the demise of planning control it is necessary to build up from broad population movements, through shifts in public expenditure and housing policy, to changes in the administration of planning, and finally on to politics and the mire of pressure group lobbying and polemic.

Changes in the distribution of population in post-war Britain

There have been substantial geographical shifts in the location of the population of Britain over the past thirty years. As population movements are defined on the basis of where people live, there is obviously a close, though not perfect, link between changing housing requirements and population change. The link is weakened by differences in household size (moving households might be of larger or smaller size than non-movers, for instance) and because each area has a different stock of existing housing available to satisfy any given level of housing need. But generally population movements indicate the places where pressures for new residential development are greatest.

The most significant trend since 1950 has been the movement out of the traditional urban cores. At first, population loss occurred in the inner districts of the largest cities, followed later by similar losses in smaller cities. Initially net migration was to the fringes of large urban areas and to freestanding towns near them. But during the last census decade, 1971 to 1981, even the suburban fringes began to lose population whilst more distant locations gained it. This process of population decline in the traditional British urban system has been described as being like mould on an orange. Existing patches gradually intensify and spread outwards while, simultaneously, new spots break out elsewhere (Randolph and Robert 1981).

In the decades 1951–61 and 1961–71 the population of Britain increased quite rapidly, by 5 per cent for both periods, whereas during 1971–81 there was only a marginal increase. Yet the changing distribution of population meant that marked losses and gains have continued to affect areas of the country. The principal criterion distinguishing areas of loss and gain is their place in the hierarchy of urban areas. The typology of areas gaining or losing population can be classified in a number of different ways (see table 8.1) but each classification shows that in the period 1971–81 the population shift was to smaller towns and villages. The absolute magnitude of the change is dramatic: Greater London lost ¾ million population between 1971 and 1981 whilst non-metropolitan, non-city districts gained 1¾ million.

The effect on the number of households and, hence, dwellings in the traditional urban areas has not been quite so dramatic, as a sharp drop in average size of households meant that the number of households in Britain increased by over 10 per cent between 1971 and 1981 (*Social Trends*, 1982 estimates). The growth was principally in one-person households of which cities have a disproportionate share.[3] The changing spatial distribution of population has not been identical, consequently, to that of households. The resultant spatial changes in housing needs are far from clear but a broad generalization is that in urban areas there has been a growing mismatch between households' housing needs and the existing and frequently decaying housing stock, whilst in the non-city areas of population growth there is a strong demand for new housing and pressure is often highest for family or retirement accommodation. Overall the number of areas where there is strong pressure for new

Table 8.1   Population changes in England and Wales, 1971–81, classified in two ways

Classification A[1]	Population 1981 000s	Population change 1971–81	
		000s	%
England and Wales	49,011	262	0.5
Greater London	6,696	−756	−10.1
Inner London	2,497	−535	−17.7
Outer London	4,199	−221	−5.0
Metropolitan districts	11,235	−546	−4.6
Principal cities	3,486	−386	−10.0
Others	7,749	−160	−2.0
Non-metropolitan districts	31,080	−1,564	5.3
Large cities	2,763	−149	−5.1
Smaller cities	1,687	−55	−3.2
Others	26,630	1,768	7.1

Source: R. Drewett, 'A comparative analysis of urban problems in Europe', *The Planner*, November/December 1981

[1] Groups of areas defined within the *1981 Census Preliminary Reports: England and Wales*, HMSO

Classification B	Population change 1971–81 000s	%
London	−756	−10.1
Other metropolitan areas	−546	−4.6
Other cities	−204	−4.8
Other industrial areas	200	3.1
New towns	283	15.1
Resorts/Retirement	156	4.9
Accessible small areas	661	17.8
Remote small areas	468	10.3
England and Wales total	262	0.5

Sources: 1981 Census Preliminary Reports; W. Randolph and S. Robert, 'Population redistribution in Great Britain 1971–81', *Town and Country Planning*, September 1981

development has increased with population dispersion. This means that for any level of housebuilding there will be more local planning authorities confronting the problem of which land to designate for new development.

Population changes are not, of course, a cause of pressures on planning in themselves but a product of a whole series of factors influencing land-use change. Higher living standards, changing life-styles, the growth in use of the motor car and a road network to accommodate it, the decline and decentralization of manufacturing employment and a host of other reasons need to be drawn together to account for it. A long exegesis of those divergent factors is unnecessary here but it is important to note that these changes have been associated with a new phase in the spatial requirements of capital accumulation and the spatial distribution of population associated with and made possible by it.

The impact of public expenditure cuts

The population changes of the 1950s and 1960s were closely matched and aided by large-scale state expenditure on the built environment. Both figure 8.1 and table 8.2 give indications of the extent of this expenditure. The most notable feature of figure 8.1 is the meteoric growth of public non-housing expenditure up to 1970 and its equally sharp fall since then. Between 1955 and 1970 public non-housing expenditure on construction work tripled (at constant 1975 prices). Included in this category are projects associated with government administration, defence and the investment programmes of the nationalized industries which are not directly relevant to residential spatial change. Data do not exist to break down the various components of this global public non-housing category, yet it is to be expected that most does relate to the potential transformation of residential space, either directly, as in new town and urban renewal programmes, or indirectly by altering the potential residential attractiveness of particular locations through road building and other transportation expenditure or as part of residentially linked welfare programmes, such as schools, health centres and hospitals.

State expenditure on the built environment has an important effect on the profits of speculative housebuilders. When public works take place at non-inner city locations they help to create a huge pool of potential development gain which is gradually realized by speculative builders in their housebuilding programmes. The rapid decline in public non-housing expenditure since 1970 has reduced this source of development

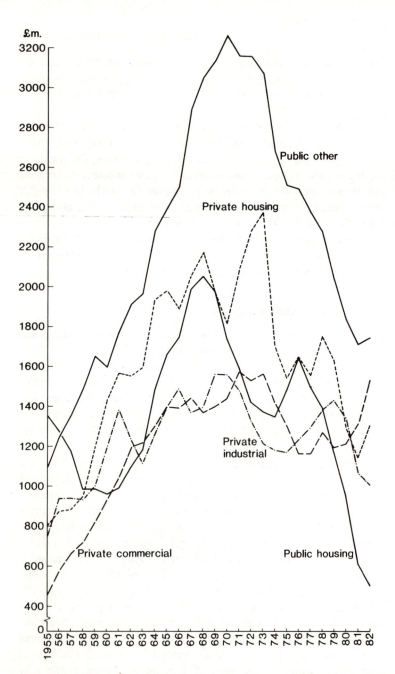

Figure 8.1 Construction output by type of work (at constant 1975 prices), 1955–82

Sources: 1955–69, DoE; 1970 onwards, HCS

Table 8.2 The rise and fall of production of the new built environment, 1955–82

| | Selected indices 1975=100 | | | | | |
| | Construction | | Housing | | | |
	New orders: all new work	Output: all work	Starts	Com- pletions	Slum clearance demolished or closed	Government- assisted dwelling renovations
1955	—	58.0	96.9	101.4	55.4	10.2
1956	—	63.0	86.3	96.0	73.6	23.5
1957	82.3	64.2	85.5	96.2	89.9	22.7
1958	80.6	65.4	80.2	87.4	104.7	21.7
1959	98.9	73.1	98.4	88.4	112.5	49.2
1960	114.9	78.5	95.7	95.1	109.2	79.8
1961	115.6	84.7	96.7	94.6	116.9	78.3
1962	118.4	87.5	100.2	97.6	117.3	67.9
1963	130.1	89.0	114.0	95.5	116.6	73.4
1964	148.9	100.0	131.9	119.4	119.9	74.8
1965	141.1'	105.8	121.6	122.1	120.9	76.5
1966	134.2	108.1	117.5	123.2	132.3	68.2
1967	154.4	115.3	138.6	129.2	143.1	71.2
1968	142.4	117.9	122.1	132.2	143.3	75.6
1969	133.7	117.4	106.4	117.2	138.1	69.0
1970	131.1	115.2	98.8	111.9	133.9	83.2
1971	139.2	116.2	106.6	112.0	143.7	116.4
1972	146.3	117.4	108.7	102.0	134.2	169.9
1973	141.4	118.1	101.7	94.0	132.1	231.3
1974	101.1	106.0	78.1	86.1	87.4	219.5
1975	100.0	100.0	100.0	100.0	100.0	100.0
1976	105.5	98.5	100.8	100.7	91.9	101.0
1977	94.6	98.0	82.7	93.9	75.4	109.2
1978	98.5	104.6	82.0	86.6	61.6	113.6
1979	88.1	100.8	68.9	75.0	58.7	125.7
1980	74.2	95.7	47.5	72.0		125.6
1981	79.7	51.2	47.5	63.7		
1982	83.0	53.2	59.8	54.6		

Source: HCS

gain. The decline is not necessarily an immediate one as public infrastructural projects take time to complete, and the land they open up for residential development is only used up over a number of years. But the decline could not have come at a worse time for speculative housebuilders, as the 1970s heralded a period of housing market instability, rising building costs and squeezed profit margins.

So the overall trend of development profitability was falling when the number of potentially attractive residential development sites created by public infrastructural expenditure began to dry up. It is in this broad context that housebuilders' complaints about land availability must be understood. The loss of state-expenditure induced land gains intensified throughout the decade and led to mounting conflict between speculative builders and planning authorities as both tried to adjust to the changed circumstances.

For planners the new period of economic austerity in public expenditure on the built environment reinforced the logic of the previous policies of constrained nodal suburban and rural development. All new residential development generates demands for additional public expenditure, so planning authorities have tried to restrict development to places where little additional expenditure is required or where it can be incorporated in schemes fulfilling wider objectives at the same time. Different implications derive from welfare and infrastructure related public expenditure. The first type is a necessary consequence of any development as the following comment by a practising planner makes clear:

> In the current economic climate, the problem worrying authorities is that . . . an authority can be seriously embarrassed when houses are erected and the developer has moved away, if there are no resources to provide for public buildings or, indeed, staff to run services, such as schools, libraries and social centres.
>
> (*The Planner*, March 1979, p. 51)

Basic infrastructural investment, on the other hand, opens up new residential development sites. Another planner gives an example of where infrastructural expenditure at particular locations can fulfil residential and other objectives simultaneously (and, hence, cheaply):

> In Berkshire . . . the County Council wished to promote a mixed industry and warehousing scheme as part of an overall package which included the provision of a major road which would in turn unlock an area with a longstanding commitment to residential development.
>
> (*The Planner*, January 1979, p. 22)

The result has been that residential land allocated in county structure plans tends to be highly locationally specific, much to the chagrin of speculative builders who want land elsewhere. For example, complaints by builders about the abundance of land in Wigan and shortages in the attractive districts bordering Cheshire were the main outcome of an official study of land availability in Manchester published in 1979 (see below, pp. 267–9). Elsewhere complaints have been made about new towns:

> For example, it is proposed that more than 50 per cent of the expected new dwellings in Shropshire will be concentrated in a single town, i.e. Telford; the same is true of Buckinghamshire where 64 per cent of new dwellings will be in Milton Keynes.
>
> (*The Housebuilder*, January 1980, p. 440)

The economic requirements of housebuilders

Private housebuilders have been engaged in a long-term attack on existing planning policies because of the growing conflict between their economic requirements and the land allocation strategies of the planning system. The most obvious problem, already mentioned, is that a lack of public infrastructural expenditure can hit housebuilders severely by denying them hoped-for development gains. Marchwiel (the holding company of Sir Alfred MacAlpine), for instance, diversified into private housebuilding through the acquisition of a Welsh housebuilder, Price Homes, but profitability has been limited by the fact that much of the company's land bank cannot be developed until a proposed coast road is built (Grieveson, Grant & Co. 1981). So builders have become even more interested in land adjacent to already existing infrastructure which is frozen in its current use by green belt or other planning restrictions.

The trend towards greater enthusiasm for land not designated for development is heightened by changes in the housing market itself. It was shown in chapter 4 that there has been a noticeable move up-market by housebuilders, with shifts back to first-time buyers during market downturns. With greater market diversity builders want a wider selection of sites in their land portfolios. The move up-market has exacerbated conflict over planning policy because up-market developments by their nature are more successful in 'desirable' areas of traditional planning restraint. In addition, the growing instability of the market has made housebuilders less willing to wait for the lengthy procedures of planning application and appeal. The time it takes to get planning permission has economically become more significant to them. Finally the squeeze on profits has encouraged builders to try to get extra development profit by reversing existing policies on specific sites through planning appeals. By being unexpected such policy changes are less likely to affect the purchase price of land (as its hope value would have been low), leaving most of the development gain in the hands of the builder rather than the initial landowner.

Such economic needs, however, only explain why there has been sustained political lobbying by housebuilders against existing planning policies. It is also necessary to examine the political conditions which make them successful.

The administration of land-use planning

The two administrative upheavals of land-use planning, associated with the introduction of structure planning after the 1968 Act and the reform of local government in 1974, created sufficient confusion over the implementation of planning that its administrative procedures became an arena for political debate. The official enquiries that resulted, like those of Dobry (1974) and the Expenditure Committee of the House of Commons (in 1976 and 1977), gave housebuilders influential places in which to state their grievances, and the reports of those committees also helped to legitimate them.

The exercise of structure planning was claimed by its initiators to increase the flexibility and responsiveness of planning to changes in economic and social circumstances, whilst enabling a greater degree of local public participation in plan formulation. These advantages arose as structure plans are only schematic, broad policy statement as opposed to the rigid, detailed land-use demarcation of earlier development plans. The history of the introduction of structure planning, however, enabled a different light to be cast upon it. The long delays in the introduction of structure plans, stretching well over ten years, helped to give credence to the view of the cumbersome, bureaucratic, insensitive nature of land-use planning at variance with the dynamic, rapid market response of the private developer.

The strategy statements of structure plans, giving broad statements of residential development policy in terms of housing numbers and locations, also gave housebuilders firm evidence of planning policies previously lacking in published planning material. Structure planning gives the appearance of a formal exactness to policies on residential development. For its own legitimation structure planning has to play down the difficulty of controlling, or even predicting, privately initiated development. The reality of planning practice, where large tracts of land are developed for residential purposes even though they are never formally designated as such in planning terms, is assumed not to exist in the structure planning exercise. This means that housebuilders can argue against structure plan numbers as though they are a fixed planning dictat.

The mystification adopted by the planning profession that structure planning is an exact science (or pretty close to one) consequently is turned against them by builders in a critique of their land allocation policies. Using structure plan data the House Builders Federation has been able to produce calculations leading to highly politically effective conclusions as in the following:

> structure plan policies will have a devastating effect on housebuilding output. Out of a total of 46 counties, only five are planning for increased levels of housebuilding. Of the remaining 41, 36 counties are deliberately planning for a reduction in dwelling completions, whilst in five counties no comparison can be made.
>
> (*The Housebuilder*, January 1980)

Like most calculations of this sort its basis is highly contentious (particularly as it is based on all housing completions not just owner-occupied ones), but politically it helps to give housebuilders a considered advantage: the planners are on the defensive not the builders, who have all the advantage of an apparently legitimate grievance against over-restrictive, insensitive, bureaucratic muddling.

The introduction of structure planning also administratively formalized the division between plan formulation (the glamorous 'intellectual' part of the planning process) and development control (the routinized, 'Cinderella' of the planning profession). It also diverted resources away from development control to the new activities thrown up by structure planning. By the early 1970s many development control offices were regarded as poorly staffed. The 'Cinderella' view of development control was widely accepted in the planning profession:

> A job in which 'failed' planners ended up, along with those who did not have the qualifications to take them into the 'mainstream' of planning. It has been an area in which planners have been underpaid, undervalued, undertrained and overworked.   (Vickery 1978, p. 24)

Administrative difficulties with development control, in other words, were recognized by the planning profession as well as outside it.

The timing of these problems, the early 1970s, was important. They occurred precisely when there was an unprecedented increase in planning applications associated with the 1972–3 land boom. The administrative problems of development control in such circumstances proved a fertile context for anti-planning campaigning.[4] Development control produced such 'delay' in the length of time taken to process planning applications, it was claimed, that worthwhile developments were badly delayed or even scrapped. In the ensuing debate 'planning delay' has been shown not to be so restrictive nor simply a problem of delay, involving as it does necessary consultation of divergent interests (cf. JURUE 1977). But such sophistications have received short-shrift in recent government directives aimed at reformulating development control towards the role of a 'service to developers' at the expense of its other objectives (Underwood 1981).

At the end of the 1960s the main onslaughts on the planning system concerned its apparent obsession with never-ending road building and its crude environmental functionalism as exhibited in many large-scale inner city development programmes of the 1950s and 1960s (cf. Ravetz 1980). 'Community needs' and 'community action' became catchwords of many as a means of articulating such criticisms. Structure planning and the 'enhanced' public participation associated with it were in part a response to that political agitation. In the changed circumstances of a decade or more later, the principal victim of the planner no longer seems to be the 'community' but the private housebuilder. So in attempting to accommodate one form of social protest the planning system helped lay itself open to a more fundamental threat.

The other area of administrative change that has influenced political debate over planning policies has been the reform of the state agencies involved in provision of the built environment. The most significant was the organizational reform of local government (in 1974 for England and Wales) into a two-tier structure of counties and districts. In part the aim was to improve administrative efficiency, and other public agencies were reconstituted on similar, intellectually fashionable, 'spatial efficiency' lines: regional water authorities and health authorities were set up, for instance. The form which the reorganization took and its subsequent history, however, gave further evidence to complaints about administrative inefficiency and political pressure against new suburban development, casting the developer again in the role of the injured party.

The administrative difficulties stemmed partly from the period of reorganization and the delays resulting from it, but more importantly from the separation of planning functions between the tiers of local government and between them and the statutory authorities like those concerned with water and sewerage. Planning applications could be subject to severe delay when county and district disagreed, and sometimes neither would know whether the available public facilities would exist to cope with new development. These difficulties were made much of by the development lobby as evidence of the problems of planning, and the 1980 Local Government Planning and Land Act did much to clarify the planning functions of counties and districts.

A growth of anti-development political influence had emerged in 1974 from the nature of local government reform which deliberately gave political weight to 'county' anti-urban interests. Counties draw up the structure plans to which the districts must conform, and emphasis in them has frequently been placed on rural conservation. A number of counties have reacted against the decentralization of population into their areas by trying to limit severely new residential development. (Cheshire and outmigration from Manchester and Liverpool is a well-known instance.) The lack of a national or regional strategy on residential location inherent in county-level structure planning with its parochial, beggar-my-neighbour bias has enabled developers to appear in the populist guise of champions of the have-nots, lobbying central government to reverse such policies by giving them more development land:

> Unless there is a radical change of attitude towards positive planning at local level – and this involves a reversal of the use of the planning system by the 'haves' to keep out the 'have-nots', and the need for housebuilders to rely on appeal for the implementation of statutory government policy – the prospects of providing the homes required are gloomy indeed.
> (Sir Peter Trench, Chairperson of National House Building Council
> and Y. J. Lovell (a housebuilder), at the 1981 Building Societies
> Association Annual Conference)

Politics and planning policy

So far only the relation between housebuilders and the planning system as an agency of the state has been considered. It needs to be placed in the wider context of other political forces operating on land-use planning. Many pressures on planning are concerned with the land uses which planning tries to influence rather than directly with planning itself. Owner occupation, unemployment and the 'inner city' problem as political issues, for example, all have consequences for planning policy even though their effect may not be immediately apparent. Moreover the way those issues affect planning frequently is mediated through the technicist ideology of the professional planner. This makes recognition of wider forces operating on planning difficult, as most empirical material concerns only the immediately relevant state agents (e.g. Ministers, local councils and their planning committees and individual planners) and the lobbies trying to influence them, like developers, conservationists and local action groups.

The nature of the empirical material has led to two distinct strands of analysis in the literature on planning. The first, associated with the urban managerial perspective (see the survey in Bassett and Short 1980), *de*politicizes planning by emphasizing its institutional position and the consequent role of the planner in the distribution and allocation of resources. The second *over*politicizes planning by seeing it purely in the immediate terms in which local or national politics affect it (usually in a Weberian-style conflict perspective into which managerialism can be inserted if need be). Reasons for change in planning policy consequently are derived from these immediate political effects alone; say, for example, the impact of a local pressure group on policy or the planning directives of specific governments. Such politicism ignores less direct political forces and also the need for wider conditions to exist to make any political influence effective.

The importance of these wider effects was clearly seen in the evolution of the planning system described in the previous chapter. Yet, because of the pervasiveness of immediate political explanations in the literature about more recent periods, it may seem that only history affords such luxuries. Many on the Left, for example, have dismissed the political significance of land-use planning because of case studies indicating the success of middle-class pressure groups in manipulating local planning policies. Strictures like the following by Saunders are commonplace:

Because British Marxists have tended to ignore or dismiss as unimportant the real economic divisions between groups such as the suburban middle class and other less fortunate sections of the population, they have too readily fallen into the trap of supporting middle class community action against a local authority on the assumption that they are thereby aiding a popular struggle against a business dominated local state. (Saunders 1979, p. 272)

To treat the politics of planning in such parochial terms, however, denies the spatial consequences of social change. Land-use planning is a specific intervention by the state in the built environment and its spatial organization. Local political struggles over planning policy are part of much wider conflicts over the nature of society and who controls it. For example, the political struggles over aspects of the built environment at the turn of the century and after the Second World War, described in chapter 7, were part of a much wider transformation of British society and its spatial context. A central role in these transformations was played by the development of particular structures of housing provision, like the one associated with owner occupation. Owner-occupied housing provision has had since the 1920s a profound effect on the political possibility and importance of land-use planning and the form it could take (cf. the situation in the 1930s with that in the 1940s). Housing policy concerned with owner occupation, and the political process from which it was derived, at those times had effects on spatial structures and the politics of land-use planning, and the politics of land-use planning affected the development of owner occupation in a mutually determining way. The same is true today. The impact of the growth of owner occupation consequently is far greater than just a housing issue; it affects the spatial structure of the built environment, how different activities interrelate spatially, the need for state intervention and the form that intervention can take.

This argument suggests why the longer-term growth of owner occupation and the mounting problems faced by housebuilders have had such a significant impact on the weakening of planning controls. Once governments and local authorities are committed to the expansion of this tenure in its current form, necessary consequences stem for planning policy. So, although the Thatcher government is being instrumental in the changes that are taking place and is ideologically committed to an anti-planning rhetoric, it is difficult to see how any government could have acted very differently unless it attempted to alter the current nature of owner-occupied housing provision.

The Council for the Protection of Rural England pamphlet (CPRE 1981) mentioned earlier, which documents clearly how changes in planning policies in suburban and rural areas are to the advantage of developers, does mistakenly see the new emphasis as simply the product of one person's actions, the then Secretary of State for the Environment, Michael Heseltine. (Its own political credentials presumably stop the CPRE from extending its critique to the Tory Party as a whole.) The implication is that another individual minister could simply reverse the policy in isolation from other governmental programmes. That is not true. Even if the Secretary of State does have the formal power within the planning system to reject 'restrictive' structure plans, send out advisory circulars to local authorities, and to formulate legislation, what conditions have enabled him to favour suburban developers so directly? The trend moreover grew to dominate planning policy under the 1974–9 Labour government during whose administration much of the 'damning' evidence on the planning system was formulated.

There have been sharp contradictions in the land-use planning policies of all governments over the past fifteen years because they have been trying to coax market forces which they do not want to take over or control. So coherent government strategies cannot be isolated to explain the weakening of the planning system. The years since 1973, in particular, have seen sharp changes in government policy over the financial gains from land development. Some changes have been the result of differences between Conservative and Labour administrations, particularly over the introduction of the Community Land Act in 1976 and its subsequent repeal in 1980.[5] Yet in the main they have not so much been a result of differences between governments as of contradictions within each government's own policies.

In respect of planning and housing policies two central contradictions can be highlighted in the programmes of successive governments. The first concerns suburban and rural areas, where all post-war administrations have wanted to avoid suburban sprawl like that of the 1930s, yet they have all supported the expansion of owner occupation. Despite the widespread publicity given to a handful of inner city owner-occupied housing schemes, suburban expansion is the only way to get large numbers of owner-occupied houses built by speculative builders. So programmes towards urban expansion (or its containment) are frustrated by support for owner occupation. And the importance of this contradiction must be seen in the context of the rapid decentralization of population and the cutbacks in public expenditure mentioned earlier. The second contradiction concerns the continued support for urban renewal, albeit increasingly posed in the modern, less physically determinist guise of inner city regeneration. High unemployment rates and the obvious poverty and growing radicalization of many inner city dwellers have markedly increased the political significance of this issue over the past decade, especially after the widespread street rioting of 1981. Yet each successive government has presided over drastic cuts in public expenditure and state involvement on the built environment in these areas, even though such state intervention must be a necessary element of any urban renewal programme (including the improvement and renovation of existing housing).

These two contradictions in state policy are closely linked by the trend towards exclusive support for owner occupation in an era of state expenditure cuts on the built environment. Both Labour and Conservative governments have championed private development as a means of filling the gaps left by state expenditure cuts. In a number of respects this strategy is quite simply impossible, as only the state can compulsorily purchase sites, clear them adequately where necessary, and build the large-scale public facilities, like roads, necessary for any development, public or private. In addition, by trying to encourage private development successive governments have had to accept the logic of the conditions economically necessary for such development. Developers must be able to produce schemes that will sell at a profit or, in their phraseology, be marketable. So the absence of public expenditure that helps to create those conditions where previously they did not exist has put pressure on areas where they are already present. In inner city areas this has involved the takeover by private housebuilders of the best, but only the best, sites previously cleared and serviced for council housing. In suburban and rural areas pressure is on cherished open space in well-situated and serviced localities.

The changes in planning policy have mainly been implemented by central government. Opposition to them by local authorities, however, has been muted. Owner occupation has been actively supported by many Conservative-controlled councils and some Labour ones for a long time. The relative independence of local authorities from central government enabled them to develop land-use initiatives over owner occupation prior to central government directive. There has been confusion and vacillation by other Labour councils worried about the impact on votes of anti-owner-occupation housing and planning policies. Some councils became well known for their innovatory deals with speculative housebuilders. Projects introducing speculative builders to slum clearance sites in Liverpool and Nottingham, for example, were well publicized in the late 1970s whilst their councils were controlled by Liberals and Conservatives respectively. Labour councils in Swindon (an expanding town) and Norwich pioneered greenfield council/developer partnerships on their urban fringes.

The desire to encourage owner occupation has led to a convergence of housing and housing-related planning policies at a local level. This convergence was aided by a change in the rules by which local authorities could get central government subsidies for housing in their areas, introduced in 1977 under the Housing Investment Programme scheme. It directed funding away from council housing to projects in other tenures as local councils could bid for finance from central government for any housing tenure, whereas previously most funding was directly related to council housebuilding (see HPCD 1977 and Direct Labour Collective 1978). Added to which the continuous reductions in central government subsidies for new council housing since 1975 left owner occupation as one of the few means open to local authorities to create new or improved housing in their areas, either through area improvement programmes or by encouraging housebuilders to build on redevelopment sites.

Owner occupation, planning and the inner city

For many local authorities in inner city areas local housing strategies by the late 1970s had become focused on ways of encouraging owner occupation, through political choice or necessity.[6] So, encouraged by central government, many councils set up schemes in conjunction with speculative builders. Inner city build-for-sale and partnership projects, where speculative builders enter into arrangements to build on local authority land, received greatest publicity. But, in addition, land throughout urban areas originally purchased for council housing was sold off to private builders. This process was aided by the requirement introduced in 1980 that local authorities, along with other public bodies, should publish registers of their land holdings. Private builders therefore are now able to initiate the purchase of council land rather than waiting for a council to offer it for sale.

The political advantages to councils and central government of encouraging new owner-occupied schemes in inner city areas are clear. It saves on public expenditure and gives the appearance that something is being done about housing problems in those areas, whilst the possibility such schemes offer of cheap entry to owner occupation appeals politically to certain strata of the working class. Despite the publicity given to these projects, however, their aggregate effect has been derisorily small.

The House of Commons Environment Committee estimated that in aggregate all the schemes where central or local government helped to create additional owner-occupier housing, such as improvement for sale, building for sale and homesteading, were creating only a maximum annualized rate of 4000 new and improved houses (HCEC 1981a). This represents only a minute percentage of total private housing output.

The reason for the low output of these schemes, of course, is the usual one in a private market: they happen only when the private builders involved can make a profit, and the inner city areas where such schemes are proposed generally do not provide sufficient scope. Even when the site is cleared and the basic infrastructure provided by the local authority, making the sites similar to greenfield ones, the potential development profit is small or non-existent. Builders as a result are prepared to pay little or nothing for the land or even want a state subsidy to build (euphemistically called a 'negative land value').[7]

These schemes represent a classic example of where the needs of the developer, related as they are to profitability, contradict the social requirements embodied in state housing provision. There is little or no accountability or political control over the projects: the local authorities do not know the profits made by the builder, they have to accept the housing quality and standards provided by the builder, and only occasionally can they have any effective control over the type of household buying or the selling price (a control which anyway is simply a distributional issue between the builder and first purchaser, as the latter can sell the house at its full market price in a few years when they get a discount). At the same time, these schemes necessitate large-scale public expenditure on site acquisition and clearance, new infrastructure and administration. One case of a 4-acre site in Nottingham in the late 1970s, for example, cost the council £200,000 to acquire, demolish and infill, but could only be sold to private housebuilders for £50,000 (interview). Private housebuilders cannot undertake inner city redevelopment projects alone, given such costs. It is not that the local authority is the major landowner in inner cities that makes their intervention so important for such schemes. Instead their involvement enables large-scale building subsidies to be channelled to these projects and removes the risk element for the speculative builder (as the council will often guarantee to buy unsold houses where necessary for use as council housing).

Build-for-sale schemes and other inner city housing partnerships between local authorities and speculative developers in summary are an expensive irrelevance to the housing problems of such areas, diverting attention, and much needed land and public finance, away from the real housing problems of inner city dwellers who can never afford owner occupation. The expense and irrelevance stem from the fact that private market criteria are never questioned by them. Studies like a much publicized one produced by the Department of Land Economy at Cambridge (Nicholls *et al.* 1980), which reached the earth-shattering conclusion that housebuilders will build in inner city areas as long as they make a profit, quite simply miss the point: at whose expense is the profit made?

Similar comments can be made about the land sales in recent years made by public authorities from land banks built up at inner city and suburban locations. They were expensive for the local councils to build up and prepare for future housing and development programmes, yet private builders will only pay prices which give them substantial development profits, and so only take the pick of public authorities' land holdings at rock-bottom prices. From the scattered evidence available it would seem that the resultant development profits have been very good. Bryant's Chairman's Report 1980 comments favourably on the considerable amount of good public sector land bought by them in that year. Barratt Developments in a record-breaking profit year bought half of its land in the year to June 1981 from local authorities (Savory Milln 1982). Fairview Estates even changed the location of its market after, commented a stockbroker, 'buying sites which local authorities in London boroughs have been forced to sell' (Savory Milln 1982). With the 1000 or so plots acquired Fairview shifted from being a home counties to a Metropolitan housebuilder. Later in 1982 the company announced it was to pull out of housebuilding, after it had taken the development profits from this non-repeatable source of land supply. Land sales can have devastating longer-term effects on council housebuilding programmes in the larger urban areas as they may remove almost all the potentially available housebuilding land.

Political support for the current structure of owner-occupied housing provision can be seen therefore to impose severe limits on land use and its planning in inner city areas. Support necessitates accepting the criteria of speculative housebuilders, which means satisfying their needs for profitability. The ability of land-use planning to respond to the needs and political demands for land uses which contradict this basic market criterion is thereby severely limited. Land uses, furthermore, are not autonomous but interrelate over quite wide areas. The allocation of land to speculative builders therefore not only fixes the land in one use, design style and estate layout, but also affects the use to which adjacent or nearby land can be put, with long-term repercussions for the future of whole areas as a result.

Marketability as a planning problem

The need for profit by speculative housebuilders is a structural fact of their social location as capitalist producers. Yet such profit requirements are more politically palatable if they are ideologically rephrased as a problem of 'marketability', that is, of being able to sell houses at a profitable price. Housebuilders have been successful in gaining widespread political acceptance of this marketing problem. The spatial location of their housebuilding operations can only be shifted to places desired by planners if a marketable environment for selling houses can be created at the new locations. In an economic situation of limited state expenditure on the built environment and a changing housing market of the type described in chapter 4, the location requirements of profitable new developments for housebuilders and locational preferences articulated through the planning system diverge considerably. Developers will, in particular, want to build in well-situated suburban areas, precisely the place where the most articulate and sensitive political reaction to new development is found. Because of such local political pressure and the need to regenerate rundown urban areas at minimal state expense, planners, on the other hand, have tended to try to force new building into existing urban areas.

Housebuilders have used the criterion of site marketability to launch a successful attack on planning land allocation policies. Its success has led to a much greater release of suburban development land. Marketability has enabled them to question the location of land designated for development in land-use plans. By doing so they have been able to defeat planners' arguments that enough land is available for owner-occupied housing development. It might be available on the planning criterion of physical designation, retort the housebuilders, but much of what is allocated does not create houses that will sell. The dichotomy between planning criteria and market criteria is sharply drawn in this debate. The success of the housebuilders' campaign has meant that since 1980 their criteria of marketability have been of foremost importance.

Again this shift in the planning system was the product of trends initiated prior to the Thatcher government. The first, experimental, joint study between local authorities and housebuilders of land availability in Greater Manchester was initiated in 1978 by the previous Labour government (although its results were not published until late in 1979) under terms of reference that were highly favourable to housebuilders. The success of this study in demonstrating shortages of 'marketable' land in that area then became a justification for setting up joint studies in other areas and for creating a more favourable framework for house-builders in the planning appeals procedure. As a consequence, local authorities are now formally required to consult housebuilders on suburban land availability and ensure that there is an adequate supply of 'marketable' development land available, with the speculative builders deciding what is marketable. This procedure is the antithesis of the aims of the 1947 legislation which wanted to go against the market criteria of the speculative builder. At the same time large additions of new housing land have had to be written into county structure plans in order to get their statutory approval by the Secretary of State for the Environment. All this procedure is prior to that of the piecemeal granting of planning permission for sites not previously designated for residential develop-ment, which has been such an important component of the planning system in the past.

The threat to planning

The growing emphasis on the market obviously corresponds to the ideology of the Thatcher government. Yet that ideology is not new, nor are directives by Conservative Ministers to force local planning auth-orities to release more land for private housebuilding. What has changed is that previously the dominance of national planning criteria, overriding an unrestrained market, was accepted as essential by all post-war govern-ments. Earlier directives to release more land were essentially treated as resolving temporary shortages during periods of high housing output. Now output is low and the changes have been designed to ensure the dominance of marketability in land release.

The choice of potential development sites open to speculative builders as a result of planning and housing strategy changes is now likely to be far greater than at any time since the war. The increased choice is a relative one given the present, historically low, level of housebuilding which has helped to hide the radical reduction in the ability of the plan-ning system to direct and contain speculative development. Once the political changes within planning become entrenched, however, their effect could hold for virtually any level of housing output. Land-use planning is slowly reverting to its role as a cosmetic exercise; the most beautiful parts of the countryside may remain sacrosanct against large-scale development but the fate of much of the rest increasingly depends on its potential profitability for residential development. At the moment, a combination of economic crisis and rising transport costs are likely to be more effective guardians of the countryside than the whole planning apparatus. Unless there are radical shifts in housing and land policies, the current political emphasis on the dominance of private development interests will survive the demise of the Thatcher government. Like the period after the First World War, once again there is a 'containment' of urban planning to the advantage of the speculative builder.

A case study of Greater Manchester

The arguments of the Greater Manchester Study (HBF/DoE 1979) highlight the differences between traditional planning criteria and the land requirements of speculative housebuilders. As mentioned earlier, this study of land availability was jointly set up by the Department of the Environment and the House Builders Federation (HBF), the trade organization of speculative builders. Its existence derived from earlier political debates over housing land availability. Representatives from the House Builders Federation and local housebuilders met separately with officials from the ten local district councils of Greater Manchester County to discuss the land allocated by planners in each district over a 3½-year period from 1978. Aggregating over the ten districts gave a land availability picture for Greater Manchester as a whole, and the results showed that 25 per cent of the land designated as available could not be built on for marketing and other reasons. The HBF claimed that this shortfall was leading to housing production rates of less than two-thirds of 'demand'. In retrospect those demand forecasts were wildly optimistic. Yet the Study had the desired effect for the HBF of politically validating their claims of land shortages throughout the country to which the Conservative government readily responded. Detailed examination of the Study's report, however, shows that it was the location of the land planners had designated for private housebuilding in the Greater Manchester area that was the main source of contention rather than the total allocation.

The issue is an example of how pressures on the planning system operate at the local level. Decentralization of population out of central Manchester and the reaction of politicians and planners in the post-1974 framework of local government, faced as they were with public expenditure cuts and the problem of urban renewal, provided the setting for the structure plan's attempts to direct private housing development.

Local government reorganization tightly constrained the administrative area of south Manchester, so that a significant proportion of the City's wealthy suburban hinterland remained under the jurisdiction of Cheshire County Council. Not surprisingly, in light of vociferous pressure from well-heeled existing residents, Cheshire severely restricted further housing development. The Greater Manchester structure plan (GMC 1979) similarly continued this policy within its southern fringe, allocating most new development to northern districts instead. This strategy achieved structure plan objectives whilst minimizing the political reaction from local residents. The suburbs of southern Manchester were protected with their remaining green belt, and new infrastructural works could be concentrated in districts suffering from industrial decay and residential decline. In the main the district councils (who actually allocate sites and give planning permission) concurred with the County's objectives. This was shown in the land availability study where the four 'southern' districts (Stockport, Manchester City, Trafford and Salford) had half the county's existing population but only a quarter of the available development sites.

An implicit assumption of the structure plan is that housebuilders will shift their operations to the localities designated by the planning process. The Manchester Land Availability Study was a statement of the reluctance of housebuilders meekly to conform, especially with regard to localities suffering from industrial and urban decay. 'Positive' planning and its chosen tool, the private developer, were at odds; so housebuilders used the Land Availability Study as a vehicle to get publicity and political action on their grievances.

The terms of reference of the Study took advantage of the separation of powers within local government. The disagreement was essentially between the aspirations of builders and the structure planning strategy, yet the discussions excluded the county-based structure unit, dealing directly with the individual districts instead. Districts do not have the same commitment to the structure plan, in the sense that they did not draw it up and are not concerned with county-wide matters. They are more likely, therefore, to acquiesce to builders' objections that sites are unavailable.

Housebuilders clarified their objections to the location of designated land by distinguishing three sectors of the new housing market – lower, middle and upper range (details are given in table 5.1). The principal problem of the structure plan strategy, they argued, was its lack of middle to upper market sites, which could not be created in the depressed areas the structure plan hoped to rejuvenate. Most objections centred on two areas, Wigan and the Mossley area of Tameside: 80 per cent of the land the Study reclassified as unavailable for the whole of Greater Manchester came from these two districts. It was argued that they provided sites for only very low-priced houses. Of Wigan there were 'too many sites in unattractive parts of the town and at best suitable only for very cheap housing which is unviable to a builder' (HBF/DoE 1979, p. 21) and of Mossley:

> The area was socially deprived, visually unattractive and physically difficult. Demand for housing in the area was very low and if it were to be improved at all (which they [the builders] doubted) there would need to be substantial and long term public investment in infrastructure and social facilities.   (HBF/DoE 1979, Appendix, p. 71)

To overcome the shortages of land in the medium and upper ranges the builders suggested the release of green belt land.

Conflict, élitism and acquiescence

The Manchester Study shows that both builders and planners are right on their own criteria. The problem is they do not correspond: a point not made within the traditional debate as both parties refuse to accept the possibility of such a contradiction. Self-interested housebuilders are hardly likely to point it out, and the ideology of planners will not let them articulate such a possibility either. Planners, according to the profession's dominant ideology, are land-use technicians. They convert conflicting political pressures, social and economic problems and physical constraints into land-use planning policies to the greater benefit of society. These policies are then realized using the self-interest of individual actors to get them to achieve the aims of the plan with a minimum of actual direction. Possible antagonism to the plan's strategy is recognized only in a limited redistributive sense (both spatially and across social groups). Planning in this ideology is concerned with a spatial redistribution of land uses, independent of the internal workings of the uses themselves.

If one major participant in the process of land-use change says the land-use plan is impossible, this technically neutral view of planning has no real answer. It can only do one of two things. First, it can say the objection is wrong. With regard to private housebuilders' complaints that means saying that builders can make profits where they say they cannot (which is reminiscent of the Uthwatt Committee's claim that planning only shifts land values), in other words, planners know the housing market better than builders. The other response is to incorporate the demands of the builders. But that denies any autonomy to planning goals. The politics of planning in Britain over the past five years have seen the planning system adopt both responses whilst ignoring the existence of the inherent contradiction.

Other potential political conflicts with developers have also been diffused. The main places where housebuilders want to build and the planning system tries to stop them are suburban and rural areas. The primary objectors to new development are the people already living in those districts, their political representatives and the lobbies they support. Given the social geography of the housing market such localities are inhabited disproportionately by upper- and middle-class households from whose ranks the leading local anti-development activists are likely to come. As objectors, 'conservation' is their catchword, and politically they will ally with the broader conservationist movement. The social composition of such local protest makes it comparatively easy for developers and their political allies to take advantage of populist sentiments and parody protest as the cries of a social élite out to defend its patch. In doing so the development lobby can combine with broader right-wing populism in its anti-planning, anti-corporatist views. The recent years of Conservative government have provided a fertile political context for this perspective, so conservationists' protests have been curtailed amidst general public resignation or acceptance of the weakening of planning control.

What is surprising is the lack of protest by the planning profession. Many leading planners, in fact, have welcomed the shift towards 'entrepreneurial' planning compared with the conservationist protests that dominated the early 1970s. The chief adviser to the House of Commons Expenditure Committee's investigations of development control echoed this sentiment clearly:

The early 1970s was a period of resistance to change: motorway protest groups, community action against slum clearance, opponents of town centre redevelopment. Strategic and entrepreneurial interests were sacrificed to local community pressures. Conceivably, however, we are moving into a different period in which the entrepreneurial role becomes dominant. . . . We need the entrepreneurial approach to planning.   (Davies 1980, p. 23)

This sentiment ignores the fundamental attack that is occurring on the ideals of the post-1947 planning system on which the role of the professional planner is based. Yet, despite the size and organization of the planning profession and its institutions, there has been virtually no response by the planning establishment to the rundown of planning control and planning departments. The cynic would say that this is because planners can still make plans even if they will not be implemented. Yet more important is the ideology and structural position of the profession. Ideologically, planners tend to see themselves as central pillars of a corporatist state directing conflicting interests to the benefit of the social whole, whilst ignoring the structural (i.e. capitalist) nature of that society. At times of economic crisis the economic needs of society become paramount, which to the corporatist means the direct and immediate interests of capital. Hence 'we need the entrepreneurial approach to planning'. By refusing to recognize that its intervention is a political one into an arena of class struggle rather than an exercise in social harmony, this ideology of planning converts planners into an élite operating in the interests of dominant social forces.

So professional élitism has produced planners' mute acceptance of their current lot. As Davies said of development control: 'Its role, as I see it, is to provide the forum within which we balance the competing claims of developers . . . and communities' (Davies 1980, p. 24). Who are the 'we' is never specified. Most of the planning literature assigns that role to the 'objective' planner. The claimed unity of God's will and the thought of the feudal king or queen led to the royal 'we'. Under this new absolutism the voice of the planner is now claimed to be at unity with the will of society. Planning is not in crisis because the notion of a planned society has failed in practice; instead an élitist form of planning is collapsing under its own contradictions. The total abolition of land-use planning would only create the spatial chaos of the 1930s. The current framework of planning must change but a new, democratic form of planning needs to take its place.

9

Owner occupiers:
a social category and
its consequences

When examining the structure of owner-occupied housing provision the situation of households living in the tenure obviously has to be considered. What is important when trying to understand the contemporary political and economic consequences of owner occupation is the nature of owner occupiers as social agents. Do owner occupiers, for instance, constitute a specific social grouping with a coherent economic status, ideology or political impact? What are the immediate costs to households of owner occupation, and what is the significance of money gains from house ownership? This chapter considers these issues. First of all the socio-economic characteristics of owner occupiers are described by comparing them with households in other tenures.

The present-day owner occupier

As the majority of households in Britain are owner occupiers a broad cross-section of the population lives in the tenure. Yet, not surprisingly, there is a tendency for the better-off members of the population to be owners; so there is a dispersion of households around a more prosperous mean than is the case with other tenures. This characteristic is shown most clearly when average incomes are compared. In 1978, the Family Expenditure Survey showed that the mean head-of-household income for owner occupiers was £4453 whereas it was only £2935 for council tenants.

For most comparisons between tenures it is useful to distinguish between those owners who own outright and those with a mortgage, who will be called mortgagors. As table 9.1 shows, over a fifth of all households owned their house outright, roughly a third each were mortgagors and council tenants and most of the rest rented private unfurnished accommodation. Much of the following discussion is concerned only with owner occupiers and council tenants, who together represent over 85 per cent of all households.

There are sharp differences in the dispersions of household characteristics for outright owners, mortgagors and council tenants; table 9.2 summarizes some of these differences. Figure 9.1 looks at the data from the opposite perspective of the tenure location of particular types of household.

Outright owners are mainly older households: 51 per cent are over 65 years old and 77 per cent are over 55. They consist, therefore, mainly of households who are retired or approaching retirement. The social composition of this group is quite wide, reflecting the characteristics of those who bought houses in earlier decades. The popular image of the outright owner being a middle-class professional is generally not true: only a fifth of these households were from the top two socio-economic groups (table 9.2); 37 per cent were manual workers.

Owner occupiers with mortgages are distinct from outright owners. In the first place virtually all, 93 per cent, are in full-time employment and the mean household income is over 50 per cent higher (figure 9.2). Mortgagors, not surprisingly, are mainly in their peak earning years: 83 per cent are aged between 25 and 54. Childless couples and small families predominate, constituting almost 60 per cent of mortgagor households (table 9.2); one-parent families conversely are under-represented. The socio-economic group location of mortgagors varies widely but figure 9.1 shows that semi-skilled and unskilled manual

Table 9.1   Household tenure (%)

			Tenants	
Outright owners[1]	*Owner-occupied mortgagors*[1]	*Local authority*	*Private unfurnished*[2]	*Private furnished*
22	30	34	12	2

Source: General Household Survey 1980

[1] Sample slightly understates % of owner occupiers in the population
[2] Includes housing associations

workers are under-represented. 31 per cent of mortgagors are professionals, employers or managers, 22 per cent other non-manual and 32 per cent skilled manual. In other words, 86 per cent of mortgagors were in the higher echelons of economically active households. Data in figure 9.1, in fact, show that house-ownership with a mortgage is by far the usual means of housing consumption for those groups. Only for skilled

Table 9.2 Social characteristics of owner occupiers and council tenants, December 1977

	Owned outright %	Owned with mortgage or loan %	Rented from council %	All households (including other tenures) %
Type of household				
One person aged under 60	4.5	4.0	4.9	6.3
Small adult household	9.2	22.9	10.9	15.3
Small family	6.7	37.0	17.9	20.6
Large family	3.4	12.8	12.1	9.2
Large adult household	16.8	18.4	18.5	17.1
Older small household	35.7	3.8	17.5	17.2
One person aged 60+	23.6	1.1	18.2	14.3
Age of head of household				
Under 25	0.5	4.5	3.8	4.7
25–34	2.8	32.4	13.4	17.8
35–44	6.5	28.5	12.9	16.0
45–54	13.4	22.0	18.0	17.3
55–64	26.0	10.1	21.8	18.0
65+	50.9	2.5	30.1	26.2
Employment status of head of household				
Employed full-time	38.4	93.2	52.6	62.4
Employed part-time	5.0	1.2	3.6	3.2
Unemployed	2.3	1.6	5.9	3.5
Wholly retired	36.1	2.1	20.8	18.2
Housewife	15.4	1.2	12.7	9.4
Other	2.7	0.9	4.3	3.3
Socio-economic group of head of household				
Professional	4.9	10.4	0.6	5.1
Employers and managers	17.9	21.2	4.3	13.8
Other non-manual	18.9	22.3	11.1	17.8
Skilled manual	24.9	32.4	35.9	30.3
Semi-skilled manual	9.6	8.0	19.5	12.9
Unskilled manual	3.1	1.8	9.6	5.0
Other	20.8	4.0	19.1	15.1
Totals	100	100	100	100

Source: National Dwelling and Household Survey 1978, HMSO

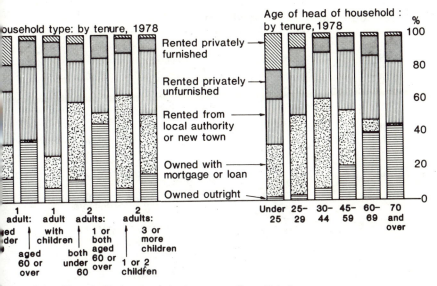

Rented privately
furnished

Rented privately
unfurnished

Rented from
local authority
or new town

Owned with
mortgage or loan

Owned outright

Figure 9.1a Household characteristics by tenure, Great Britain

Source: General Household Survey 1978

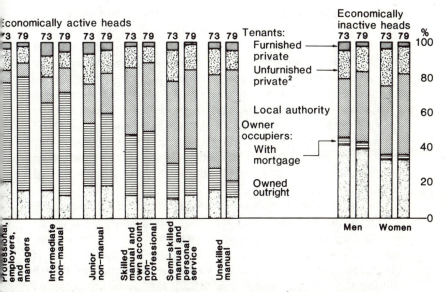

Figure 9.1b Tenure of households: by socio-economic group[1] of head of household, Great Britain

Source: Social Trends, 1980, 1981

Notes: [1] Excluding Armed Forces, full-time students and those who have never worked
[2] Including those renting from a housing association, and those renting with a job or business

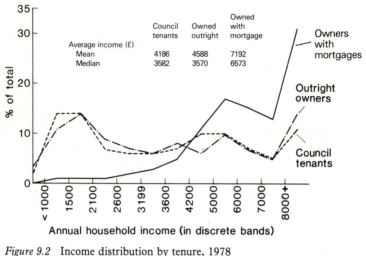

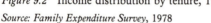

Figure 9.2 Income distribution by tenure, 1978

Source: Family Expenditure Survey, 1978

manual workers does any other form of housing consumption, council housing, come close in comparative importance. This pattern contrasts with the other socio-economic groups: those who are semi-skilled or unskilled manual workers or who are economically inactive. For them owner occupation with a mortgage is comparatively rare. There is therefore a cascade effect to the propensity to buy a house with a mortgage between socio-economic groups: the greater the ability to earn a higher current income the greater is the likelihood of being a mortgagor. Given the socio-economic composition of mortgagors, it is not surprising that the distribution of income is different from the other tenures: most owners have substantially higher incomes than the average council tenant, for instance (figure 9.2).

When looking at changes in the housing tenure of socio-economic groups during the 1970s some interesting results emerge. For all the economically active groups, bar unskilled manual workers, the percentage of owner occupancy rose between 1973 and 1979 (figure 9.1). One of the sharpest increases was amongst semi-skilled workers, even though still less than half are owners. It is difficult to treat owner occupation simply as the 'middle class' tenure. It is quite likely, for example, that the majority of trade unionists are owner occupiers. There is some evidence, moreover, that even more skilled and semi-skilled workers would like to be owners (cf. HCEC 1981b). The economically weakest groups, however, have had a different experience: the proportion of owner occupiers has actually fallen for unskilled manual workers and for those economically inactive. Part of the latter decline is attributable to the greater provision of sheltered housing for the elderly by local authorities. But the principal reason must be the rising costs of ownership.

The evidence of a growing social polarization between owner occupation and council housing is shown clearly when the data for mortgagors are compared with those for council housing. Only just over half of the heads of council households are in full-time employment. Many are retired, 21 per cent are in the last working years of their lives: 52 per cent were 55 years or older (table 9.2). The bi-modal distribution of income in council housing, shown in figure 9.2, highlights the importance of the distinction between economically active and inactive households in the council sector. Amongst the former, manual workers were the major group, and even the incomes of those in employment are generally much lower than for mortgagors. There is a growing trend for council housing to be the tenure of those excluded from owner occupation, either because they are too old to embark on the mortgage cycle or because they cannot afford to. A recent Labour party pamphlet points out that:

> The median household income of council tenants, for example, fell from 83 per cent of that of home buyers in 1953 to 58 per cent in 1978; the number of council tenant households with economically inactive heads rose from 25 per cent in 1968 to 35 per cent in 1976, and apparently to about 40 per cent in 1977, and the proportion of all Supplementary Benefit claimants who are in the council sector rose from 49 to 59 per cent between 1970 and 1979.
>
> (Labour Party 1981, p. 51)

Throughout the 1970s council housing has increasingly become a welfare net, the residual housing tenure. The result is that a choice of housing tenure for most households is now non-existent. Other tenures, apart from these two major ones, have virtually disappeared in most regions of the country. And between owner occupation and council housing there is little effective choice. If you can afford to buy, you have to buy; if you cannot you can only hope to be housed by the council in one of the better-quality dwellings in its highly variable stock. Moreover, council housing is very expensive comparatively for those not eligible for rent rebates (or for those who do not take up the rebates they could claim).

It is quite likely that owner occupation is now the majority tenure for the 'economically active' sectors of the working class (on virtually any definition of that class). If the 'middle class' are included, with its contradictory social location (see Wright 1978), most groups which have a major power to push up wages at the expense of profits are substantially affected by the cost of owner occupation. The shift of housing subsidies away from council housing towards owner occupation has consequently been a shift away from a housing tenure whose households have increasingly been economically marginalized (either through not being 'economically active' or being one of the weaker ones of those that are) towards the tenure of those with greater economic power to undermine the profitability of capital. The political significance of this point will be developed further in chapter 12.

The economics of house purchase

Most owner occupiers, given the existing alternatives, are hardly likely to regret the lack of tenure choice they have. Owner occupation for them is by far the cheapest housing tenure and offers personal privacy and house types not available elsewhere. Precise comparisons of housing costs either between households in different tenures, or even between households in owner occupation, are impossible because of the wide temporal and spatial variations in personal housing costs.

Most owners enter the tenure by taking out a mortgage of around 75–80 per cent of the purchase price of their house. Their initial outgoings consequently are high. They have to finance the unmortgaged part of the purchase price (almost £4000 on average in 1981) and to pay high transactions costs associated with the mortgage deed, legal conveyancing and stamp duty taxation. Over time their housing costs then fall quite quickly, as long as there is inflation. This results from the so-called 'front-loading' effect of mortgage repayments. The monetary value of the mortgage debt is fixed, so in inflationary periods, when money incomes are rising, the impact of mortgage repayments on household income declines in relative importance over time.

Outright owners obviously face no mortgage costs, they have only outgoings on repair and maintenance (which can be high for the third of the owner-occupied stock built before 1914). All existing owner occupiers, however, enjoy the financial benefits from house price inflation. With inflation the price of houses rises, which increases the money value of owner occupiers' wealth. The financial benefits to individual home-owners of house price rises and their aggregate effect unfortunately are very complex. It is difficult to avoid double-counting but these financial gains must be examined because of the real benefits they provide to home-owners and their effects on short-run house price fluctuations.

The difference between the initial purchase price and the resale price of a house is defined here as money gain. When a house is not actually being sold the owner has only property with potential money gain. At the time of sale the money gain may be *transferred* by using it to help purchase a similar or more expensive dwelling: the latter is known as 'trading up'. Trading up necessitates an additional commitment of funds by households, usually through taking out a higher mortgage. This extra expense is made easier for existing households when there is inflation because of the temporal impact of mortgage front-loading described earlier. It means that households can trade-up by readjusting the real level of mortgage repayments towards the old income-mortgage ratio. Money gains may alternatively be *realized* by taking them out of the housing sphere as personal revenue. Realization obviously occurs when a household dissolves or moves to another tenure or country. Existing owners, however, may also realize part of their money gains by trading down to cheaper dwellings, or they may realize some of their gain by taking out a higher mortgage than otherwise would be necessary. The benefit to the individual in the last case, however, is limited to the cheaper post-tax cost of mortgage interest compared to some other form of finance. Buying a car by taking out a higher house mortgage, for example, may be cheaper than buying it on hire purchase.

Existing owners do seem to cash in some of their money gains when moving to take advantage of the tax benefits of borrowing for house purchase. Empirical studies in the earlier 1970s found that on average half of owners' money gains were realized at the time of sale (see Building Statistical Services 1973 and Ball and Kirwan 1975). The proportion of money gain realized is likely to vary depending on the absolute size of the money gain, the household's financial commitments and income, and the state of the housing market. It has been suggested that much of the money gain realized at the time of moving, however, is used to finance the high transactions costs of moving (cf. Stow Report 1979).

The aggregate difference between transferred and realized money gain is important as the first involves no extra commitment of funds to owner occupation whereas realized money gain does. Realized money gain, therefore, is a financial 'leakage' out of the housing system. Yet the total sum of owner occupiers' wealth represented by their houses does not have to be balanced off by an equivalent outstanding debt elsewhere in the economic system because most of it is not realized. It is fallacious, therefore, to look for losers in the economy to balance out owner occupiers' new-found wealth: for example to search for an implicit cross-subsidy by new owners of all existing owners' money gains. Such redistributive effects have been suggested (e.g. in Saunders 1979, see below, pp. 290–2) but they simply do not exist. The wealth of existing owners is a paper one, fictitious until it is realized.

This explains why it is has been possible for owner-occupied housing to become such a large component of personal wealth without generating huge shifts in economic resources that would bring the rest of the economy crashing down. Some authors have suggested such monetarist-style crowding out effects (cf. Kilroy 1978) yet principally they are based on confusing real economic flows with paper revaluations. By the second half of the 1970s owner-occupied housing had grown to nearly a half of net personal wealth, compared to only 20 per cent in 1960 (see *Economic Trends, passim*). But that calculation is based simply on multiplying the size of existing owner-occupied stock by the current average house price and then subtracting the total outstanding mortgage debt. Its potential as a command over non-housing economic resources obviously depends on taking that wealth out of the housing sphere. A housing market 'leakage' is required, and the magnitude of the leakages is far less than the estimated net value of the total owner-occupied stock. Most personal wealth consequently is now necessarily tied up in housing. This does not deny the importance of money gain tied up in owner-occupied housing, but it does question whether the apparent substantial redistribution of wealth in Britain implied by the existence of a large owner-occupied housing stock is actually a redistribution of real economic power in society.

This raises the broader question of the general economic status of owner occupiers. Ownership of real property makes owner occupation crucially different from other forms of housing consumption. Recognition of this difference, however, does not enable its effects to be stated because they depend on the wider social context within which house ownership exists. Yet often the effect of owner occupation on individual behaviour is regarded as virtually axiomatic, rather than open to analysis. One common approach is to treat owner occupiers as a combination of house users (with consumption preferences) and housing capitalists (with investment motives). Housing is an 'investment' into which individuals sink 'capital' or 'equity' and end up making 'capital gains' from price rises. This type of terminology is deliberately avoided here. Money (in its role as a measure of value) is used to denote owners' financial gains from house price rises, so those gains are called 'money' gains rather than 'capital' gains. This is because the purchase of houses by owner occupiers does not make them capitalists, for it does not involve them in the social relations implied by the existence of capital.

The point is not simply terminological, it highlights the distinctiveness of the owner-occupied market. Households are undoubtedly aware of the financial advantages of owner occupation, and their financial calculations influence their choice of tenure and how much housing they buy. Yet housing is a means of shelter, it cannot be opted out of by a household. Capitalists, on the other hand, are intent upon expanding an amount of money capital through making a profit. In fact, their money only remains as capital if it is used to make profits. Capitalists, therefore, cannot opt out of turning their money capital over, i.e. maintaining a positive cash flow.

Previous owner occupiers and capitalist suppliers have different economic reactions to changing market circumstances. Capitalist housebuilders, for example, obviously stop being housebuilders when they no longer supply houses. Slumps in the market put them in crisis as they cannot make sufficient profit. If they stop housebuilding completely they are either on the verge of bankruptcy or have switched to another more profitable activity. The same is not true for owner occupiers. Their ability to sell their houses might be impaired in a market downturn, but their ability to consume housing is not. They can hold their houses off the market if necessary until the market improves, even if this gradually creates increasing inconvenience and cost. The two types of supplier consequently respond in completely different ways to changing market conditions and should not be conflated by treating both as capital. Housebuilders, after all, have to live with market conditions in which the majority of suppliers are not other capitalists but owner occupiers.

The criticisms just made concern particular ways of looking at differences in the personal incidence of housing costs. They are not an assertion that those differences do not matter. They are criticisms of approaches which emphasize individual effects at the expense of the wider systemic effects of housing cost differences and in doing so end up mis-specifying what those individual effects are. Consumers do not live as the isolated recipients of commodities with pre-given incomes and preferences, but as constituents of a social system that determines their lives, including their income, preferences and their housing situation.

Commodity purchasers, however, are not automatons buying the correct quantities of commodities at pre-determined prices in accordance with the dictates of the social system and the individual's place within it. Capitalism is based on the freedom of individuals to own and exchange commodities; freedoms that are created and maintained at the expense of other freedoms. Most fundamentally workers have to work and be exploited at work so that others can enjoy the freedoms engendered by ownership of economic wealth derived from capital accumulation. Even so the freedoms associated with commodity ownership are one of the real bases for the widespread acceptance and enthusiasm for owner occupation: it gives the appearance of enabling one important facet of life, housing, to be under the direct control of the individual household. House ownership does mean individual control over a unit of that commodity, and purchase is undertaken on terms and conditions that individuals find acceptable subject to the relevant constraints and alternatives (colloquially known as 'a fair purchase in the current circumstances given what I can afford'). There is thus a degree of autonomy for individuals over decisions to purchase commodities. So what is purchased depends on aspects of individual ideologically derived consciousness.

Individual autonomy over commodity purchase is none the less relative, not absolute, as wider social forces influence the conditions within which it can be exercised. Actions based on the characteristics associated with consumers' purchases of commodities therefore are not invariant. As the situation changes within which decisions are made, so will those decisions (and so will consumers' preferences, expectations and their resultant political demands). The recognition of a difference like housing tenure within a social class consequently does not enable any statement to be made of the effects of that difference; for those effects are not constant but dependent on changes within society as a whole and within structures of housing provision. This conclusion forms a basis for the examination of ideology and political consciousness amongst owner occupiers.

Owner occupation and ideology

A central concern of much radical literature on owner occupation is its effect on the ideological consciousness of working-class homeowners. The question asked is: 'does the current structure of owner-occupied housing provision lead to any generally accepted values or beliefs?' Saunders (1979) in his survey of the literature notes that most Marxist analyses of owner occupation have made ideology the central core of their explanations of the current role of owner occupation. Owner occupation is frequently argued to be part of a process of ideological incorporation of important strata of the working class into the dominant ideology of capitalism. 'Owner occupation is seen as an important means whereby capitalist interests, aided and abetted by the state, have divided the working class and strengthened their [the capitalists'] control' (Saunders 1979, p. 81). There are two issues to be considered here: has this been the case and is it a necessary result of this tenure form? To an extent the former question can be answered by considering the second one first.

It has long been recognized that forms of housing provision influence the behaviour and beliefs of those consuming housing. The interrelation between the home and 'morality' was well recognized in the nineteenth century: in subtle ways in middle-class homes (Burnett 1978) and in the explicit attempts at reforming working-class life-styles to fit middle-class morals via 'self-help' in the 5 per cent philanthropy movement (Tarn 1973) and the industrial garden villages like Bourneville and Port Sunlight. Politicians also have long made extravagant statements about the effect on working-class consciousness of housing reform. Lloyd George, after the First World War, claimed that the 1919 Addison Housing Act would provide 'Homes for Heroes' that would 'keep the Bolsheviks at bay'. More recently the advantages of owner occupation in giving everyone a stake in the system through property ownership and the personal wealth it implies have been on the lips of many a politician, not to mention building society managers and others with vested interests in promoting home ownership. What is surprising is that such statements have been taken as proof of the ideological effects of certain housing tenures and owner occupation in particular by many who would treat most utterances by such politicians with considerable cynicism. It seems to be far easier to accept the truthfulness or effectivity of Machiavellian designs on working-class consciousness via housing policy than, for instance, claims that the First World War was 'the war to end all wars'.

Owner occupation undoubtedly influences individual beliefs but ideological formation cannot be deduced solely from one particular commodity transaction. Whether, for example, a house is thought of by its owner occupier as a mini landed estate which enables him or her to identify with the highest in the land or as a millstone made of unrelenting mortgage debt depends on much wider questions of the formation of ideology than simply on housing tenure. Most forms of housing provision in capitalist societies, moreover, reproduce the individual, private, family-orientated nature of housing consumption and the need to pay regular, large amounts of money to have a home to live in (Cowley 1979 and Gray 1979). These characteristics therefore cannot be attributed to owner occupation alone.

To attribute to ideological incorporation the sole reason for the existence of widespread owner occupation amongst the working class is both to mis-specify ideologies held by owner occupiers and, as Saunders has suggested, to adopt a crudely instrumentalist theory of ideology.

> The owner occupier is seen as the unreflexive receptable of ruling class ideology, an individual whose horizons are limited to the defence of his own little patch and who cannot see that in reality he has nothing to defend. He has been bought off by a mere trifle, and the extension of house ownership since the war has been one of a variety of ways in which he and others like him have been hoodwinked, divided, re-pressed and moulded for continuing exploitation.
>
> (Saunders 1979, pp. 82–3; gender as in original)

What the ideology of incorporation tends to confuse is a lack of organized political protest about particular aspects of housing provision amongst owner occupiers with ideological acceptance of the political *status quo*. Rent strikes, protests about repairs, organized tenants' move-ments and other features of direct local action, though sporadic, are well documented amongst local authority tenants whereas their equivalent amongst owner occupiers hardly exists. Unity between local authority tenants is always likely to be stronger than between owner occupiers because of the greater class homogeneity in the council sector. But as important is the fact that political protest by owner occupiers takes different forms because it occurs within a distinct structure of housing provision.

The incorporation view has moreover to extend beyond housing issues. The owner occupier has to exhibit greater passivity on all politi-cal issues: a proposition which it would be hard to demonstrate. It is interesting to note in this light one historical case to the contrary: that of the South Wales miners, one of the most militant working-class groups in the 1920s and 1930s. Mardy in the Rhondda was founded as a working-class mining village at the end of the nineteenth century and much of its housing was built in *owner-occupier* schemes. This did not stop the village from being dubbed 'Little Moscow' for the militancy of many of its inhabitants and their widespread support for the Communist

Party. In another area of the coalfield, the Anthracite District, it has been suggested that home ownership was one source of strength of the resistance during the 1926 lockout: 'Many miners also owned their own houses and would "eat them" before being starved back to work' (Francis and Smith 1980, p. 247, and their ch. 5 for the Mardy example).

Central to the understanding of ideological formation and housing provision is the recognition that all forms of housing in Britain are provided on the basis of commodity exchange. This general economic character of housing provision forms part of the framework within which ideologies are created and, therefore, influences the determination of those ideologies. Within ideologies related to housing provision are general pressures to accept the inevitability of commodity exchange, all the disciplines that go with it (like regular payments), and generalized notions of 'fairness' within that exchange. Invariably there are widespread attempts to present the issues in terms of their relation to some exchange norm: frequently a mythical, unfettered, unsubsidized, private market. Even so, attempts at the formation of particular beliefs about housing provision depend on historically specific problems concerning structures of housing provision, and housing ideologies cannot be distinguished from the struggles of which they are part.

Aspects of council housing provision can be used to illustrate this point. Council housing is provided as a commodity and it involves state expenditure (so it is widely believed that all council housing is subsidized). The meaning of the notion of subsidy, however, is not so much tied up with the money earmarked for local authority housing by the Treasury but instead by comparison with a hypothetical private market. If council housing is subsidized by the state its price must be less than a private market rent, it is often claimed. But this need not be the case, nor does it seem immediately relevant. Yet all rent policies with respect to council housing have been couched in those ideological terms, and so have most struggles against those policies. Take, for example, three periods when the principles of rent fixing were a major issue. In the first one it was claimed by the contemporary government that as council housing is subsidized it is only fair that rents are pooled so that all may share in the subsidy (the issue *circa* 1935). In the second it was suggested that council housing is subsidized too much, so all rents should be raised to a 'fair' level, defined to be a notional free market rent minus an equally notional 'scarcity' element (the issue *circa* 1972). In the final one, the government said that rents should be decided by local councils but they should be neither too high nor too low but 'reasonable' (the issue *circa* 1975). Each of these approaches to council rents took place during attempts to reduce state expenditure on council housing in specific political circumstances. Their predominant ideological effect was to centre discussion on the level of rent and not on the structure of provision that determined the cost of council housing, nor on the need to pay rent at all.

The overwhelming dominance of a 'fair-exchange' ideology in housing can be seen in the role of council housing in the post-1945 welfare state. Council housing is supposed to be one of the pillars of the welfare state – along with the National Health Service, Education and the National Insurance/Social Security systems – yet it is the only one in which there has been no questioning of the need for direct payment. Commodity provision is accepted as virtually axiomatic and debate is over the nature and forms of subsidy – not over the nature of provision as is the case with the other constituents of that welfare state apparatus.

In owner occupation the ideological implications of commodity exchange are different, as ownership is being exchanged as well as possession of the commodity. There is moreover no unitary supplier (the council) against whom to direct political protest but instead an anonymous market consisting mainly of other owner occupiers. The material conditions in which ideologies are formed therefore are different from the rental case. Ownership also results in different economic interests, which have to be protected, and the interpretation of what those interests are also takes ideological forms. Owners' economic interests have varied over time as have the threats to them. Since the early 1970s the major threat has been to the mortgage interest tax relief and to money gains from house price inflation via proposed reforms of the tax system (see chapter 12). The ideology of the virtuous 'self-help' owner occupier in contrast to the subsidized, 'welfare state scrounger' council tenant has evaporated in the face of rapid rises in the cost to the Exchequer of mortgage tax relief and arguments over the need to tax money gains. The result, however, is that the owners' perceived economic position takes on a changed ideological form: political resistance to threats to those interests do not go away once one justification appears to falter under changing events. The link between economic interest, ideology and political demands is a complex and continually changing one. It is, not surprisingly, impossible therefore to identify an ideology of owner occupation in general.

There is another dimension to housing consumption that also produces an ideological effect via the process of commodity exchange.

Reproduction of social classes or specific strata within them has a spatial dimension related to proximity to work and to demands for territorial exclusivity; the latter is the case especially for those in the higher class strata. The development of this pressure for spatial exclusion is a comparatively recent phenomenon in Britain, associated with the separation of home and workplace with the advent of capitalism and the squalor of many towns in the eighteenth and nineteenth centuries. The separation has generally been highly successful. All neighbourhoods in most towns can be classified with ease by their current social composition and by their past ones, often just by walking up and down a couple of streets for a few minutes. The creation of socially uniform mini-societies with their own sets of ideological beliefs within class societies has been achieved in this way and is well documented by urban sociology. The principal means by which this territorial separation is achieved occurs through the market and a concomitant spatial division of house types and house prices/rents. This took place when private landlordism was dominant as well as with present-day owner occupation.

This spatial price mechanism, although highly effective, does not always achieve the desired results. In the nineteenth century it would frequently be reinforced in upper-class residential areas by guarded gateways at entrances to such housing areas, as, for example, in the case of the Bloomsbury Estate in central London. In the second half of the twentieth century, the class structure of society has changed, owner occupation become virtually universal amongst certain social groups and guarded housing estates either illegal or prohibitively expensive. Now the price mechanism is reinforced by local residents' pressure groups and by the manipulation of the planning system. Frequently territorial exclusiveness is ideologically justified in terms of the maintenance of property values in the locality.

The most famous of these cases took the ideological form of conflicts between owner occupiers and council tenants in the great controversies concerning overspill estates in the 1950s and 1960s (Cullingworth 1960). The maintenance of residential exclusivity is also the major type of locally based political protest by owner occupiers that has been documented. The existence of such conflicts, however, cannot be taken to imply that all owner occupiers have a similar ideology of snobbism and a desire for social homogeneity in their neighbourhoods. A few empirical examples cannot justify a shaky general theory. The refusal by whites in the United States to sell their houses to blacks, for instance, which created some notable victories for the Civil Rights Movement in the

1960s, cannot be used to deduce that all white home-owners are racists. Racists used the housing market for their own ends, not the other way round. In Britain, with the demise of council overspill schemes in the 1970s the main pressures on prosperous suburbia have come from the threat of new low-priced owner-occupier estates in the locality. Defence of the cherished green belt is as strong here as against the earlier council housing schemes. Potential ideological divisions between owner occupiers are more politically important than any notion of an ideologically subdued home-owning proletariat.

Discussion of ideology and owner-occupied housing provision, furthermore, should not be limited to the occupiers themselves. The other agents within the structure of provision also have their own ideological beliefs and act accordingly. In contrast to the wide literature on owner occupiers, little material is available on them. Ford (1975) has looked at building society managers but the wider building society movement is still uncharted territory. Speculative builders have particular views of events through which they interpret changes in the housing market, interpretations which have considerable political weight. But again notions of ideology have not formed part of the analysis of their actions. Perhaps as capitalists it might be felt that there is a one-to-one correspondence between ideological beliefs and direct economic interests. That might be the case, though not necessarily, but even if it were true their understanding of the workings of the housing market will be the basis on which those interests are justified politically. It was argued in earlier chapters that their beliefs about the market are wrong, not through Machiavellian design but because they are formulated on the narrow perspective of the problems facing individual housebuilders rather than on the nature of the structure of owner-occupied provision as a whole.

Owner occupiers as a political grouping

The discussion of ideology has implications for conclusions about political action over housing. There is no necessary reason why owner occupiers should act as a political unity, even over housing issues. Yet frequently the tenure is said to have the effect of creating an inevitable political unity amongst owners that can be deduced directly from the tenure's inherent characteristics. Owner occupation is taken out of its historical context and given universal consequences that unfortunately close off further enquiry into the relation between owner occupation and political consciousness and, hence, limit discussion of potential forms of political action and alliance. To keep open the possibility of such an analysis this section presents a critique of some widely held views about owner occupation and political action.

When examining individuals located in any particular social class it is necessary to avoid reducing their political actions solely to immediate objective class economic interests. Location in a particular class, for instance, is obviously not the sole determinant of the ideology within which any individual understands and reacts to the world around them. Nor does class location produce a unique political reaction. The need to distinguish individual political action and class location is brought out clearly when housing provision is examined. Economic differences alone suggest the importance of avoiding simplistic class-level analysis because individuals from the same class may well live in different tenures and so face different costs and shares of those costs borne directly by the consumer and the state through subsidies. Not surprisingly, therefore, how individuals' political demands over housing are formulated and how they relate to individuals' class locations has been an area of considerable interest especially for those concerned with the analysis of community action.

Most discussion has centred on the effect of owner occupation on working-class political consciousness. At best the conclusion is neutral, although empirically incorrect, when it is suggested that owners do not actually get the financial gains they are frequently said to, or at least the working-class ones do not (cf. the discussion over council house sales in *Critical Social Policy*, 1, 2, 1981). More common, however, is the view that owner occupation has adverse political effects because it gives workers privileges that weaken their militancy, or, in the case of discussions that are limited to housing alone, groups who have common political housing demands are divided on the basis of housing tenure rather than class. Moreover, those housing tenure based political divisions are said to be in conflict. The debate has tended to remain at a conceptual level but the political implications of these conclusions are presumably that traditional support by the Left for those who rent should be maintained but it must be recognized that this support is generally against the interests of owner occupiers. Hostility or indifference is thus the general political attitude towards owner occupation.

This type of argument implies that forms of housing consumption necessarily generate particular political actions. The principal weakness is that its conclusions about political action are not products of analysis but of the way in which tenures initially are defined. Dunleavy (1979) for instance examines the voting patterns of individuals in different social locations. Social location is defined on the dimensions of social class and whether the individual lives in public or private housing and travels by public or private transport. His data are derived from a Gallup opinion poll undertaken in 1974 so his classification had to be limited to the information provided by the poll. He discovered that there was a correlation between voting intentions for the Labour and Conservative parties and whether the individual lived in or travelled by the private or public modes. This correlation was found to exist across class lines. He concluded that 'individualized consumption locations in housing and transport clearly influence alignments towards the right, and involvement in collective modes influences alignment to the left. . . . Overall these effects are comparable to social grade' (Dunleavy 1979, p. 443).

The statistical dubiousness of the generalization in Dunleavy's conclusion is outweighed by the transformation of the data used to derive its content. The conclusion is tautologically based on the initial definitions of the principles on which the two tenures operate rather than the results of the survey. Owner occupation is said to be individualized consumption whereas council housing is said to be collective consumption. But it is difficult to see how the data lend support to that conceptualization. If the data used do accurately measure housing-related political differentiations, all that can be concluded is that in 1974 owner occupiers and council tenants were more likely to vote Conservative and Labour respectively. At the time, this was not surprising as both tenures were in crisis and each party was offering different policies. Council tenants had been highly politicized as a result of the Conservatives' 1972 Housing Finance Act which forced up rents sharply and tried to reduce state subsidies. Owner occupation was in crisis as the result of the housing boom in 1972–3 and then the sudden collapse of the market. Whatever the reasons, the political positions were an historical product. Yet Dunleavy chose to ignore history and to base his conclusion on supposed permanent principles which necessarily lead to an opposition between tenures. The inevitability of political opposition between households in different tenures, however, is not justified but simply asserted. This prematurely closes off any possibility of investigating the feasibility of creating a political unity across tenure boundaries over housing matters. So the significance of the point goes beyond academic dispute.

Another influential study in recent years suggesting that owner occupiers constitute an identifiable political grouping with respect to housing has been that of Saunders (1979). Like Dunleavy he concludes that 'the two major patterns of housing tenure − ownership and renting − are important determinants of the real political divisions which are constituted in housing struggles' (p. 102). His argument is based first on an elaboration of owner occupation as a housing class within a Weberian sociological perspective, followed by a critique of certain crucial parts of Weberian notions of housing classes, then finally an attempt to amalgamate the post-critique remnants with an interpretation of Marxism. A step-by-step account and critique of the argument would consequently be lengthy and is unnecessary here. What is important is that the claimed political distinctiveness of owner occupation is not made on the basis of some supposed ideologically created political unity. The ideological approach is forcefully rejected and replaced by a division based on economic criteria alone. The economic differentiation of owner occupation is made by defining renting as a means by which housing is consumed whereas owner occupation is a means of accumulation as well. The political conclusions about the two tenures follow directly from these definitions and so it again constitutes a clear example of political differentiation by definition, rather than through an analysis of the historical situation.

Owner occupiers, according to Saunders, can accumulate wealth because of house price inflation, low rates of interest on mortgages, and tax relief. The political actions of owners consequently are said to be based on defending those sources of wealth (presumably in national politics) and at a local level defending property values against incursions by undesirable land uses and rate rises. He searches for the source of the wealth in the economic flows associated with owner-occupied housing finance, finding it in the losses of others. Owners' gains, therefore, have to be defended against others who lose:

> it is clear from the analysis in the previous section that relations of exploitation can be established − i.e. that the sources of returns for owner occupiers can be identified. They derive their increasing wealth from highly specific sectors of the population (aspiring home owners and building society investors), or from virtually the entire population (through tax relief).   (Saunders 1979, p. 96)

The argument that all owner occupiers' money gains are a redistribution of wealth was criticized earlier. Moreover, any redistribution of wealth that actually does occur is simply a redistribution not a mechanism of exploitation. Saunders' particular definitions of exploitation and capital are at variance with Marxist notions, yet this is never made explicit within Saunders's discourse. Marxist notions of capital and class are based on exploitation within the process of production. Saunders's definition of owner occupation as a means of accumulation, where owners exploit non-owners, is a definition of capital based solely on a distribution of wealth (a Weberian 'command over resources') and of exploitation as an unfair one (through a form of 'unequal exchange' in owner-occupied housing finance).

The point here is not to plead for theoretical purity; instead it is to suggest that the argument is a dead end for housing analysis. Discussion by Saunders is erroneously limited to a competition between individual consumers for limited resources. Yet there is no theoretical basis for such a narrow distributional perspective. Links exist, for example, between wage levels and housing costs; similarly state subsidies to housing tenures arise from much wider forces than the immediate demands of groups living in particular tenures. The analysis by Saunders of owner occupiers' economic interests is too narrow. Yet why should there be such a direct correspondence between economic interest and political action in the first place? Saunders is highly critical of reductionist approaches to owner occupation that deduce its political effects from notions of a unitary ideological consciousness engendered by living in the tenure. Yet he seems to have fallen into a similar position through simply replacing 'ideological consciousness' with 'economic interest'. Individuals' political responses can therefore be read off immediately from their economic interests. No concrete analysis of the conditions in which political actions are formulated consequently is required. Within Saunders's approach it is impossible to examine the changing and contradictory role that owner occupation has played in British society and continues to play. Historical change has been frozen out in an attempt to find one immediate politically effective economic interest.

Owner occupation and political consciousness

The problem with all attempts to deduce a unified political response by owner occupiers from a single cause is that the tenure has to be seen as conferring a common and universal consequence on everyone living there. The obvious variety in the 55 per cent of the population who are owner occupiers, therefore, has to be ignored in the process of generalization. The result is that claims about owner occupiers' political consciousness derived from either a political passivity produced by ideological effects or through a unity of economic interest can easily be dismissed by reference to empirical cases that do not conform to the claimed generalization. The diversity of economic circumstances of owner occupiers is clear from the descriptive data of the characteristics of owners given earlier in the chapter and from the subsequent description of the nature of owner-occupied finance. Some owner occupiers have no mortgage costs, some have cripplingly high ones; some realize large money gains, others see the price of their houses drop; repair costs for some low-income owners are so burdensome that they have to live in a physically deteriorating home whose price falls through neglect, whereas others prosper from buying up dilapidated houses and improving them. General market trends can be discerned but the individual incidence of them obviously varies.

The methodological difficulty suffered by the studies of political consciousness referred to above is that they all try to derive political action by starting at the level of the individual. They ask the question 'what does owner occupation do to the political consciousness of the owner occupier?' Once that question is answered, it seems that the mystery will be uncovered of the success of owner occupation in structuring the housing policy of the state. Yet a more successful approach is likely to be one that takes a broader systemic view of the problem: 'what are the conditions that have enabled widespread differences in individuals' economic circumstances and political beliefs to be channelled into broad, popular political support for one housing tenure, owner occupation?' Analysis should be concerned, therefore, with examining the conditions under which the structure of owner-occupied housing provision gained political hegemony over state housing policy and sustained its dominant position.

Three implications follow from the latter approach. The first is that although individual consumers are still important they are only one group of social agents involved in the political process. There are other political forces involved in the structure of provision and yet others who are affected by it. It is the ability to accommodate a sufficient breadth of such disparate interests that has created the political success of owner occupation. The second implication contrasts with the earlier reductionist views of owner occupiers' housing situations. One of the conditions likely to have helped the current form of owner occupation succeed in generating popular support is precisely the wide differences in housing costs faced by individual home-owners. The economic burdens of living in the tenure are not spread evenly across all households living there and their incidence for particular households varies over time. Not all home-owners have to bear the costs of the growing crisis of the tenure; consequently, some might even gain from it by being able to realize higher money gains. The final point concerns the dynamic implications of achieving political hegemony. Once support for owner occupation is sufficient for the tenure to dominate the housing policy options of the state, that dominance has to be sustained. This means there is a need to adapt to changed circumstances. Once, for example, the tenure avoided the need for state expenditure on housing, now it requires a growing and uncontrolled amount of it. Owner occupation required little direct state expenditure during the inter-war years as few people enjoyed mortgage interest tax relief and all owners also paid an imputed rental tax (i.e. schedule A tax). But by the 1980s the state subsidy to the tenure was huge.

What is so interesting about owner-occupied housing provision in the early 1980s is that as soon as it gained such political hegemony in the late 1970s that the future path of housing policy in Britain seemed to be leading towards owner occupation as the sole tenure (Kemeny 1980), the conditions sustaining its dominance began to crack apart. A housing crisis is emerging the like of which has not been seen since the political stalemate over state intervention into housing provision in the years prior to 1914. The social agents who dominate the structure of owner provision have similarly been plunged into crisis by the forces that are breaking up the hegemony.

10

Building societies and the changing mortgage market

Building societies have exerted a considerable influence on the growth of owner occupation in Britain. They have provided most of the mortgage credit for home ownership by successfully tapping an expanding share of personal savings. In addition they have been at the forefront of the propaganda machine exhorting the benefits of the current form of owner occupation: undertaking extensive lobby campaigns, providing supportive material to the media and anyone else they can interest, and maintaining close links with government and the financial establishment.

The societies have prospered with the growth of owner occupation but are now facing mounting problems as the expansion of owner occupation begins to falter and other financial institutions, particularly the clearing banks, take a growing interest in the mortgage market. Building societies are having difficulties both in getting sufficient money deposited with them and in lending it out as additional mortgages. In consequence they are becoming increasingly unhappy about being tied to owner occupation and are beginning to push for legal changes in their status to facilitate substantial restructuring. A number of leading building society managers are now predicting inevitable major changes over the next few years. Alan Cumming, chairman of the Building Societies Association, has predicted a bleak future for the societies without substantial legislative change (*Guardian*, 27 January 1982). A divide is opening up between building societies who want to reinforce their links with housing property, by becoming property owners and dealers in housing to rent and for sale, and other societies who want to reinforce their links with the financial community by expanding their banking functions (cf. *Financial Times* survey of banking, 27 September 1982). Whatever the direction taken, the cosy life building societies had as monopoly providers of owner-occupied mortgages seems gone forever.

To understand the problems faced by building societies their economic role and recent history has to be examined.

Building societies as economic agents

Building societies are capitalist enterprises, albeit of a special kind, out to accumulate and expand the size of their capital. They are a type of merchant capital, known as money-dealing capital, that accumulates by lending money at a higher rate of interest than they pay to those from whom they borrow. They specialize in borrowing money on short-term deposit and lending it as mortgages to owner occupiers. Nearly all of the money invested in building societies comes from private individuals. Building societies consequently operate with funds generated within and channelled to the personal sector of the economy.[1]

Even though they are capitalist enterprises building societies' activities are not solely influenced by profit maximization. Formally they are Friendly Societies, with their actions limited by legislation and overseen by the Chief Registrar of Friendly Societies. They are not legally able to charge a profit on their financial intermediation activities but can charge only administrative costs and roll over surpluses into their reserves. Expansion of assets and income, nevertheless, is still necessary to their rationale as enterprises. Building society managements tend to be closed oligarchies subject to little control by the membership of a society, so a society's objectives become closely linked to the importance for senior management of high salaries, status, perks and power: all of which are associated with increasing size in terms of turnover, branches, etc. There are also economic reasons for increases in size associated with the ease of funding and turning over long-term mortgage debt in the face of short-term deposits, which help the societies to confound the maxim against borrowing short and lending long. The substantial scale economies in mortgage activities account for the rapid decline in the number of societies and the large proportion of total assets held by the top ones: the top two, for example, have 35 per cent of all building society assets (table 10.1).

Their special legal status means that building societies are prohibited from owning land except for the purposes of conducting their businesses; cannot invest in other organizations or directly buy or sell property; have to have first claim on mortgaged property and can only lend a limited percentage of their funds as large advances or to non-owner occupiers. Building societies consequently are unable to expand into other areas of housing provision or to diversify out of housing altogether; they are locked into lending to owner occupiers. About 80 per cent of societies' assets are mortgage advances to owner occupiers; most of the rest is held in liquid investments to provide working capital and to cover any short-term excess of withdrawals and lending over deposits (the ratio of liquid assets to all assets is called the liquidity ratio).

The specialism of building societies in lending for owner occupation was finally determined by the growth of mass home ownership in the inter-war years. Legislative restrictions on societies' activities had gradually evolved from the nineteenth century and their effect was to help societies dominate this sector, following and reinforcing trends within the movement rather than redirecting them. The legislation was particularly

Table 10.1   Building societies: societies, branches and assets

Number of societies

1900	1930	1940	1950	1960	1970	1981
2,286	1,026	952	819	726	481	251

Number of branches

1968	1970	1972	1974	1976	1978	1980	1981
1,662	2,016	2,552	3,099	3,696	4,595	5,716	6,203

Number of borrowers (m.)

1940	1950	1960	1970	1975	1981
1,503	1,508	2,349	3,655	4,397	5,484

Mortgages outstanding (£m.)

	1960	1970	1973	1975	1977	1979	1981
Actual	2,647	8,752	14,532	18,802	26,427	36,801	48,854
Deflated to 1975 prices (deflator: RPI)	7,272	16,207	21,122	18,802	19,590	22,900	22,400

% of building society assets held in 1981 by:

Top 2	Top 5	Top 10
35	55	71

Sources: BSA Bulletin, Wilson Committee (1980), BSA (1981a)

helpful in bolstering public confidence in the movement as a whole in the face of periodic crises of confidence (cf. the 1939 Building Societies Act referred to in chapter 2). Successive governments, furthermore, have excluded building societies from the restraints of monetary policies, although recently this position has begun to alter.

The isolated and specialist position of the societies for most of the post-war years has been a great advantage to them. For many years building societies had a free rein in the owner-occupied market, tapping the personal savings of households in the face of until recently muted competition from financial institutions like the clearing banks. In the 1980s, however, it is clear that the societies are facing serious problems in sustaining their expansion. Although the difficulties they face are interconnected, the easiest way of describing them is to divide them on the basis of the two sides of their money-dealing function, the uses and sources of their funds. The societies face a demand problem of finding new outlets for mortgage advances at viable rates of interest and an income problem of short-term variations in the inflow of funds. Both are linked to the new instability of the owner-occupied market.

Finding new mortgage outlets

Building society expansion can be measured by the rate of increase in mortgages outstanding with them. In money terms each post-war year has been a record one. But since 1973 much of the expansion has simply reflected general price inflation. In real terms the record growth of earlier years has stopped. From the early 1970s onwards the real rate of change of outstanding mortgages became closely correlated with the booms and slumps of the housing market. During downturns the real level of mortgage redemptions outweighed new advances, causing the real value of building society mortgage assets to fall. They would then pick up again during the next surge in the housing market (table 10.1). Steady real growth, therefore, could no longer be relied upon from the owner-occupied market so societies began to look around for new potential outlets. The problem for the societies has been that every potential option they have threatens the advantages of their special and isolated status. Certain of them question the justifiability of their privileged legal position, whilst others have woken the previously dormant competition of the banks.

The dilemmas faced by building societies are considerable. Initially they had hoped that expansion could take place by moving into Western

Europe following Britain's entry into the Common Market in 1973. Legal complexities and the perilous state of many EEC housing markets has to date limited this approach to a few fact-finding tours and conferences. On the home front, the potential mortgage glut could possibly be overcome by relaxing the societies' restrictive lending policies. To an extent this has happened but major business can only be drummed up through more generous responses to existing owners' additional loan requests. But to do this creates problems as such loans indirectly finance consumers' non-housing expenditure rather than additional housing expenditure by enabling owners to 'cash in' some of their housing money gains as mortgage refinancing, bringing with it the danger of effective monetary control in place of the present weak supervision.

The public sector has provided one successful area of expansion of the societies' mortgage business. Conservative housing policies, particularly over council house sales, have proved to be a substantial, if temporary, source of new mortgage business. Over £2000m. was raised by local authorities from council house sales in 1979–81, and a significant but unknown amount of it was mortgaged to building societies (*BSA Bulletin* 31). In addition, the societies began negotiating successfully with a number of local councils to take over their outstanding mortgages (*Guardian* 28 August 1982), which in 1982 stood in total at £3900m.

The other option building societies have had is to diversify out of owner-occupier mortgage lending altogether. This option can take a number of forms. First they can move into other property activities. In general, this requires legal changes but Abbey, the second largest society, has already set up some pilot rented housing schemes through taking advantage of loopholes in existing legislation. Second, they can provide a wider range of consumer services, on the lines suggested in the following report:

Building societies may start selling holidays, theatre tickets and postage stamps if the deputy chairman of the Building Societies Association has his way.

Speaking at the annual conference in London yesterday, Mr Herbert Walden said that the constraints of the Building Societies Act are 'too restrictive'.

While Mr Walden wants to retain the 'mutual' origins of societies, he called for wider powers for societies on both the savings and housing sides: 'In these days householder durables are sometimes difficult to distinguish from fixtures. Should we not be able to lend for washing machines, refrigerators, fitted carpets, etc.? We may need to finance the provision of the computer linked in to the cable TV facility we are to expect.'

Building societies are facing competition for both savings and mortgages and Mr Walden wants new powers to be able to compete. This includes the power to hold land for housing development and to hold property for letting. He would use the branch network to provide services, even competing with the post office to pay pensions and sell savings certificates.

Mr Walden acknowledged that some people want even more far reaching changes, in which case he would want a 'very detailed investigation' into the consequences.　(*Guardian*, 20 May 1982)

Finally, societies can expand their retail banking functions. Some societies have already begun to enter this field by announcing in 1982 plans to set up cheque account facilities and external cash dispensers. Halifax, the largest society, has started experiments alone whereas others have made agreements with banks (Leicester with Citibank and Abbey with the Co-op Bank). Others are said to be only waiting for the advent of electronic banking systems over the next few years to avoid the costly business of cheque clearing.

Retail banking brings building societies into competition with the clearing banks. Competition with the banks has been the most publicized aspect of building societies' mortgage problems in the 1980s. From being a relatively insignificant source of mortgages in the mid-1970s the banks aggressively took renewed interest in the owner-occupied mortgage market from 1978 onwards (see table 10.2). The clearing banks took almost 40 per cent of net advances in the last quarter of 1981 and even more in 1982. Startled by their success and sensitive to government grumblings about the effects on monetary policy, the banks announced plans to curtail the growth of their activities in the summer of 1982. But they look set to stay a major force in the mortgage market.

In part the movement by banks into owner occupation reflected the decline of traditional borrowers from banks brought about by the economic slump of the early 1980s. Traditionally banks have been geared to institutional borrowers and have tended to ignore the personal sector. In addition, government restrictions on bank lending were lifted in 1980 which enabled them to compete more effectively for mortgage business. But the principal cause of banks moving into owner-occupier mortgages has been a response by them to a long-term attempt by building societies to overcome mounting problems caused by fluctuations in their income. The nature of those problems can be seen by looking at the sources of building society income.

Building societies' sources of funds

Building societies have three principal sources of finance for mortgages:

1 *Repayments of mortgage debt* by existing borrowers. In part they are regular mortgage repayments but mainly they are mortgage redemptions at the time of moving which usually will be replaced by a new loan. Many redemptions, therefore, imply a prior commitment to another mortgage loan, so gross mortgage advances are much higher than the net advances which are not matched by redeemed mortgages (see figure 10.1).

2 *Interest credited to investors' accounts.* Most people do not withdraw the interest earned on their investment accounts. About two-thirds of interest is retained. The relative competitiveness of building society interest rates does not seem to influence this proportion, although the level of interest rates will affect the sums involved: higher rates mean obviously more money credited.

3 *Net receipts.* Building societies get new money from the net inflow of additional savings over withdrawals from investment accounts.

Table 10.2   Competition in the mortgage market, 1973–81

	Net advances per annum						
	Building societies *£m.*	*%*	*Local authorities £m.*	*Insurance companies and pension funds £m.*	*Banks £m.*	*TSBs £m.*	*Other public sector £m.*	*Total £m.*
1973	1999	69	355	183	310		46	2893
1974	1490	61	557	189	90		113	2439
1975	2768	74	619	150	60		133	3730
1976	3618	92	67	103	80		60	3928
1977	4100	94	4	119	120	1	18	4362
1978	5115	92	−43	166	270	5	17	5530
1979	5271	80	293	357	590	7	74	6592
1980	5722	77	461	376	490	93	247	7389
1981	6207	67	250	239	2200	182	348	9426
1982								

Source: BSA Bulletin 31

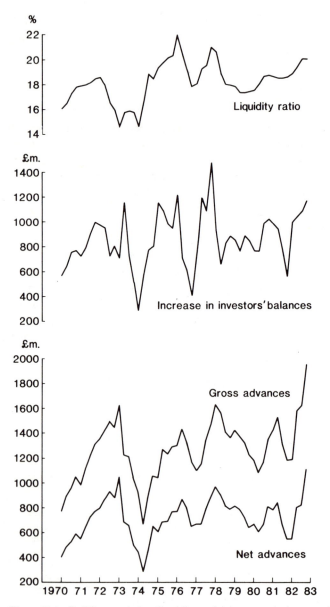

Figure 10.1 Building societies: liquidity ratio, increases in investors' balances, gross advances and net advances, 1970–82, seasonally adjusted (1975 prices)

Source: HCS, deflated by RPI

The factors influencing net receipts depend on the size of those receipts. Small sums (£500 or less) are fairly insensitive to interest rate variations. They depend instead on seasonal factors (like holiday and Christmas expenditure) and on the spread of the branch network. Large sums, on the other hand, are highly sensitive to interest rate competitiveness. The size range from £500 to £2000 tends to be balanced out by withdrawals of money to purchase consumer durables, so the band of importance is the £2000+ category of deposit. Empirical research has shown that the stock of deposited money tends not to switch between potential outlets depending on the relative rates of interest offered (HPTV II), presumably because the short-term transactions costs are too high. Instead it is the flow of new savings which is most sensitive to relative rates of interest.

The proportion of net receipts derived from small and large sums varies sharply. In 1977, 69 per cent of societies' net receipts were large sums and 31 per cent small ones. In 1979 the proportion was almost exactly the reverse, 33 and 67 per cent respectively, owing to a rise in small savings and a decline in large ones (Stow Report 1979). The differential rate of interest used to measure competitiveness is the local authority three-month deposit rate. Regression results have found this to be a better proxy than banks' deposit rates (HPTV II). Little money is actually invested in such debt, instead it is said to be a good proxy for general short-term money market rates. The societies' actual main competitors for liquid funds are banks and national savings (table 10.3).

Overall the building societies have been very successful in attracting personal sector liquid savings. In the early 1960s they had only about 20 per cent of this market. Yet, as table 10.3 shows, by 1970 they had a 35 per cent share and they continued rapidly to increase their share until 1977. But since then their share has stagnated and even declined slightly. National savings were the main loser to the societies' expansion in the 1960s and 1970s. An expansion of building society branches helped to increase their market share. There were almost four times as many branches in 1981 as compared with 1968. (Additional branches also help to drum up more mortgage business.) Branch expansion, however, has diminishing returns. The year in which building societies reached their peak market savings share, 1977, was also the peak year for savings accounts per branch (5457), after which time the number per branch gradually fell. Meanwhile management expenses rose as the decline in real average administrative cost per shareholder and borrower of the early 1970s was reversed (BSA 1981, tables 6.11 and 6.9). The era of branch expansion now seems over.

What is interesting about building society income is the secular changes that have taken place. In the first decade after 1944 building societies were able to take advantage of their sheltered position, as in the 1930s, to charge comparatively high mortgage interest rates and so could pay, in turn, high rates to investors. This enabled the movement to expand rapidly on the net receipts attracted. Table 10.4 compares the average interest paid to building society investors with bank rate/minimum lending rate for the post-war years. The societies' huge interest rate advantage was gradually eroded by the second half of the 1960s. The rapid expansion in branches and share accounts dates from this period. After 1965 the movement rapidly expanded to attract over 1 million new accounts each year. But then in the 1970s the instability of net receipts became an increasing problem (Gough 1975a). The societies began to offer additional rates of interest to large sums held on a fixed-term basis in an attempt to stabilize their income. These term shares increase from only 2 per cent of funds invested in 1974 to 15 per cent in 1980.

Expansion by building societies in the post-war years, in other words, has been associated with three distinct phases marked by different

Table 10.3   Competition in the market for personal liquid assets, 1970–81

	National Savings %	Local authority temporary debt %	Deposits with banks %	Deposits with savings banks %	Deposits with building societies %	Other %	Total %
1970	22.8	1.1	34.6	6.1	34.5	1.0	100.0
1971	21.4	0.8	33.8	6.2	36.8	1.0	100.0
1972	19.9	0.8	34.5	6.3	37.8	0.7	100.0
1973	17.5	0.9	37.7	5.9	37.7	0.4	100.0
1974	15.7	0.7	40.0	5.4	38.0	0.2	100.0
1975	15.0	0.6	36.5	5.3	42.4	0.1	100.0
1976	14.4	0.4	35.4	5.6	44.0	0.2	100.0
1977	14.5	0.3	31.8	5.8	47.4	0.1	100.0
1978	14.6	0.3	31.7	5.8	47.4	0.2	100.0
1979	11.7	0.3	33.7	7.6	46.4	0.2	100.0
1980	11.4	0.3	34.9	7.1	46.1	0.3	100.0
1981	14.8	0.3	33.6	5.0	46.0	0.2	100.0

Source: BSA Bulletin

strategies for winning new funds. They are a progression through three alternative approaches open to a market monopolist: charge high prices until the competition catches on (first phase), non-price market expansion (the second 'branch' phase) and market segmentation based on differences in price elasticities (the third 'interest differential' phase). Each of them has built upon the gains created by the previous one. The recommended rate system of the Building Societies Association cartel helped to sustain the initial high interest levels and created a tradition of generous margins between investment and mortgage interest rates. The lack of pressure to cut administrative expenses then enabled the societies to enter into areas of non-price competition such as branch expansion. Branch expansion, in turn, helped to increase the number of investors to the point where small and large savers could be treated discriminately.

Branch expansion illustrates the unique advantages building societies have had in the financial environment of post-war Britain. A much higher proportion of adults do not have bank accounts in Britain than in most other advanced capitalist countries: only 61 per cent had accounts in 1981 (*Financial Times*, 27 September 1982). Building societies have been able to capitalize on this gap in the retail banking market. Advertising campaigns by building societies try to cultivate an image of probity combined with convenience and informality to attract people put off by the cost and complexities of current account banking, yet who need somewhere safe to put their cash. The societies' expanding branch network has been able to provide a competitive service to current account banking and to other savings institutions. High Street branches are convenient, withdrawals are easy, the accounts pay interest, and the interest is net of tax. The latter plus the cash nature of most counter transactions, in addition, makes it a haven for money from the so-called informal economy. Savers, therefore, tend to accept relatively lower rates of interest from building societies and that enables the societies'

Table 10.4   Average interest rates (%), 1945–79

	1945–9	1950–4	1955–9	1960–4	1965–9	1970–4	1975–9
Grossed up[1] building society share rate	4.07	4.34	5.51	5.77	7.21	8.94	10.84
Bank rate/MLR	2.00	2.97	4.81	5.01	6.87	8.17	10.78
Difference	+2.07	+1.37	+0.70	+0.76	+0.34	+0.77	+0.06

Source: 'Mortgage Finance in the 1980s' (Stow Report) (1979)

[1] 'Grossed up' is share rate grossed up by standard rate of tax

own lending rates to be comparatively low. Throughout the first half of the 1970s, this enabled mortgage rates to be comparatively low. The banks consequently had no chance then of competing for mortgage business.

The tax position of building societies further increases their comparative advantage. Building society share interest is not liable to basic rate tax because it is taxed at source under a composite tax arrangement with the Inland Revenue. The composite rate is the estimated average basic rate of tax of all investors and, as not all investors are liable to income tax, the composite rate is lower than the basic rate. In 1979–80 it was 21 per cent compared with a 30 per cent basic rate. As the Wilson Committee concluded:

> The effect of this arrangement is to enable building societies either to pay higher rates to their depositors than would otherwise be possible, thereby giving them a competitive advantage over other deposit-taking institutions, or to charge lower rates for mortgages, thereby increasing the demand for them, or some combination of the two. From 1 December 1979, for example, the recommended building society share rate has been 10.5 per cent net of tax, equivalent to 15.0 per cent for a basic rate taxpayer before tax. But the cost to the societies including their composite tax payment is only 13.3 per cent. Were they to offer this amount to their depositors gross they might benefit from having a more attractive rate to offer to non-taxpayers, who are at present not entitled to reclaim any part of the tax deducted under the composite rate arrangements. But depositors liable to tax account for the greater proportion of building society funds and are likely to be more responsive to interest rate differentials than those who pay no tax. Without the composite arrangement the building societies would therefore almost certainly have to pay a higher gross rate. This would seem likely to be reflected mainly in the rate charged for mortgages, which at present are held below the market clearing level, rather than in a reduction in their margins.   (Wilson Committee 1980, para. 696)

The nature of building society advantages and operations in the savings market has been a key aspect of their dominance of mortgage finance up to the 1980s. Their comparative advantage sheltered the societies from general money market competition. Interest rates could be set depending on the level of activity the societies wanted. In particular, mortgage interest rates could be pitched slightly below general market rates to exclude potential competition from other financial institutions and to encourage excess demand for mortgage funds. Excess demand enabled the societies to minimize the risk of mortgage default and was useful politically as it could be claimed to demonstrate their popularity.

The new volatility of the housing market and societies' net receipts broke this successful formula in the mid-1970s. They forced the societies to raise their interest rate structure towards levels which enabled competition to break down their control. To offset potential shortfalls in income, term shares were introduced. The extra interest offered on them would, for the crucial interest rate sensitive portion of investors, help to reverse the societies' long-term loss of interest rate competitiveness. Furthermore societies wanted to start tapping the wholesale money markets to cover marginal shortfalls of funds. Both term shares and intervention in the wholesale money markets could only be financed by raising mortgage interest rates. The alternative of squeezing the interest differential on which their money-dealing activities are based was obviously unattractive. The societies tried to raise mortgage rates in ways which would least affect the demand for mortgages.

In the absence of alternative sources of mortgage finance, larger mortgages are likely to be less price elastic than smaller ones. So the importance of differential mortgage interest rates grew, whereby progressively higher rates of interest were charged on larger mortgages. Differentials had the additional advantage of avoiding the political opprobrium of declaring higher basic mortgage interest rates.

The attempt by building societies to stabilize their receipts consequently gave the clearing banks a chance to re-enter the mortgage market. The building society monopoly had lasted so long that they seemed to have forgotten about the threat of competition. The differential mortgage rate strategy projected precisely the most attractive sector of the mortgage market into the grasp of the banks. Whether banks remain a significant force in the mortgage market will depend on whether building societies can rediscover a formula for collecting sufficient cheap retail savings to enable them to isolate the owner-occupied mortgage market again.

In 1982 the state of play was that building societies did not know what to do. Movements in interest rates had put them at a temporary advantage but rationalization and diversification were the order of the day. They are being squeezed on both sides with problems in expanding, or even sustaining, their mortgage business and in generating enough income. The new competition in the mortgage market, however, has added a political dimension to their problems. Disquiet over the potential destabilizing effects of the mortgage market on monetary control has grown and the advantageous tax position of societies seems less justifiable the more they diversify into non-housing areas. A sizeable proportion of the £10 billion lent to owner occupiers in 1981, for example, indirectly financed non-housing consumer spending, according to the *Bank of England Quarterly Bulletin* (September 1982). Not only does this raise the spectre of monetary control of housing finance but also threatens the tax relief on mortgage interest. The justification for tax relief on owner-occupied mortgage interest payments alone declines when much of that borrowing finances non-housing expenditure.

As building societies diversify, the more difficult it will be for them to justify their own special tax arrangements. Someone may dust off the Wilson Committee's recommendations for the abolition of their composite tax arrangements and the special low corporation tax rate they pay (40 per cent against the usual 52 per cent). The Wilson Committee commented on the composite arrangements that

> The Treasury have told us that they would be unlikely to favour such a system now if they were to be starting afresh, and the Inland Revenue have resisted the extension of similar arrangements to other institutions such as the Trustee Savings Banks.
>
> (Wilson Committee 1980, para. 697)

Building societies, therefore, are not only facing economic problems but political problems as they are being forced to confront strong, entrenched interests as they try to overcome their current difficulties. The special tax status of the mortgage market and of building societies, in particular, is likely to be increasingly criticized by some influential opponents as building societies diversify. Other types of financial capital will resist the threat generated by the current problems of the societies. The political power of the City to influence government policy as a result might well be added to the growing chorus of demands for the reform of housing finance.

Building societies and the housing market

The building societies' problems started to mount in the early 1970s at the same time as the owner-occupied market became more unstable. This is not surprising because the mortgage market and the housing market are closely interlinked.

The most obvious question to ask concerning the interlinkages is the effect of mortgage finance on the level of activity and rate of price change in the housing market. Monetarist-style claims that increases in the availability of mortgage credit precede and cause house price inflation have a long pedigree (cf. Gough 1975a and Mayes 1979).[2] Empirically this view was contradicted by the end of severe mortgage rationing in 1980–1. House prices did not rise despite plentiful mortgage finance and a historically low house price/income ratio by the end of the period. Theoretically, the argument ignores many of the interlinkages between the two markets and conflates correlation with cause.

There is a close correlation between increases in savings balances with building societies (net receipts and interest credited) and the level of mortgage advances by building societies, as can be seen in figure 10.1. There is a lagged relationship because the societies cannot instantly lend money. If changes in competitiveness were also plotted for the period shown in figure 10.1 it would be seen that increases in balances are closely related to the state of building society interest rate competitiveness. These correlations could be read as implying a simple causality running from competitiveness through increases in balances to mortgage advances. But the matter is more complicated. The relationship between balances and advances is fairly fixed (the only other thing building societies can do with their money is to increase their liquid balances). So it is feasible for the causality to be in the reverse direction: building societies fix interest rates to investors to provide a level of funds commensurate with the demand for advances: in which case advances determine balances. The causality of balances determining advances, in other words, assumes that there is always an excess demand for mortgages at the prevailing rate of interest, which is not necessarily true. The sharp fall in advances in 1974, for example, could have been a product of the slump in the housing market instead of the slump being caused by a shortage of mortgages.

The importance of not assuming a simple causality of mortgage availability determining the level of housing market activity is reinforced by the fact that building society receipts are influenced by the level of activity in the housing market. As owner occupation has increased in size it has come to affect substantially the level of building society income; in particular the increased volatility of the housing market has made building society income more unstable. Table 10.5 shows the variation in the relative importance of the three main sources of building society funds from 1977 to 1981. It can be seen that the importance of each source varies considerably. Net receipts, for example, represented 53 per cent of the total in 1977 but only 27 per cent in 1981. The changes depend on the housing market as well as on competition in the liquid savings market.

Repayments of mortgage principal, the first row of table 10.5, depend on the level of sales of the existing stock. Obviously when turnover in the housing market is high the greater are repayments. In the last quarter of 1974, for example, approximately 60,000 existing dwellings were sold, whereas in the last quarter of 1977 over 180,000 were sold, three times as many (figure 11.3). Some existing owners will not take out a further mortgage with building societies, releasing funds for other borrowers. The sharp 47 per cent rise in repayments between 1980 and 1981, shown in table 10.5, presumably illustrates the extent to which existing owners were then switching to banks for their mortgage finance. Most owners, however, will take out a further mortgage with the building societies. So the extent to which building societies are involved in turning over their existing mortgage business depends on the state of the housing market. The difference between gross and net advances brings out this relation. Net advances are gross advances minus repayments of principal. The difference between gross and net advances is highest during booms, whilst during downturns it narrows as sales of existing houses drop the fastest. This effect is shown clearly in the quarterly data for building society mortgage advances from 1970 to 1981 given in figure 10.1.

It can also be seen in figure 10.1 that there has been a secular increase in the difference between gross advances and net advances. This means that the level of building society business is becoming more and more associated with funding the moves of existing owners. As the ability of existing owners to move depends on the existence of an active housing

Table 10.5   Sources of building society mortgage funds, 1977–81

		1977	*1978*	*1979*	*1980*	*1981*
Mortgage capital	£m.	2,789	3,619	3,832	3,892	5,708
repayments	%	31	43	40	35	45
Interest credited to	£m.	1,377	1,512	2,254	3,343	3,585
investors' accounts	%	15	18	23	30	28
Net receipts	£m.	4,722	3,310	3,515	3,816	3,474
	%	53	39	37	35	27
Totals	£m.	8,888	8,441	9,601	11,051	12,767
	%	100	100	100	100	100

Source: BSA Bulletin

market, building society business becomes increasingly locked into the cycle of boom and slump in the housing market.

The impact on net receipts of housing market fluctuations is more complicated. As was shown earlier it is the larger sums of money (£2000+) that are most sensitive to comparative interest rates. New savings funds of such magnitudes are unlikely to be derived from savings from income but instead from legacies, maturing insurance policies and the sale of personal assets. A signficant element of those sales will be of houses. 46 per cent of personal net wealth in 1976 was represented by the value of dwellings, so much of the flow of funds for which building societies compete is derived from the housing market itself. In order to be saved elsewhere the money realized has to be withdrawn from the housing market. A large proportion of sales of existing houses do lead to such withdrawals. Of the estimated 1 million house sales in 1979, for example, 270,000 involved sales resulting from deaths, household dissolutions, moves to other tenancies or sales of rented property. Each of them releases funds from owner occupation that are available for saving. The Stow Report (1979) calculated that these last-time sellers realized £4500m. from the owner-occupied market in 1979 (75 per cent up on the 1975 figure because of intervening house price inflation). Trading down by elderly owners constitutes an additional source of housing-based funds available for saving. The funds flowing out of the housing market are the main source of the large sums of personal sector liquid assets available for investment. In contrast, maturing life policies, the other principal source, yielded an estimated £2500m. in 1979: only 55 per cent of the housing market sum. The building societies reckon that they have attracted up to a third of the proceeds of house sales in the past (Stow Report 1979). Net receipts, therefore, are to an extent dependent on the housing market.

The interdependence of net receipts and sales of existing housing has grown as owner-occupied housing has expanded as a source of personal wealth. Moreover, the ability of last-time sellers to sell their houses and the prices they can get depend on the cycle of boom and slump in the housing market. More money therefore will be drawn out of owner occupation in this way during booms than during slumps. So whilst interest rate competitiveness influences the proportion of funds that are invested with building societies, the amount of funds available for investment from which that proportion is derived is dependent on the level of activity in the housing market. Building society mortgage advances help to finance the movement of funds out of house ownership (possibly via a chain of sales), so a part of building society net receipts paradoxically are financed by their own net advances.

There is a multiplier effect to building society lending with one loan helping to generate the conditions for further loans. The size of the multiplier, however, is highly variable depending as it does on interest rate competitiveness, the number of last-time sellers and the amount of money they manage to realize.

A conclusion of the analysis of building society income consequently is that lending money for house purchase helps to create the funds that can be borrowed for house purchase. This has not only helped to sustain the expansion of owner occupation but also has exacerbated the instability of the housing market, as the sources of funds for building societies are dependent on the level of activity in the housing market itself. To an extent, therefore, booms generate their own sources of mortgage finance which help fund the price rises and feed the expectation that those price rises will continue. Similarly, slumps themselves contribute to the shortage of mortgage funds that can perpetuate those slumps. Sources of income for building societies cannot be seen solely as an external constraint imposed upon the housing market; a prey to credit squeezes and interest rate increases. The latter are obviously important but are not the end of the story. Building society finance is instead part of a housing system that generates its own volatility.

Conclusion

Building societies have prospered and expanded with the growth of owner occupation. Through their preferential political treatment with respect to monetary policy and taxation they were able successfully to isolate the mortgage market from the rest of the financial system. They succeeded, in other words, because they did not threaten the economic interests of other financial institutions. Mortgage business was not very attractive to other forms of money-dealing capital nor did the building societies syphon funds away from other institutions. Instead it helped to tap new sources of personal savings by appealing to the non-banking sections of the population and by recycling funds within the owner-occupied market. The government might have lost some potential revenue from a loss of national savings but for most of the post-war era that loss has been small compared to the policy gains of privately funded housing finance.

This conclusion is important as it puts a political dimension on the debate over whether owner-occupied housing finance 'crowds out' alternative investment outlets and forces up the personal savings ratio (as wrongly claimed, for example, by Kilroy 1978). Increases in saving with the building societies do not necessarily imply increased saving in the personal sector as a whole, because saving by one person with a building society usually will be matched by expenditure from another person via a mortgage advance from a society funded by that saving. Building societies, in other words, deal in flows of funds in the personal sector.

The flows of funds in which they deal, moreover, have generally been overwhelmingly restricted to transactions associated with owner occupation. Building societies usually get about two-thirds of their income from the owner-occupied market (Stow Report 1979). So they are money dealers who deal predominantly with owner occupiers' and ex-owner occupiers' housing costs and money gains. In general, they do not divert funds to housing but deal in the costs and gains of owner-occupied housing. As no other form of financial capital has been crowded out by these activities of the building societies, they politically have had an easy life to date.

The mounting crisis of the structure of owner-occupied housing provision, however, has broken down the isolated position of owner-occupied housing finance and brought the building societies into direct conflict with other forms of financial capital. The future of mortgage finance in Britain is now in the melting pot. The importance of this change is likely to be far more important politically than in direct economic terms. The political consensus, or at least an acquiescence to the *status quo*, that has characterized the financial communities' attitude towards owner-occupied housing mortgage finance has gone. The direct economic consequences are less certain as mortgage credit has been supportive and reactive to changes in the housing market rather than a direct influence. Money dealers have taken advantage of housing costs associated with owner occupation rather than caused them.

11

The unstable housing market

What has been stressed in previous chapters about the owner-occupied housing market is how unstable it has become since the early 1970s. Periods of house price explosion are followed by periods of stagnant prices. The level of sales shows similar volatility although it does not closely correspond to movements in prices as will be shown later. Many features of housing market transactions, in fact, illustrate this instability: housing starts and completions (chapter 4), for instance, or building society receipts (chapter 10) and the house price/average earnings ratio (figure 11.1). Their causes obviously are interconnected and this chapter argues that they are related to the growing importance of existing owner occupiers in the housing market, which creates a destabilizing bunching of transactions.

House price movements obviously are an important element of housing market change. Short-run price formation is a product of the interaction of supply and demand. The characteristics of the housing market affecting short-run price change have been well documented elsewhere (HPTV II and BSA 1981a). Housing is a durable, expensive commodity which everyone must consume in one form or another, so demographic factors have an important effect on the overall level of demand. The level of real incomes similarly influences how much people can afford to pay, whilst mortgage credit must be available to finance desired purchases. The cost of owner occupation relative to other tenure forms plays a part in determining which tenure is chosen (at least, for those with an effective choice). On the supply side new supply is a small proportion of the total

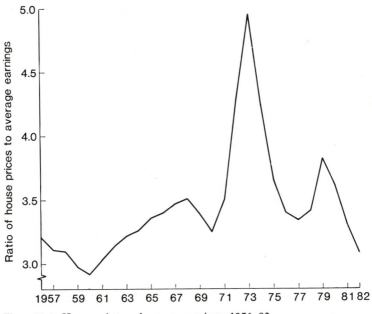

Figure 11.1 House prices and average earnings, 1956–82

Source: BSA Bulletin

stock (1 to 2 per cent) and it is unresponsive in the short term to price movements because of production lags, whilst over the longer term its price is gradually rising (see chapter 4). Quite how all these elements link together, however, is not so clearly understood, leaving much room for housing market pundits to predict the time of the onset of the next bout of house price inflation.

Market volatility is a characteristic feature of markets for durable commodities where there is a large existing stock in use. The markets for such commodities are the places where adjustments to stock demand and supply take place. Those adjustments may be small in terms of the total existing stock but in terms of the commodities actually traded they can have a substantial numerical impact. Less than 10 per cent of the total owner-occupied stock is traded in each year, so an imbalance between total housing stock demand and supply has a significant effect in the market. These imbalances can be created through adjustments to longer-term trends, like changing household formation or rising real incomes, but superimposed upon them are much shorter-term variations. Because of the slow responsiveness of housing supply most of these short-term variations are demand based. The most reliable correlation with house price movements econometric studies have found, for example, is the contemporary rate of change of real disposable income (BSA 1981b).

The owner-occupied housing market is different from other durable commodity markets in a number of crucial respects which add to its volatility. A large proportion of demand and supply is combined in the moves of existing owner occupiers. By the earlier 1980s (when over half the total UK stock was owner occupied) about half of the houses sold each year were put up for sale by moving owner occupiers. Another quarter or more sales involved household dissolutions. They again could lead to new purchases by reconstituted households or, as the last chapter showed, to increases in the pool of mortgage credit. The extent of the interconnectedness of demand and supply is unique to owner occupation. Another feature which adds to this market's instability is the heterogeneity of supply: no two houses are exactly identical in dwelling characteristics and location. This makes the market balancing act particularly precarious.

Both characteristics of the housing market are illustrated in the notion of a chain of purchases and sales. Most owner occupied transactions are part of interconnected chains where a series of buyers and sellers have to make successful moves before any one of the transactions can be completed. The length of a chain depends on the number of existing owner households and houses within it. The shortest chain has a single link, when a first-time purchaser not dependent on funding from a house sale buys a new house. The longest chains involve successive purchasers and sales by existing owners. Within chains of sales moving owner occupiers may be 'trading up' to a more expensive dwelling or 'trading down' to a cheaper one. This requires additional balancing relations. Households can only trade up when new higher-quality dwellings are available for them to do so or when other owner households trade down or leave the tenure, vacating part of the higher-quality stock.

Exploring the impact of the interconnections in the owner-occupied market in causing its instability is a major theme of this chapter. In doing so the role of first-time buyers and existing owner occupiers has to be distinguished. This necessitates looking at their relative housing costs and incomes, and the importance of money gain in facilitating existing owner moves.

Transactions in the owner-occupied housing market

The types of household and dwellings involved in owner-occupied transactions, and the flows of houses and households implied by those transactions, are summarized diagrammatically in figure 11.2. The resemblance to a Heath Robinson type plumbing system is deliberate. The owner-occupied housing market constitutes the mechanism that both enables and regulates the flow of houses and households into and out of owner occupation. Transactions in it, therefore, reproduce and replenish a stock of dwellings and their aggregate money value and also sustain a body of owner-occupier households and their willingness to accept the housing conditions and financial payments involved. Whilst transactions each year represent only a small part of the whole stock, their smooth flow is essential to the economic and social reproduction of that total stock. The analogy to a plumbing system highlights this flow/stock relationship and enables the complex ways in which flows in the market relate to the stock to be expressed in a simple diagram.

Treating the owner-occupied market metaphorically as a set of pipes through which transactions flow also reveals certain of its other features. Owner-occupier households and houses are discrete entities. Once built, the stock of houses is physically fixed; it never flows anywhere. Households move but only when they acquire or lose title to an owner-occupied house. The market consequently does not involve physical flows, it just enables them to happen; subsequently as household moves, or previously as housebuilding or renovation. What flows through the market are legal titles to house ownership, the prices households are prepared to pay for them, and the revenues derived from those transactions. They are all discrete entities which flow, continuing with the metaphor, because they are held in suspension in a fluid of money credit and household savings.

Within the market there are a series of agencies that regulate and assist the flow of transactions (or, as in the case of conveyancing monopoly, impose themselves within the system). Newspapers, magazines and estate agents spread information about transactions. Solicitors undertake conveyancing. Building societies, insurance companies, banks and mortgage brokers arrange and process mortgages. Yet others are involved: surveyors, the Land Registry and insurance policy dealers to name a few. Not all are essential to each transaction, but all regulate and sustain the aggregate flow; an important role in a market where transactions are linked in chains and where failure of one means the collapse of all transactions in the chain. These agencies can be regarded as the valves of the market system: filtering out potential transactors who cannot fulfil certain criteria, enabling flows to be directed in a harmonious way, smoothing out surges of supply and demand by creating queues which

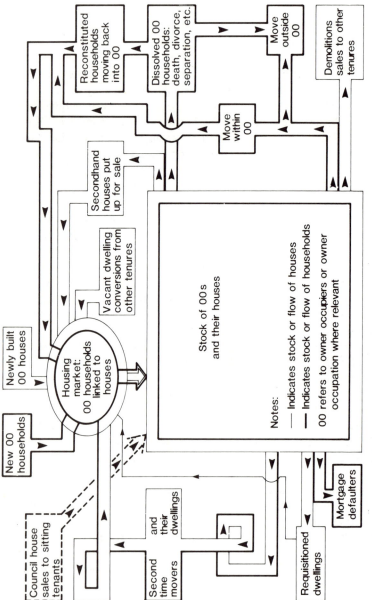

Figure 11.2 Transactions in the owner-occupied housing market

Newly built OO houses

New OO households

Council house sales to sitting tenants

Reconstituted households moving back into OO

Dissolved OO households: death, divorce, separation, etc.

Move outside OO

Demolitions sales to other tenures

Move within OO

Secondhand houses put up for sale

Vacant dwelling conversions from other tenures

Housing market: OO households linked to houses

Stock of OOs and their houses

Notes:
— Indicates stock or flow of houses
▬ Indicates stock or flow of households
OO refers to owner occupiers or owner occupation where relevant

Second time movers

and their dwellings

Requisitioned dwellings

Mortgage defaulters

otherwise would cause prices to oscillate frantically, ensuring the legality of transactions, and so on. Yet they are not only advantageous. Like most valves they are expensive, of varying efficiency and effectiveness, tend to jam up when overloaded, and it takes time to pass through them. House purchase is not only the most expensive transaction that most households undertake but must also be one of the longest. In an age of instant electronic information, house transactions are inefficient and slow. Just one stage, mortgage approval to mortgage completion, generally takes 2–3 months. The time it takes to pass through the housing market can lead to transactions being negated, most spectacularly through 'gazumping', and to chains of transactions being broken. These 'valves' can consequently heighten fluctuations (especially of prices) as well as smooth them out.

Three aspects of transactions in the housing market must be examined to see why demand tends to bunch into periods of boom followed by slumps. They are, first, the means by which transactions are financed; second, the role of moving owners' money gains in their moves, and, finally, the subsequent ongoing mortgage costs faced by first-time buyers and moving owners.

Financing house purchase

The purchase of any commodity obviously involves a transfer of a sum of money in one form or another from the purchaser to the seller equal to the commodity's price. If markets are aggregated across the economy as a whole it can be seen that purchases involve a circular flow of funds: firms, for instance, hire workers who use part of their incomes to buy the commodities they produce. This type of interrelation forms the basis for Keynesian-style explorations of effective demand. Yet for most commodities there are insignificant feedback effects at the level of the individual market. Bread workers, for example, buy only a tiny proportion of the bread they produce. Such feedback effects in the owner-occupied market, however, are substantial, occurring both with the proceeds of existing owners' house sales and in the funding of the mortgages advanced for house purchase.

Financial interconnections between transactions in the housing market mean changes in the level of market activity and house prices cannot simply be looked at by examining the behaviour and constraints faced by individual purchasers, suppliers and financial institutions. They only give a partial picture that ignores the interdependencies between each of them. The simplest way to start the explanation of these interconnections is to describe a situation where they do not exist. If a first-time buyer buys a newly built house using a mortgage, there is a flow of funds from the purchaser and the building society from which they borrow to the housebuilder. In this case the money flows directly out of the housing market as the builder uses it to finance labour and materials costs, profits, interest payments and perhaps the purchase of more land. But if the first-time buyer buys a house from an existing owner occupier who wants to buy another house, the money stays in the housing market to help finance further transactions. Here the funds used by the first-time buyer are subsequently used by the seller to pay off his/her own existing mortgage and to convert into a monetary form, at least momentarily, the wealth he/she had tied up in the house he/she sold (from his/her original downpayment on it and any money gain accruing during the period of ownership). What happens next in the chain of sales depends on the type and price of the house purchased by this existing owner and the means by which he/she finances it.

If an existing owner occupier's house is bought for the same price as the one sold, using the same sized mortgage as before, the effects are easy to trace. Looking first at the mortgage finance implications, one mortgage is paid back and another equivalent mortgage simultaneously taken out. The net effect on total mortgage advances of the move consequently is zero. The need for additional new mortgage loan finance in the illustrative example, so far, is just that of the first-time buyer; yet it has enabled two moves to take place. The example could be extended further and further: if a chain of sales extends to other similar owners who move under identical financial circumstances it could go on forever. More realistically the funds that start off a chain of sales, or subsequently enter it, eventually leak out of the housing market. This occurs for three reasons:

(i) the purchased house is new to owner occupation;
(ii) a moving owner occupier cashes in some of his/her money gain by taking out a higher than necessary mortgage;
(iii) the purchased house is sold by someone trading down or as a last-time sale caused by a household dissolution or the household moving to another tenure or emigrating.

Injections of funds by households into the housing market consequently have multiplier effects in terms of the number of transactions they facilitate. The size of the multiplier depends on how long it is before the initial financial impetus leaks out of the owner-occupied market. As every chain of purchases and sales involves a unique sequence of moves the speed of the leakage is highly variable, and consequently so is the multiplier effect. All that can be generally said is that the multiplier is likely to be greater during market upturns as the number of second-hand sales in any chain is likely to be higher.

The existence of a transactions multiplier means that not all transactions in the housing market require new sources of finance to enable them to occur. The need for injections of new funds depends as much on the types of transactions that occur as on their volume. Moreover, if there are few leakages of funds out of the owner-occupied market a large number of transactions can be financed at progressively higher price levels with quite small injections of new funds. A surfeit of mortgage advances, therefore, is not a necessary prerequisite for a house price boom.

Whether a new financial injection leads to a greater turnover at fairly constant prices or helps to spark off a house price explosion depends on the types of chain generated. Chains of sales require buyers to be matched with sellers quickly and harmoniously. But the needs of purchasers in terms of trading up or down or moves between regions, for instance, can frequently be at variance with what is available for purchase; especially where chains involve a large number of existing owners and so are interlinked in complex patterns of transactions.

The notion of financial multipliers in the housing market can be extended to the injection of funds themselves. Many building society net advances are financed by money that is withdrawn from the housing market, as the last chapter showed. So transactions in the housing market help to generate the funds that start off other chains of transactions. Again this multiplier effect is variable in size as the proportion of net advances funded from the housing market depends on how much of the deposits of the liquid assets of last-time sellers the building societies manage to attract. But, as chapter 10 concluded, the multiplier is likely to be higher during booms than slumps.

The recent state of the market will have a strong influence on the likely success of transactions at stable prices. When the housing market is coming out of a period of slump, for example, it is probable that many potential moves have been frustrated for some time and that there will be few financial leakages out of the market. So long chains and many transactions are possible at fairly stable prices. But once a few buyers cannot find the house they are looking for, or begin to feel that shortages are appearing, prices start to be bid up. As prices start to rise subsequent transactions may require additional injections of funds. Ultimately there are widespread breaks in chains and a market downturn. This is not a complete explanation of booms and slumps, but it does highlight the fact that there is no tendency towards an equilibrium, market clearing price for owner-occupied housing. Oscillations of varying temporal amplitude between market ceilings and floors are endemic.

The variation in the level of housing market transactions is substantial. Accurate data are not available but, for the 1970s at least, building society mortgage data are a good approximation because they funded most transactions. In the early 1980s with the large-scale entry of the banks their data are less accurate. Figure 11.3 shows quarterly data for building society advances from 1970 to 1981. Advances to existing dwellings (which mainly are already in owner occupation) and new dwellings are given separately. There are two opposite trends over the period for existing and new dwellings. If rough trend lines are drawn through their two frequency distributions they show a doubling of purchases of existing dwellings and a halving of purchases of new ones over the period. The fall in new house sales shows the extent of the crisis of housing production, whereas the growth of existing dwelling sales has greatly exceeded the overall growth in the owner-occupied stock itself (figure 4.1). Owner-occupied houses, in other words, were being turned over at a faster rate over the eleven-year period.

Advances to existing dwellings are also volatile around their trend change. The first boom, which peaked in 1972, and the subsequent slump show the greatest deviations from trend but there are also three later noticeable cycles (figure 11.3). If the variations in turnover are compared with the rate of change of house prices given at the foot of figure 11.3, it can be seen that only in the first house price boom of the 1970s did turnover and house price change together. This has led to the erroneous belief that the volume of mortgage advances was the sole determinant of house price rises (cf. Gough 1975a) but data from the rest of the decade discount that view. In 1975 and 1976 market activity rose well above the levels of the 1972–3 house price boom, but without sparking off a new price explosion. Conversely, the rapid increase in house prices during 1978–9 took place in a period after the turnover of existing dwellings had peaked and was falling rapidly (inclusion of bank advances, however, may conteract that fall to an extent).

One reason for the sharp variations in purchases of existing houses is that a failure to move by an existing owner occupier leads to the withdrawal of both a purchase and a sale. If owners cannot get a satisfactory

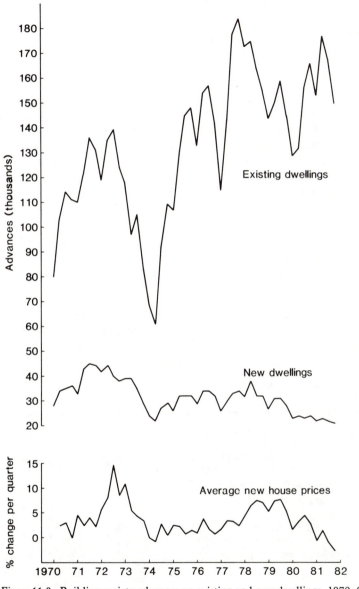

Figure 11.3 Building society advances on existing and new dwellings, 1970–81

Source: BSA Bulletin

price for their existing houses they hold off from the market until they can. Much of demand and supply, therefore, is balanced off. Housing being sold by those leaving owner occupation, trading down or acting for legatees, is also likely to be held off the market during temporary downturns, as the whole object of such sales is to get a 'realistic' price for the house. The market cannot freeze up forever but such holding operations can be sustained for long time periods as the cost of the alternative (massive price cutting) is too great. When market conditions improve a whole series of previously frustrated moves consequently take place, resulting in greatly increased levels of turnover.

Most moves by existing owner occupiers are 'voluntary' in the sense that employment relocation or changed financial circumstances do not force a move. A suvey by the Nationwide Building Society of people who borrowed mortgages from it in 1981, for example, found that 83 per cent of existing owners only moved 25 miles or less (60 per cent as little as 5 miles or less), and that changed employment location accounted for only 16 per cent of owner moves (table 11.1). Existing owners on the whole, therefore, do have some flexibility over the timing of their moves and they will move when market conditions make it most attractive.

Table 11.1   Existing owner occupiers' reasons for moving, 1981

Reason for moving		% of moving owner occupiers
Change in personal circumstances:		33.6
Getting married	2.5	
Change in family size	9.0	
Change in income	6.4	
Change in job/work location	15.7	
To obtain better or more suitable property:		42.5
Present accommodation too small	25.3	
Present accommodation too large	3.8	
To obtain more modern property	9.3	
To obtain garage or garden	4.1	
Location of property:		17.0
To be nearer work	5.1	
To be nearer amenities	1.8	
To be nearer friends/relatives	2.4	
To move to better neighbourhood	7.7	
Other reasons:		6.9
Total		100.0

Source: House Buyers Moving, Nationwide Building Society Survey, September 1982

For many existing owners the time of moving is important because inflation affects the real costs of 'trading up'. After a few years of ownership the front-loading effect of mortgage repayments with general price inflation means that a household can trade up by raising their mortgage/income ratio back towards its earlier level. They can also transfer from their previous dwelling the money gain derived from the intervening house price inflation. Both characteristics of inflation tend to encourage the bunching of owner moves. The front-loading aspect is reactive: once moves are bunched it tends to reproduce the bunching again in the future, as profiles of the real incidence of mortgage costs on incomes becomes similar for the large number of people who move during a particular boom. The temporal incidence of money gain similarly depends on when a house is bought. But its size depends on the rate of house price inflation rather than on general price inflation. When house prices rise slowly it obviously takes longer to build up a substantial money gain than when they rise fast. A hypothetical example illustrates how great the variation can be. If someone bought the 'average' house with an 80 per cent mortgage advance and sold it after four years the net money sum realized on sale would have varied during the 1970s from 50 per cent to as much as 150 per cent of the initial purchase price (see table 11.2). The potential aggregate impact of money gains on the housing market consequently varies quite sharply over time, and it is related to the time of booms and slumps reinforcing the bunched effect of existing owner moves.

Mover owners transfer much of their money gain to their new dwellings, so examination of variations in the percentage mortgage advance to former owners shows how important is the variation in the temporal incidence of money gain in the housing market. As figure 11.4 shows, the drop in the percentage mortgage advance to former owners has been substantial, falling on average from two-thirds of purchase price to a half in only ten years, during which time house prices have more than trebled. Yet, because of the importance of the timing of purchase just described, there has been a ratchet-type effect within the overall decline. There are sharp falls in the mortgage percentage during house price booms and their immediate aftermath (that is, 1971–3 and 1978–9) which are then followed by slight rises. The reason is simple. During the booms, money gains rise fast so that, even though prices were doing the same, the proportion of purchase price financed by mortgage falls. The proportion continues to fall initially in the subsequent slump because those second-time buyers able to sell their properties in the depressed market still have money gains acquired in the previous boom. The longer the period after the boom, however, the smaller is the relative number of previous owners with large money gains who are involved in housing transactions. All those who transacted during or after the boom would either not move again for a while, or have realized part of their money gain, or be recent owner occupiers whose money gain is low. The result is that owner movers with small money gains gradually form an increasing proportion of previous owner-occupier transactors after the initial onset of a market slump, raising the percentage mortgage advance. The percentage advance does not rise to the pre-boom level, none the less, because much of the overall money gain remains in housing and thereby still influences transactions. So the increase in mortgage percentage levels off at a new lower-than-pre-boom level. The same process happens again during the next boom and its aftermath.

The ability of previous owner occupiers to use money gains to purchase houses shows in part why house price booms can be self-sustaining, and why long periods of sluggish price rises result once a boom is over. Once house prices start to rise, existing owners' money gains enable them to purchase at the increased prices without incurring an equivalent increase in mortgage debt. Once the boom is over there

Table 11.2   Money realized by selling a house, 1970–81[1]

	Purchase price 4 years previously £	80% mortgage advance £	Sale price £	Money realized £	Money realized as proportion of initial house price
1970	3,850	3,080	5,000	1,920	0.50
1971	4,080	3,264	5,650	2,386	0.58
1972	4,340	3,472	7,420	3,948	0.91
1973	4,660	3,728	10,020	6,292	1.35
1974	5,000	4,000	11,100	7,100	1.42
1975	5,650	4,520	11,945	7,425	1.31
1976	7,420	5,936	12,759	6,823	0.92
1977	10,020	8,016	13,712	5,696	0.57
1978	11,100	8,880	15,674	6,794	0.61
1979	11,945	9,556	21,047	11,491	0.96
1980	12,759	10,207	24,307	14,100	1.11
1981	13,712	10,970	24,810	13,840	1.01

Source: BSA Bulletin and own estimates

[1] Hypothetical examples; see text for method

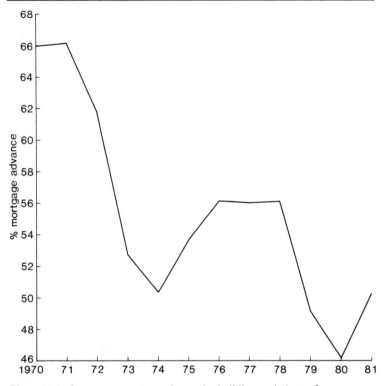

Figure 11.4 Average percentage advance by building societies to former owner occupiers, 1970–81

Source: BSA Bulletin

will be little money gain around to help fund price rises, possibly for a number of years. After a time, however, enough money gain will exist to help sustain another boom.

Income and house prices

So far emphasis has been placed on the economic forces that help to generate instability in the housing market. The need for some form of housing and the level of personal incomes puts limits on the magnitude of the fluctuations associated with this instability. Incomes act as a limit, stopping house prices from escalating for ever, and also help to create the conditions for an improvement in demand in the depths of a market slump. The effect is different for first-time buyers and previous owners as the former have no money gains from ownership so their house deposits will consist solely of personal savings.

First time buyers

Table 11.3 shows mortgage and income data for the 'average' first-time purchaser buying the 'average' first-time purchaser's house. Income distributions are skewed so most new purchasers have less than the average income and hence tend to be in a worse financial position than the 'average', but average data do give some indication of the temporal variation in costs. The table shows that mortgage repayments, not surprisingly, varied over the 1972–9 period in relation to the mortgage

Table 11.3   Average male earnings and first-time purchasers' initial mortgage outgoings, 1972–9

		1972	1973	1974	1975	1976	1977	1978	1979
Average recorded income of first-time purchasers[1]	(£)	2281	2734	3231	3753	4285	4800	5283	6290
Average male earnings: gross[2]: manual	(£)	1710	1980	2270	2900	3390	3720	4200	4840
non-manual	(£)	2260	2500	2830	3560	4240	4620	5240	5880
net[3]: manual	(£)	1430	1800	2070	2510	2880	3260	3690	4390
non-manual	(£)	1770	2160	2440	2940	3440	3860	4380	5120
Gross mortgage outgoings[4]	(£)	552	700	890	942	972	1035	1088	1534
Mortgage interest rate[5]	(%)	8	9½	11	11	10½	10½	9¾	11¾
First-time buyers' income as proportion of: manual		1.33	1.38	1.42	1.29	1.26	1.29	1.26	1.30
non-manual		1.07	1.09	1.14	1.05	1.01	1.04	1.01	1.07
Gross mortgage outgoings as % of average net income of: manual		39	39	43	38	34	32	29	35
non-manual		31	32	36	32	28	27	25	30

Source: Department of Environment

[1] *Source:* DoE/BSA 5% sample survey of building society mortgages. Recorded income is the income the building society takes into account in assessing the mortgage application; this may include income (e.g. wife's earnings) other than the basic income of the applicant

[2] *Source: Department of Employment Gazette;* New Earnings Survey estimates (April) – full-time men over 21, all industries and services

[3] Calculated net of income tax (married man's allowance, no children) and tax relief on mortgage interest in first year of mortgage

[4] First year of 25-year mortgage for 80% of average first-time purchasers' dwelling price at the given interest rate

[5] BSA recommended rate as at 30 June

interest rate and the rate of house price rises. If the standard 'ideal' of a first-time purchaser (i.e. male, married and no children) had had an income equivalent to the average income of a male manual worker throughout that period, mortgage repayments initially would have cost between 32 and 43 per cent of disposable income; average non-manual incomes fared 4–8 per cent better. These high percentage outgoings indicate the drain on disposable income involved in the initial years of house purchase and the need to rely on joint earnings, especially for manual workers. The period after a boom when both house prices and interest rates are relatively high is the worst, viz. 1974 and 1980 (although the latter is not shown in table 11.3). During 1980 initial repayments reached the astronomical height of 45 per cent of average earnings before tax (*BSA Bulletin* 27).

Booms and their aftermaths, however, do not simply alter the incidence of housing costs but appear also to change the social composition of first-time buyers. The first row of table 11.3 gives average recorded income of first-time buyers, that is the income that the building society takes into account when assessing the mortgage application. It is likely to exclude or understate, therefore, income which the society does not regard as regular and reliable (e.g. overtime payments, women's earnings, etc.). It can be seen none the less that the size of this income in relation to both average manual and non-manual incomes varies considerably over time, increasing during a boom and its aftermath, then falling back. This indicates that lower-income households are squeezed out of house purchase during booms.

It would appear, therefore, that aggregate demand by first-time buyers is counter-cyclical, falling during house price booms but building up again in the ensuing slump. This effect helps to bring booms to an end by choking off part of demand, and also creates the preconditions for a new boom during the slack intervening period. The cyclical variation in demand by first-time buyers has been associated with a fairly static longer-term trend. There was not an overall upward shift in first-time buyer demand in the late 1970s as expected by the Housing Policy Green Paper (HPTV I, p. 129); if anything, demand remained below the levels of the early 1970s. This undoubtedly reflected the impact of house price increases and the financial difficulties imposed on households of lower means whose shifts to the tenure are necessary if owner occupation is to expand proportionately as a tenure. This lack of an increase in first-time buyers has helped to reinforce the relative importance of previous owner occupiers in the market.

Previous owner occupiers

Previous owners tend to take out higher mortgages when buying their new house than do first-time buyers, despite the money they already have tied up in housing (see table 11.4). Previous owners, however, tend to have higher incomes so their mortgage advance/income ratios are lower on average than those of first-time buyers (table 11.4). The ratios indicate, none the less, that repayments still represent quite high proportions of income and that they also vary with the house price cycle.

Higher mortgages plus money gains mean that previous owners are on average dealing in the more expensive sectors of the market. It is noticeable that the difference between the average purchase price for previous and first-time buyers has increased substantially over time: this can be seen for 1970–80 in the last column of table 11.4. The increasing difference has not been smooth, accelerating during house price booms and falling back slightly during downturns. The widening price gap is the result of differing rates of price inflation for the houses bought by each group as well as from a growing ability of existing owners to trade up. Comparison of the increase in average purchase price for new and previous owners (table 11.4) shows that the rate of price increases always tends to be higher for previous owners, and that their house prices accelerate considerably faster during booms than those for first-time buyers. The price series are not 'quality deflated' so both relative price and quality (i.e. trading up) effects are taking place at these times. Only in 1974, the bottom of the mid-1970s slump, did house prices increase faster for first-time buyers than for previous owners, doubtless reflecting earlier high pricing in the sectors of the market dominated by former owners. Mortgage advances to former owners also increase as house prices rise, so that during booms these repayments take an increasing proportion of their income. Booms eventually are choked off by the rising mortgage repayment ratios of the previous owners who move. The mortgage cost effect, therefore, is counter-cyclical like that for first-time buyers.

The interrelationship between money gain, house price rises and previous owners' mortgage/income ratios again shows that money gain cannot be seen solely as an unqualified benefit for its owners. If it is not taken out of the housing market it does enable some physical improvement in housing quality through trading up. But it also fuels house price inflation so that those trading up incur additional mortgage outgoings. As with tax relief on mortgage repayments the secondary effect of money gain on house prices helps to negate the apparent initial advantage.

Table 11.4 Average house prices, mortgage advances and incomes: first-time buyers and former owner occupiers, 1970–80[1]

	First-time buyers					Former owner occupiers					Ratio of average house prices former and new owners
	Average dwelling price (£)	% house price increase p.a.	Average advance (£)	Average income (£)	Average advance/ average income	Average dwelling price (£)	% house price increase p.a.	Average advance (£)	Average income (£)	Average advance/ average income	
1970	4,330	5.7	3,464	1,766	1.96	5,838	7.8	3,854	2,168	1.78	1.34
1971	4,838	11.7	3,914	1,996	1.96	6,666	14.2	4,407	2,466	1.79	1.38
1972	6,085	25.8	4,954	2,281	2.17	8,965	34.5	5,538	2,748	2.02	1.47
1973	7,908	30.0	6,115	2,734	2.24	11,900	32.7	6,273	3,118	2.01	1.50
1974	9,037	14.3	6,568	3,231	2.03	13,049	9.7	6,577	3,618	1.82	1.44
1975	9,549	5.7	7,292	3,753	1.94	13,813	5.9	7,409	4,299	1.72	1.45
1976	10,181	6.6	8,073	4,285	1.88	15,160	9.8	8,509	4,997	1.70	1.49
1977	10,857	6.6	8,515	4,800	1.77	16,246	7.2	9,101	5,558	1.64	1.50
1978	12,023	10.7	9,602	5,283	1.82	18,792	15.7	10,611	6,161	1.72	1.56
1979	14,918	24.1	11,286	6,290	1.79	24,074	28.1	11,837	7,101	1.67	1.61
1980	17,533	17.5	12,946	7,749	1.67	28,959	20.3	13,359	8,688	1.54	1.65

Source: BSA Bulletin

[1] Data for 1981–2 give a misleading impression as they exclude bank mortgage lending

The anatomy of house price rises

Having considered the various elements of the demand side of the owner-occupied housing market, it is now possible to summarize the overall determinants of the level of market activity and the rate of house price increase. From what has been said in earlier parts of the chapter there is no simple or singular answer. House price rises may be caused by a variety of reasons none of which by itself needs necessarily result in price rises. Whether they do or not depends on their impact on the inter-linkages in the market. Earlier sections showed that to a great extent the level of market activity is dependent on the pre-existing level of activity: slumps sustain slumps and booms sustain booms. Underlying each phase is the ability of new and existing owner households to pay the contemporary level of mortgage costs. The rate of price increase then depends on what is going on within chains of purchases and sales, and on the ability of purchasers and sellers to enter into them.

House purchasers want particular types of dwelling (e.g. price, size and location) and suppliers are also offering specific house types as well. When chains of sales are possible that balance off demands and supplies, prices are likely to remain fairly static. When households cannot find the house they want or fear that prices will rise because of a high level of market activity, this induces price rises and multiplier effects rapidly escalate any such initial imbalance. The role of transactions agents is important as they both may slow down transactions (for example, when mortgages get short) and have an incentive to create the impression that prices are about to rocket to ensure that purchasers buy now. When chains of sales break because suppliers cannot find purchasers, the reverse process is likely to happen, with prices stagnating.

There are certain temporal conditions that are likely to create rising, steady or stagnant price changes. In a period of rapidly rising house prices, for example, existing owners who move are able to use the money gains resulting from the inflation to feed further the inflation process, enabling price rises to occur without necessitating proportional increases in mortgage lending. Leakages of such gains also increase the deposit base of building societies so that lending itself can be raised. Rising prices, however, do eventually choke off demand, first from new entrants, then from moving owner occupiers. Mortgage finance becomes scarce, mortgage repayment costs rise and the transactions agents (the 'valves' of the plumbing system) cannot cope with the increased level of transactions. Individual buyers and sellers can be affected directly or because they are involved in a chain of sales in which one of the links gets 'broken'. During downturns, the preconditions for another boom are created. Money incomes, for instance, begin to rise faster than mortgage repayments. There is a gradual build up of money gain by potential transactors, and so on. This is why bursts of rapid price increase are relatively short, lasting from one to two years, and why the subsequent downturns tend to last longer, during which the preconditions for a new boom build up.

There are, however, longer-term trends that show why the market is becoming more volatile. One is that the general cycle of economic activity as a whole has become more pronounced since the early 1970s, with slumps in 1974, then again even more sharply in 1980. But there are two important factors within the housing market indicating that the fluctuations of the 1970s and 1980s are not simply the product of external shocks. There is no indication that the housing market is oscillating round a stable equilibrium path; if anything the opposite is occurring. Instability is being generated by the growing importance of existing owners within the market, and by the long-term decline in housing output.

Figure 11.5 distinguishes mortgage advances to new and existing owners. It can be seen that former owners rapidly increased in market significance over the period from 1970 to 1981, although the graph understates their impact from 1979 onwards because it does not include bank advances. The growing preponderance of moving owners is also dependent on a high level of transactions, so their role varied across the three transactions cycles that occurred during those years.

During the first cycle, first-time buyers dominated the market, although there was a notable rise in moves by former owner occupiers. In the second cycle, despite building societies' avowed preference for first-time buyers during the period, the number of first-time buyers was below that of the earlier cycle and the number of second-time buyers markedly higher. Advances to both groups followed consequently a similar numerical pattern. In the final cycle, advances to both groups initially rose rapidly but first-time buyers peaked earlier and fell off rapidly. A distinct gap opened, therefore, between first-time buyers and former owners, one that is understated by looking at building society data alone. This gap is a new, and possibly permanent, feature of the owner-occupied market. Its size varies but it reflects the simple arithmetic of the rise of owner occupation plus the increased propensity of owners to move.

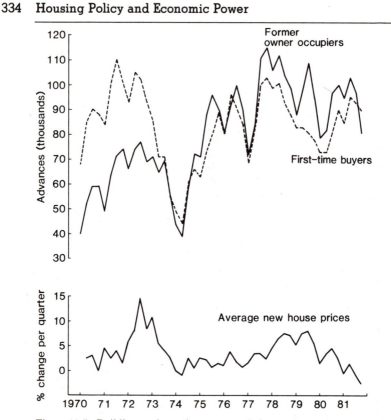

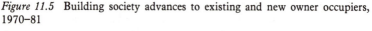

Figure 11.5 Building society advances to existing and new owner occupiers, 1970–81

Source: BSA

An owner-occupied market dominated by existing owners requires longer chains of sales in it than one in which existing owners do not dominate. The longer the chain has to be, the more likely it will not be achieved, as more purchases and sales have to be matched. The problem is compounded further by existing owners using moves to trade up and by the bunching of such moves. Households can only trade up if housing is available for them to do so. That requires other households to trade down or for new high-quality housing to be built. But new housebuilding has fallen substantially over the years so significant shortages of more expensive, middle- and up-market housing appear during booms.

So the problem goes back again to the lack of housing supply discussed in chapter 4. The housing market is kept at such tension because of a lack of new supply. New housing output is too small a proportion of the total stock and it takes too long to build for it to have a significant impact on short-run changes in house prices. But a lack of new housing output has helped create the conditions in which excess demand, fuelled by the money gains of existing owners, can push up prices dramatically. Because of the time it takes to build a house, this lack of new output must be seen as a longer-term effect across short-term price booms and stagnations. House prices during the 1970s had to rise by such large percentages to induce the new housing output that was produced, and prices did not rise fast enough to stop output dropping dramatically.

12

The political economy
of owner occupation

Approaching the housing policy of the state

Analysis of the structure of owner-occupied housing provision in the previous chapters makes it possible now to examine recent trends in housing policy. There are two related but distinct questions to be considered, pertaining to the present situation and to future possibilities. The first, about the present, is why has owner occupation in its existing form come to dominate the housing policy of the state? As chapter 1 pointed out, part of this domination is actually a paralysation of policy initiatives: a fear by successive governments of undertaking any fundamental reform that includes owner occupation because of the perceived threat of a political backlash. This fear paradoxically has led to a depoliticization of housing issues by closing off debate on the central question of the nature of owner occupation. Political dispute, for example, may rage between the Conservative and Labour parties over other housing tenures, but it is all remarkably abstract and contradictory, and hence politically marginalized, because policies towards other tenures are not integrated with reform programmes embracing owner occupation. Political stalemate consequently has led to political inaction.

The second question about housing policy relates to change: is it possible to create a new structure of owner-occupied housing provision which enables the tenure to become part of a socially progressive programme of housing reform? The possibility of mass support is at

The political economy of owner occupation 337

issue here, so any proposed programme must appeal to a wide section of owner occupiers as well as to those living in other tenures. The possibility of housing reforms which break the current political impasse is the really interesting question in the context of which the current situation must be evaluated. Ultimately the answer to it is one of political judgement rather than of analysis alone. But any political calculation needs obviously to weigh up the balance of social forces over an issue before a final judgement is made.

This chapter argues that a broad group of economic interests came together in the post-war era to support the growth of owner occupation and the subordination of other forms of housing provision to it. The current economic problems of the structure of owner-occupied provision outlined in earlier chapters, however, are now breaking up this support. In order to keep the discussion down to a reasonable length emphasis is concentrated on the period since the late 1960s and on broad themes rather than detailed policies or political manoeuvrings. Much of the chapter examines the situation of particular groups of social agents in relation to owner occupation and to the principal current alternative associated with council housing.

The argument presented here suggests that economic interests and changes in those interests are important influences on the housing policy of the state, which is why this chapter has been called the political economy of owner occupation. This does not mean, however, that there is a perfect correspondence between objective economic interests and political demands over state policy, far from it. Reasons why this is the case are explained where appropriate. What is important is that the connections between economic processes and political action are explored. Neither should be reduced to being simply a component or expression of the other.

The politics of housing in Britain and elsewhere have been strongly influenced by contemporary ideas about the nature of housing problems. The consumption-orientation of most housing analysis, criticized in chapter 1, has had its effect on political debate over housing provision. Part of the reason for the political stalemate over owner occupation is that it has been misunderstood as a form of housing provision by both the political Right and the Left. Ideas about housing consequently have ended up having a substantial influence on the politics of housing and in creating a housing crisis. Housing tenure has been conflated with the particular structures of provision associated with those tenures. This type of political discourse is called here the politics of tenure.

On the Right owner occupation has been regarded as a privatized back-to-the-market solution to housing provision, whereas, as will be shown, it has involved large and uncontrollable amounts of state expenditure and market failure in the sense of an inability to satisfy housing needs cheaply or adequately. The Left, on the other hand, by unquestioningly accepting the back-to-the-market thesis of owner occupation has had the political frustration of trying to develop an alternative housing strategy around council housing in its current form whilst tacitly accepting the *status quo* in owner occupation. The lack of electoral appeal in the approach has meant that the obvious political advantages to be gained from highlighting the current housing crisis have been lost. The lack of an effective, popular alternative housing policy has forced housing issues out of the political limelight. As far as the Right is concerned housing provision is back in the market place where it should be, except for a small state sector for those in real need, so they do not want to highlight the inadequacies of the current situation. The Left, on the other hand, can offer no viable alternative. The only prominent political role given to housing is the easily ideologically acceptable Keynesian one of more housebuilding to reflate demand in an alternative economic strategy. This characteristic exposes, once again, the failure of the Left, both inside and outside the Labour party, to treat social needs as an integral part of economic policy (Rose and Rose 1982).

One interesting area which will not be explored in depth in this chapter is the increasing centralization of control over state intervention in housing. The history of state intervention has been one of a development from the local level upward, eventually to large-scale central government intervention. This history created and reproduced a tradition of local diversity and autonomy. The desirability of local political control of state intervention in housing has continued to be a principle to which successive governments have had to pay lip service. Yet, with the shift towards owner occupation and attempts to cut state expenditure on housing from the late 1960s onwards, central direction is now overwhelming, leaving little flexibility for local initiative or political diversity.

State policy and the growth of owner occupation

Before assessing political support for owner occupation it is important to examine what the state has actually done towards the tenure. There have been many pieces of legislation directed at owner-occupied housing provision stretching back into the nineteenth century. Most, however, can be classified into four main areas, each of which are reactions to changing market conditions rather than key determinants of the expansion of the tenure.

(i) *Creating the preconditions for market exchange:*
This has taken place through legislative measures affecting property rights and the ease of property transfer, and through the encouragement and control of the building society movement.

(ii) *Attempts to lower the income threshold to ownership:*
This has been a major concern of a whole series of legislative measures since the 1950s. Most of them have had primarily a cosmetic effect on the significant barriers to ownership for the lower paid: the only economically active socio-economic group to record a fall in the proportion of households in the owner-occupied tenure in the 1970s was the unskilled manual group of low-paid workers. Special reductions in mortgage interest rates, loans and grants to first-time buyers, preferential credit arrangements and mixed tenure schemes have been the main instruments used.

(iii) *Supply subsidies and land policies:*
With the expansion of owner occupation into the older parts of the housing stock the state has tried via a series of increasingly generous measures to get individual owners to rehabilitate the stock. This policy appeared to be spectacularly successful for a brief period in the early 1970s when an annual peak of 188,000 renovation grants were approved for this tenure. Since that date approvals have slumped dramatically, fluctuating between 60,000 and 90,000 for the rest of the decade, belying the earlier successes. Local authorities have also sold, at low prices, dwellings from the council rental stock and have shown a growing interest in building for sale and in other schemes with private builders. The most important influence on owner-occupied supply has been via the land-use planning system and state expenditure on the built environment.

(iv) *Tax reliefs:*
The most significant and well-known fiscal policy towards owner occupation is tax relief on mortgage interest payments. By 1981 the Treasury estimated that this relief cost £2000m. in lost tax revenue, which was equivalent to all the central and local government housing subsidies to council housing, housing associations,

new towns, rent rebates and rent allowances put together (Treasury 1981).

Unlike many other housing policies, mortgage interest tax relief was never introduced as a conscious policy to encourage home ownership. Prior to 1969 the interest payments on all personal borrowing were exempt from tax; at that date this exemption was abolished except for housing loans. Before the Second World War few households were affected by the existence of tax relief. The importance of the relief, however, grew as the number of owners grew, as house prices rose, and as the income tax net widened. Its abolition then became 'politically impossible' and the threshold above which relief could not be claimed was raised periodically to reflect house price inflation and that political impossibility. Since 1974 the mortgage ceiling for relief has been fixed at £25,000, much to the consternation of building societies and housebuilders who have argued that it should be raised to £53,000 to restore its real value at 1982 prices (*Financial Times*, 7 June 1982).*

There is also another important tax measure: households used to have to pay schedule A income tax on their imputed income from house ownership until the tax was abolished in 1963. The argument is that, as house-owners, owner occupiers implicitly derive rental income from themselves as house dwellers which like other non-money income should be taxed. Schedule A tax was imposed as part of a general system of taxing real property. Again, it was initially a fiscal measure not directed at owner occupation but one aimed at private landlordism. The owner occupier subsequently ended up falling into the same tax net. Its abolition in 1963 was a political carrot aimed at owner occupiers in the run-up to the first general election to be lost by the Conservatives for thirteen years.

The final tax measure is one from which owner occupiers have been excluded: capital gains tax. It has been argued by many that the gains made from house price rises should be subject to capital gains tax as are profits made on the stock exchange. The incidence of this tax on the individual owner occupier would roughly be 30 per cent of the difference between the original purchase and final selling prices of the house, to be paid at the time of sale. The Treasury has made a crude calculation that £2400m. of tax in 1981 was avoided in this way (roughly equivalent to a £2500 tax on every house sold).

* The March 1983 budget raised the limit to £30,000.

Capital gains tax highlights the problems of deciphering state subsidies to housing tenures. Foregone taxes conceptually are equivalent to subsidies as the financial flows through the state exchequer are altered in a similar way. But it is not easy to decide who are the ultimate beneficiaries of subsidies nor whether a tax relief/subsidy is actually being given for housing. It could be argued, for example, that mortgage interest tax relief simply sustains higher house prices than otherwise would be possible. Alternatively, a number of subsidies to council housing are not housing subsidies at all but income maintenance subsidies, most notably rent rebate schemes. Until the early 1970s many of them were channelled through the social security system instead of housing revenue accounts (Merrett 1979). In 1982 the government announced the introduction of a Housing Benefits scheme which again will be part of the social security system, so this item again should disappear from Housing Revenue Accounts. Contortions of terminology reach their heights with the capital gains tax 'subsidy', because after the 1982 Budget all capital gains taxation was put on an index-linked basis. This meant that owner occupation no longer generated this specific tax loss to the Exchequer and so as a subsidy it 'disappeared'.

The history of fiscal policies towards owner occupation can be summarized as being both haphazard and aimed generally at encouraging the tenure through tax reforms that removed adverse tax effects as they appeared whilst maintaining the beneficial ones as they grew in importance. Unlike council housing, where the whole structure of state finance is periodically reviewed and restructured, there has never been a serious attempt to alter the nature of the fiscal measures related to owner occupation (the *Housing Policy Review* of the mid-1970s carefully fudged the issue). This absence for a period of over sixty years reflects the political significance of this tenure form. Yet it indicates neither a concerted policy of encouragement of owner occupation nor unbridled attempts to subsidize the tenure, but it does suggest the existence of constraining limits on the haphazard path of state housing policy.

Notwithstanding this conclusion about the nature of state intervention within this tenure, it is still undoubtedly true that the special tax effects have facilitated the growth of the tenure. This result is not because the costs to individual households have actually been reduced in this way, as individual costs depend on the level of house prices as well as on the tax relief. Instead the tax reliefs have had an expansionary effect because they have facilitated rises in house prices. Those rising prices, in turn, have encouraged switches of the current stock from other tenures and sustained the rate of housebuilding. The existence of these tax advantages thus reflects the political power of the owner-occupied housing lobby to sustain fiscal benefits that fortuitously arose and to remove any tax burden that ended up being directed at owner occupation (e.g. schedule A tax). Moreover, because of their tax relief nature, the incidence of many state subsidies to owner occupation is automatic, so political pressure has only to be used to stop legislative change rather than the more difficult task of forcing change.

Theories of politics and structures of provision

Before undertaking the analysis of housing policy it is important to clarify the theoretical approach being adopted here towards explaining the role of the state. Theoretical positions about the state in capitalist societies obviously influence the interpretation of state policies.

If the state is regarded as some socially neutral entity with ultimate, absolute power over a generally harmonious society, state policies will tend to be viewed as attempts to correct social disorders and inefficiences. This correction process could be a long-drawn-out one, held back by institutional rigidities, prejudices, ignorance and the like, but conceiving of the state in this way does imply that there are solutions to any social problem which do not require fundamental changes in social organization. Legislation over housing therefore is seen as a process of groping towards a technical solution that provides a social optimum. Such a view has to be rejected, however, once it is recognized that a class society is being considered in which the fundamental economic interest of classes place them in antagonistic opposition to each other. The state is therefore one site for that class struggle, so any state policy must be a product of that struggle.

Marxist theories of the state, which recognize this basic feature of class societies, diverge on the implications they deduce from it. Some suggest that a general theory of the state can be derived so that state actions at a specific point in time can easily be deduced by reference to that theory. Some theories divide capitalism up into a series of stages of development, arguing that advanced capitalist societies by the 1940s had reached the stage of state monopoly capitalism. The actions of the state consequently should be considered in relation to an overriding aim of supporting the continued dominance of a handful of all-powerful monopolies. Other theories suggest that the actions of the state depend on the wider needs of capital and vary with the phases of the accumulation cycle. Both of these views deny the complexity of social relations by insisting that the state mechanically functions to the advantage of capital accumulation alone.

Other positions reject the notion of a general theory of the state, arguing for the importance of historically specific political struggles across a broad range of social groups. In some, this tends to degenerate into a pluralistic style of conflict with no framework for understanding the reasons for conflict beyond codification of antagonistic groups in terms of the immediate empirical events. Others place greater reliance on theoretical analysis of the contradictory relations between classes in capitalist societies, arguing that particular historical situations necessitate specific hegemonic strategies by classes in order to win broad support for policies and parties that favour their own interests (see the surveys in Jessop 1982 and Mingione 1981).

Mechanistic views of the state always operating in the interests of capital or a dominant fraction are of little use in understanding housing policy. They would imply that a general theory of housing tenure, for example, could be derived for particular historical epochs or phases of accumulation (for example, perhaps private renting for the competitive stage, but what then?). The variety of housing policies across different advanced capitalist countries would have to be denied; so would the considerable differences in the relative importance of tenures between them. Moreover, monopolies profiting from housing provision presumably are as likely to impose costs on non-housing monopolies (via pressures on wages and taxes) as on other fractions of capital. In other words, even if empirically a gain to a certain institution classified as 'monopoly capital' could be identified, no explanation of why that gain arises can be derived from the notion of 'monopoly' alone.

To reject such mechanistic views, however, does not explain how state policies should be considered. Talking in terms of historically specific political struggles, for instance, still enables a variety of positions to be taken over owner occupation: the ideological incorporationist and other theories of owner occupation criticized in chapter 9 would all be compatible with a 'hegemonic strategy' type of approach. A further narrowing down consequently has to be made before an adequate approach can be arrived at.

What is apparent from post-1915 housing policies in Britain is that private agencies cannot come close to meeting even the most minimal housing aspirations of the mass of the population without substantial state orchestration of their actions and direct state economic involvement.

This is as true for private tenures like owner occupation as it is for state-owned council housing. During the twentieth century, therefore, any government must have some credible strategy towards housing in the broad political and economic strategies it pursues. Support for those broad strategies from significant sections of the working class has to be won and housing policy forms a useful component in the attempt. Credibility does not, of course, mean that the strategy necessarily leads to improvements in the housing sphere. Presumably some consistency with broader strategies is required but the necessity of producing a real impact on housing provision will depend on the contemporary political importance of housing. Ideology, politics and economic interest become inexorably woven together. The housing issues are historically contingent because they depend on the contemporary forms of housing provision, the problems they create and the suggested solutions to them that have already been tried.

One of the clearest examples of the point being made is the development of housing policy in the years prior to 1919. The Liberal government after 1906 tried to build up a credible housing strategy around town planning. The acute class crisis of the pre-First World War years led to a particular strategy for its resolution by the Liberals: some reform to the benefit of the working class (particularly at the expense of landowners) whilst maintaining the full workings of the Empire, free trade and at home the market mechanism. Time could be bought and the productive capability of the working class improved in the face of economic stagnation and the restructuring of productive capital. The attempt at a credible housing strategy around self-help, town planning and land reform failed because it did not take sufficient account of the Labour Movement's opposition to private landlordism and the contemporary contradictions of that structure of provision (see chapter 7). The changed circumstances of 1915–19 forced that omission to be rectified.

Housing policy cannot just be seen in terms of broad political and economic strategies. The options open for housing policy are limited because state policies intervene and influence structures of housing provision, even if sometimes creating them as with council housing. The role of the state therefore is limited by those structures and their wider linkages. This relation between state policy and structures of housing provision is one reason why there is no single, coherent, long-run housing policy of the state, as it cannot control housing provision. The state can only intervene within the social relations constituting a structure of provision, even in the case of council housing, where the state apparatus does directly control one principal aspect of that structure. The state therefore has to operate in a situation where it has limited power, and in a situation that is constantly changing. Any housing policy of the state has to react to changes; adherence to a single pre-existing policy or plan would fail disastrously, even if one existed.

The conclusion about the limited impact of state housing policy has two implications. The first is that theories of social struggles over housing policy cannot start off from one characteristic of a tenure alone and consider its implications for relations between social classes through attempts by classes to get housing policies which rely on manipulating that characteristic to gain hegemony. This closes off supposed ideological or other innate characteristics of owner occupation as the route to understanding developments in housing policy. The second implication, however, does suggest a useful starting place. In order to understand housing policy there is a need to examine the contemporary situation of structures of housing provision, the agencies involved in them, and how they relate to wider social forces, both economically and politically.

When considering potential support for a structure of provision, rarely will political pressure over any issue derive from an homogeneous block, like the working class. Instead it will come from amorphous groupings, 'power blocs', the constituent parts of which will vary over time, and frequently those parts will not formally be linked or even recognize their common interest. Such groupings, moreover, will often coalesce around specific issues rather than broad long-term programmes. In housing, this process can be seen in the way in which political agitation becomes associated with specific tenures and particular issues within those tenures; as, for example, in the momentous, widespread agitation against the 1972 Housing Finance Act's attempt to raise council rents substantially, or in the present support for the *status quo* in owner occupation.

Groupings form around political issues because a common interest is recognized in pushing for change or in resisting it. Support is not necessarily limited to groups with a similar economic place in society. At the level of the individual, if a clearly articulated material interest is the starting-point (even though it need not necessarily be), say for example the need for decent housing at a low price by a worker or the need for higher profits by a firm, that interest has to be translated into a political demand within the ideology with which the individual conceives of their understanding of the issue. These demands in turn can be made only through some form of political representation – a political party, trade union, employers' organization, pressure group, etc. – which will itself influence the nature of those demands. There is consequently a twofold dislocation of any material interest before it is represented politically (if at all): within ideology and within the form of political representation (and its ideology). Yet from such diversity a successful 'power bloc' may form around a specific set of demands, and it might succeed in getting a favourable political response to them. The process of political persuasion could operate at a series of different levels: locally or nationally, in minor procedural amendments or in major legislative change, via smoke-filled rooms or via street demonstrations, peacefully or violently depending on the nature of the demand and the political strategy adopted.

Such political processes, and the ideological frameworks within which they operate, cannot move or be driven in any direction. They exist as part of a determinate social system that is subject both to change and to definite limits to that change. State policies usually will ensure that class struggle is contained within the general limits of the dominant mode of production. Any policy towards housing provision must be understood therefore in terms of the dominance of capitalism, and the limits imposed by it on housing policy. The political options open in housing policy are limited by the number of structures of provision in existence at one time plus the viable alternatives around which 'power blocs' have coalesced. In particular, alternatives have to be propounded before they can ever possibly gain support. And what is so astounding about the contemporary situation in Britain is that no viable alternatives are on offer. To understand developments in housing policy, therefore, the internal coherence of the structures of provision associated with council housing and owner occupation should first have to be examined. After that has been done it will then be possible to consider the nature of broader political support for these tenure forms.

Stability and change in housing provision

Associated with housing tenures are particular sets of social relations which are the product of earlier struggles over housing provision rather than the inevitable consequence of any particular tenure form. Yet their existence as established structures of provision determines the development of those tenures, how much housing in them costs, how much housing is provided and of what type. What is notable in Britain is that since the 1920s there has been little change in the social relations associated with the two new major tenures of owner occupation and council housing. This is despite sharp variations in levels of housebuilding in both tenures (little private housing was built between 1939 and 1953 for instance) and the marked shifts in the relative power of the agents involved (of which the increased power of private landowners and the land-use planning system in owner occupation since the late 1940s are two illustrations). This stability is as much a political product as an economic one, and it helps to explain the relative decline of council housing over the past few years and the political strength of owner occupation.

The structure of owner-occupied provision has been described in detail in previous chapters. A description of the agencies involved in council housing shows a different set of social relations but still ones dominated by capitalist agencies (figure 12.1). The state, in the form of

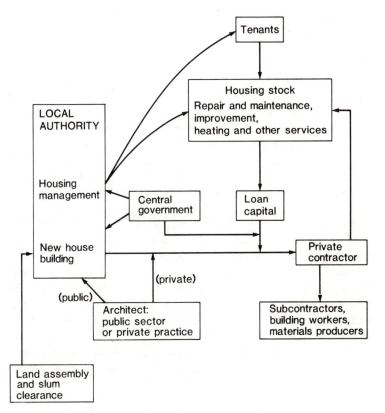

Figure 12.1 The structure of council housing provision

local authorities, legally owns council housing. Each local authority manages its existing housing stock and builds new houses on criteria fixed and overseen by central government. Generally houses are built by private contractors but sometimes by the councils' own direct labour forces. Building is financed through borrowing from loan capital over a sixty-year repayment period. The current costs of running the existing council housing stock are pooled together in a council's Housing Revenue Account: that is the expenses of new building and management bureaucracies, the interest and capital repayments on outstanding debt, and the costs of servicing, improving, repairing and maintaining the existing stock. Rent levels are set to cover these costs, taking into account subsidies received from central government and rate income. Until 1967 most subsidies were directly earmarked for new building (an immediate grant of so many pounds per year over all or part of a dwelling's assumed sixty-year life). Since 1967 under various schemes most subsidies are simply general payments into the Housing Revenue Account and so are more explicitly partial payment of the interest charges of loan capital. Subsidies, in others words, should not be seen as payments to tenants but a means by which the viability of the existing structure of council housing provision is sustained. (In somewhat simplistic terms loan capital is being subsidized, not tenants.) When subsidies begin to shrink, not surprisingly therefore so does the council housing sector.

The current financial procedures for council housing, which are a result of the existing social relations there, make the tenure highly susceptible to steep cost rises during periods of accelerating price inflation and associated increases in interest rates (the so-called 'front loading' phenomenon). They also lead to similar financial shocks when the rate of housebuilding rises or building costs increase rapidly (Merrett 1979). The costs of council housing, and the ways in which those costs change, are products of its structure of provision. So are the quality of the houses produced and the building methods used to produce them. The industrialized building programme of the 1950s and 1960s, for example, can only be explained by the interrelation between its dominant agents: state functionaries and politicians, architects and building contractors.

Contracting is different from speculative building as chapter 3 noted (see also Ball 1983). It does not involve land purchase; the labour process and hence the position of building workers are also somewhat different. The use of private contractors and private loan finance were general characteristics of local authority procurement which had developed during the nineteenth century. Council housing was introduced at the end of the nineteenth century as an extension of local authority activity in face of the shortcomings of private landlordism. Neither in content nor in form was council housing intended by the state to be a revolution in housing: the 1919 Addison Act, for example, was meant to help stop a social revolution rather than to start one. So previous local authority practices and bureaucratic procedures were reproduced in council housing with little or no opposition. Along with them came nineteenth-century charity notions of the 'deserving poor' and no means for tenants to influence management processes except by organized protest or via the ballot box, both of which are unwieldy means of expressing detailed dissension.

The increasing complexity of public housebuilding and housing management has led to the development of a large state bureaucracy which decides what are needs, whose will be satisfied, what will get built, and how it will be built and financed. There is little political control over such procedures, and tenants' organizations either get incorporated or mobilize only fragmentarily over specific and usually localized issues. These difficulties have been well aired and real attempts have been made to overcome them. Yet the current bureaucracy within council housing provision does not exist by chance, but because of the nature of the structure of provision of which it is a part. Capitalist enterprises only take part in the provision of council housing on their terms (i.e. for profit). A vast bureaucracy is required to check, process and attempt to control those private interests in order to try to get the best possible terms for the council. Similarly rents are based on the current and past costs of building; council housing therefore still takes the commodity form even if its price is 'subsidized'. Again procedures and organizational forms are required to collect rents. The collection of a rental income, whose size is determined by external cost and subsidy factors, therefore is the principal requirement of housing management in its relation to tenants. This requirement must override all others and hence ultimately management must always be an overseer of tenants rather than a potential instrument for their use. Little of the current structure of council housing provision, therefore, is open to democratic accountability or control because its major determinants (various forms of private capital) are outside the scope of potential democratic control.

The type of immediate relation council housing has to tenants and to capital has important political implications for council housing and also for owner occupation. Council housing by its nature involves a degree of collective provision as it is instigated by the state. The current nature of local authority housing finance increases this collectivist element because any individual tenant's housing costs depend on the overall expenditure and revenues associated with all of a council's housing stock. Both of these collectivist elements could result in cheap, good-quality housing and enhanced personal freedoms if they were part of a process of social co-operation based on public accountability, mutual support and need. But the existing structure of council housing provision suppresses such potential aspects of socialized forms of housing provision. This has provided a material base from which the propaganda lobby for owner occupation could launch campaigns to claim that relatively owner occupation is both financially more attractive and provides greater personal freedoms. Few would doubt their success, even if they disagree with some of the claims or dislike the methods used with their divisive play on snobbism.

There has been a long-term erosion of mass support for council housing because of the problems faced by the tenure in its current form, and the relative attractiveness and the 'marketing' of its alternative: owner occupation. The erosion of support obviously varies from area to area and particularly between social strata. There is strong, if fragmentary, evidence that it is especially sections of the middle and upper strata of the working class who no longer politically support council housing. It is amongst this section of the population that owner occupation has been growing the fastest since the early 1970s. Most of the large-scale council house sales since 1979 also have been to households from these strata. No longer is council housing seen in the imagery of politics as the tenure of the working class, because of the extension of owner occupation amongst key sections of workers. This might explain why there has been so little protest or political mobilization against the large rent increases for council housing in the early 1980s, whereas previous attempts to increase them invariably met with rent strikes and other forms of mass opposition, of which the agitation around the introduction of the Housing Finance Act 1972 is the most famous (Skinner and Langdon 1974).

Most of the agencies involved in the structure of council housing provision, in addition, have provided little political support for the tenure. Billions of pounds have been paid to building contractors, materials producers and loan capital as a result of council housing programmes, but some of them have actively campaigned against the 'wastefulness' of state expenditure on housing. This includes building contractors who, with the election of the Conservatives in 1979, actively supported the policy switch away from council housing via their employers organization, the NFBTE. This was done in part on the assumption that there would be a marked increase in private sector work. When this failed to materialize they did start to lobby vigorously for more public sector work but the pressure they unsuccessfully exerted on the Government related to public sector orders in general rather than as a defence of council housing.

All the private agencies involved in council housing provision can switch their activities elsewhere, which explains their political position. This is clearly the case for loan capital whose sphere of operations is at a world scale. Similarly, most of the larger building contractors who used to be heavily involved in council housebuilding operate across a wide variety of building activities, and so can compensate for the loss of council housebuilding work elsewhere. They are not economically tied to council housing provision in the same way as are the agencies involved in owner-occupied housing provision. The whole economic rationale of building societies, exchange professionals and speculative housebuilders is owner-occupied housing. Their political concern and coherence over the defence of owner occupation, therefore, is substantial. In contrast, the only usual political intervention made by the private interests associated with council housing is to act as a barrier to reform of the tenure's structure of provision. Successive campaigns against the use of local authority direct-labour workforces in council housing provision, for example, have been mounted (Direct Labour Collective 1978 and 1980), and the ideology of reliance on loan financing is firmly entrenched.

The only groupings within the structure of council housing provision with a strong interest in supporting its continued existence are building workers and state housing employees. Historically the political position of building workers has been weak for reasons specified in Ball (1983). State housing employees are also in a difficult position to mobilize support or to create alliances because they are the bureaucracy on whom many of the failings of council housing have been blamed. The ability for them to generate broad political support, as was achieved for instance by the National Health Service workers in 1982, is virtually impossible: nurses may have public sympathy, rent collectors do not.

The politics of tenure and ideologies of subsidy

In order to understand how particular structures of housing provision relate to wider social forces, the nature of political debate over housing issues must be considered. Debate takes place within a series of ideologies about the nature of housing problems and ways to solve them.

With virtually no exception since the end of the First World War the politics of housing in Britain have been based on housing tenure and the relative roles that each tenure should play. Political debate consequently takes place within ideologies that see housing provision as simply being about providing housing as an item of personal consumption which has to be subsidized (or not) by the state. Tenure and subsidy are the items of political struggle; who does the providing within a tenure and how they do it are just technical matters left to experts outside of mainstream political action. This type of political action will be called here the politics of tenure. It has had a profound influence on the nature of housing policy in Britain as it has forced the relations of production out of political discussion over housing provision. Tenures as a result have become associated with specific structures of housing provision.

Moreover, the form in which the discussion takes place – tenure, levels of subsidy, and the quantity, design and quality of housing – gives considerable importance to the ideology of the expert. Contradictions within structures of provision become ideologically transformed into problems of housing design and housing finance. The mass opposition to the systems building of the 1950s and 1960s, for example, brought a questioning of the infallibility of design professionals; yet conversely the escalating costs of housing finance have had the reverse effect of reinforcing the role of the financial expert. Housing revenue and expenditure have remained 'high finance' best left to the experts who consequently have considerable leeway in imposing their own politics on state housing policy in the name of objective reason.

A good example is the politics of subsidy. The level of subsidy will be the political issue, the means by which it is provided a complex technical one left to the experts. The form of subsidy, however, is vital: tax relief on mortgage interest, for example, is automatic and hence indexed to inflation and politically difficult to change; the reform of local authority housing finance in 1967 by the then Labour government, on the other hand, despite raising the level of subsidy, was organized in such a way that it guaranteed (perhaps unintentionally) that council housing costs would rocket if the rate of inflation and interest rates rose, creating a basis for widespread cuts with the onset of inflationary economic crises in the 1970s.

The politics of tenure influence beliefs about the recipients of housing subsidies. Emphasis is mistakenly put on subsidies to individual households in each tenure, ignoring the wider flows of revenue and expenditure between the agents involved in structures of housing provision. Council tenants and owner occupiers receive subsidies, so it appears from within the politics of tenure, and additional distortion is created by averaging the subsidies across all households in a tenure whereas subsidies are in fact highly selective. Mortgage interest tax relief, for example, obviously offsets only the mortgage repayments of owner occupiers with mortgages, and it is concentrated on those with high mortgage outgoings who tend to be recent movers. Almost half of owner occupiers own their houses outright and so do not receive this subsidy at all, whereas many others have only low mortgage outgoings. Yet official data invariably average this subsidy over the whole owner-occupied housing stock. (The annual housing survey in *Social Trends* is a good instance of the misleading information that can be generated by presenting meaningless averages.)

A similar effect occurs with council housing. Here it relates to the construction costs of dwellings and the types of service provided. The outstanding debt on council housing depends on the initial 'historic' costs of dwellings in the stock. So, in general, the costs of recently built or improved dwellings compared to the rents charged are much higher than is the case for older dwellings. On an historic cost basis, councils in fact make large 'profits' on their better-quality older stock and much of the subsidy is loaded onto dwellings built in the last few years. Similarly there are wide variations in construction costs over different parts of the country. Redevelopment is more expensive than greenfield site building and large cities have higher costs than smaller ones. When authorities in higher cost areas, like London, have large redevelopment building programmes the timing and building cost effects compound each other. The *Housing Policy Review* found, for instance, that the average costs of council housing in inner London boroughs were 2½ times greater than those for non-metropolitan boroughs and that the average subsidies there were 3.7 times higher (HPTV II, p. 6). A significant proportion of local authority housing also caters for special social needs, like sheltered accommodation for the elderly or specially designed accommodation for the disabled. Such accommodation, which formed a large part of council building programmes in the 1970s, also generates a disproportionate need for subsidy.

Two important conclusions can be derived from the variable effect of subsidies. The first is that tenures and subsidies to tenures are very blunt instruments with which to tackle politically significant housing issues

The politics of tenure make any more sophisticated alternatives difficult to introduce. Widespread criticisms of the distributional consequences of the present system of housing finance face little chance of making headway within its framework. To quote again the words of Peter Shore, 'we certainly do not believe that the household budgets of millions of families . . . should be overturned, in pursuit of some theoretical or academic dogma' (HPR, p. iv).

The second conclusion is that most debate takes place on the basis of poor information. Comparisons of the average state subsidies to council tenants and to owner occupiers are worse than meaningless – they are distortions of the real situation. In particular they overemphasize the subsidies going to council housing – most council houses receive much less than the 'average' subsidy – and underemphasize the subsidies going to mortgagor owner occupiers; assuming, of course, that households are the direct recipients of the subsidies.

Within the framework of the politics of tenure different social groups have different interests in the types and levels of subsidy provided by the state. The capitalist agencies involved in the structure of owner-occupied provision have every interest in subsidies which keep house prices and the level of demand high. Mortgage interest tax relief is therefore ideal for them as it is a subsidy to all movers, the size of which is in direct proportion to the size of the mortgage incurred. It is hardly surprising, therefore, that they have expressed such concern over the freezing of the mortgage ceiling for relief at £25,000 between 1974 and 1983. But others have different economic interests which the next section explores: the two most important groups to look at are capital and the working class.

Housing costs, incomes and state intervention into housing provision

How much a household can spend on housing obviously depends on its income, so as income rises more can be spent on housing. Conversely, if housing costs rise, households' standard of living falls unless incomes rise to offset the increase. The exact relation between housing costs and income is far from simple but these general directions of change are sufficient to draw out the main implications of the relationship. Not all of a worker's standard of living is financed directly from wage income, however; the state subsidizes the costs of some necessities of life and in some cases provides services free. In doing so the state alters, to an extent, the relationship between incomes and living standards.

For the working class the level of income is determined by the struggle at the workplace over wages, a struggle influenced by the contemporary profitability of capital and the strength of the working class. When profits are rising, firms are more likely to acquiesce to demands for higher wages. When profits are falling, pressures mount for capital to cut total wage costs and wage rates. The existence of state expenditure complicates the matter somewhat. However, with some qualifications that will be discussed later, some general relations between the three can be examined.

For a given level of labour productivity, a rise in any one of the three tends to depress at least one of the other two. A rise in state expenditure, for example, squeezes the profitability of capital if financed directly out of profits or squeezes living standards and puts pressure on wage levels if financed out of taxes on workers. (Borrowing to finance state expenditure produces similar effects, although the total cost is higher as interest charges have to be added and the time profile of the cost incidence is different and much longer.) Rises in housing costs, therefore, can intensify the economic conditions affecting class struggle as they may depress either the rate of profit or workers' living standards directly, or do so indirectly via their effect on state finance. Rises in the costs of the two major tenures in Britain over recent years are therefore a prime reason for an escalating crisis of housing provision. It is a crisis whose impact is not limited to housing consumption alone but extends right through the economy as a whole.

The case of a fixed level of labour productivity is obviously an oversimplification, as in a capitalist economy changes in productivity are enforced by the accumulation of capital and investment in new methods of production. Changes in labour productivity tend to cheapen commodities and thereby enable the share of the wealth produced through commodity production going as profit to be increased whilst the standard of living of workers is maintained or within limits even increased. This tends to offset conflicts between profits, wages and state expenditure but it does not remove them. The pressures are greatest during economic crises when there are sharp falls in the rate of profit. The escalation in housing costs in Britain since the early 1970s, therefore, could not have come at a worse time. The long post-war boom had come to an end and been replaced by recurrent world economic crisis and a continuous erosion of the strength of capital operating in Britain.

The position of state expenditure itself is contradictory with respect to both living standards and the rate of profit. As was argued above it can depress capitalists' profits and workers' living standards; yet some aspects of state expenditure can at the same time increase the total mass of use values produced by a given workforce. State provision of a service in the non-commodity form as a 'free good' might be the cheapest way of providing it. These state activities can either reduce the non-labour inputs of capitalist firms or raise workers' living standards and their productive potential. A good example of state expenditure cheapening non-labour inputs is the provision of an extensive road network which cuts the cost of transporting commodities. State expenditure that improves workers' living standards similarly can have a subsequent positive impact on capitalists' labour costs: first by reducing upward pressures on wages and second by raising the productivity of the labour power bought for a given wage. The national health service and the education system are examples of such possible effects. Council housing is another, but here the issue is made complex as there is also a distributional effect between types of capital (building/non-building and loan/non-loan). So, with council housing, the comparison has to be a relative one between it and other forms of housing provision. Whether council housing provides economic benefits to capital via the reproduction of the labour force consequently varies over time, depends on the contemporary characteristics of other existing forms of housing provision and relates to the type of labour force required.

When the economy is booming quite large increases in housing costs can be sustained through either higher wages or increased state subsidies. At such times increased housing expenditure might economically benefit even capitalists whose immediate profits are lowered by such expenditure (either because of higher wages or through taxes on profits to pay for the state expenditure). This is because the extra expenditure reduces barriers to additional profits and accumulation through its effects on the productivity and availability of a workforce. Low wages may produce poor housing conditions, for example, and poor housing has been well documented as a cause of ill-health and social problems. So the productivity of labour can be increased by providing better housing. Similarly, potential local labour shortages, that may put an upward pressure on wages in a locality, can be circumvented by building more houses and thereby increasing the available workforce in the area. Such factors, for instance, were of paramount importance in the boom years of the 1950s and 1960s. When the economy is in crisis, however, labour shortages no longer constitute a problem and higher housing costs may exacerbate the crisis by further reducing the overall rate of profit or the standard of living of the working class. At times when capital accumulation requires an expansion or improvement in the labour force, therefore, the quantity of housing provided is of interest to capital as well as its cost. During periods of crisis, on the other hand, cost considerations become more important.

The durability of housing means that its production need not be continuous, in the sense of regularly producing a certain amount of new stock each year, to maintain the current housing standards of most of the population. Capital as a whole might therefore find expedient a severe reduction, or abandonment, of housing production during periods of crisis. A lack of new building, however, has a cumulative effect, so shortages will gradually increase. This is where the distinction is important between expenditure by the state on new housebuilding and on subsidies which reduce the cost to households of existing housing. The switch in the 1970s away from state capital expenditure on housing towards individuals' housing cost related subsidies, outlined in chapter 1, should be seen in this context.

The basic economic relationships between the state, capital and labour obviously hold only for those in work: the sick, the elderly and the unemployed, for instance, are not involved in this direct relationship. Again, however, their housing costs depend on the structure of provision associated with the housing tenure in which they live. Elderly owners, for example, do not face large mortgage payments. The wider economic effect of state expenditure on their housing provision again depends on the phase of the accumulation cycle. The boom years of the 1950s and the 1960s made it economically easier to expand this item of state expenditure; the crisis years since the 1970s have led to pressures for its reduction.

The relationships between state expenditure, the rate of profit and wages that have just been described are functional economic relationships: if one of the three changes it tends to produce effects on the other two. This conclusion is significant when looking at housing as it highlights the importance of seeing state expenditure on housing in a broad, rather than an isolated, context. In particular, it highlights the need to maintain a class perspective on the consumers of housing. Put most simplistically: the housing costs and conditions of the employed members of the working class directly affect the overall state of the class struggle between capital and the working class. Changes in structures of housing provision are the means by which those costs and conditions can be altered. Pointing out such linkages, however, does not explain the reasons why change takes place; it simply identifies some points at which economic pressures may arise. Those pressures nevertheless are extremely important in understanding post-war developments in housing policy. So it is worthwhile summarizing the main conclusions derived from the rather complex analysis of this section.

When looking at the economic implications of particular forms of housing provision on the accumulation of capital, there is an important need to distinguish between capital that profits from housing provision and other types of capital. For most capital housing will represent only a cost: a drain on potential profits. That cost appears either in the form of taxes paid for state expenditure or wages paid to workers (whose pay in turn is taxed to help finance state expenditure). In this way any productive capital helps to fund the housing provision of its workforce and also that of other households (via state expenditure). Adequate housing for its workforce is a direct benefit to any firm and, with less force, so is the housing of potential future workers. Better housing improves the productivity and availability of labour power. Cheaper housing costs, moreover, will tend to lower the pressure to increase wages. Housing provision for other sections of the population, on the other hand, is just a cost to capital. To give it a functional name, it can be said to be a payment to maintain social harmony. Less functionally, it can be said to produce different pressures within class struggle as it does not relate to the immediate relation of production between capital and labour.

Whilst capital may have an interest in housing provision, at the moment all that can be said about the relation of accumulation in general to housing provision for the working class is that at times there is a close correspondence between working-class aspirations for decent, low-cost housing and the interests of capital, whilst at other times they are diametrically opposed. An objective of the analysis of housing provision, therefore, is to consider when and how such a correspondence or divergence emerges as it crucially affects the possibility of introducing progressive housing programmes.

Progressive changes in housing policy

The two times when there have been major progressive shifts in state housing policy, after both world wars, occurred because no other viable alternative form of mass housing provision existed at the time. Working-class political demands for particular types of housing provision could be forced onto the political agenda as a space had been created by the permanent or temporary collapse of other alternative structures of provision; whilst the interests of other social classes, including capital, forced them to acquiesce at least temporarily to those demands or to try and mould them towards their own interests.

Prior to the First World War the structure of provision associated with private landlordism collapsed. After the war, in 1919, the only possible way to expand working-class housing provision, given the lack of alternatives, was the large-scale subsidization of council housing. The interests of capital, the working class and particular state functionaries and politicians 'coalesced' to squeeze out the private landlord. It was hardly surprising that this alliance turned out only to be temporary but it introduced a new dimension into housing provision, central government funded council housing, that has existed ever since.

Similarly after the Second World War expansion of private renting was economically impossible and the structure of provision associated with the new mass inter-war tenure of owner occupation could also not build new working-class housing. To get new output, building prices had to be held down by a system of building controls. This facilitated political pressure for the expansion of council housing. For a few years, good-quality council housing was built within a framework of building controls, on land that could be compulsorily purchased at its current-use value (and after the 1947 Planning Act all development land was partially nationalized). For a while, consequently, not only the private landlord but also the speculative builder and the private landowner were politically isolated out of mass housing provision. It took until 1953–4 for the speculative builder to become a major force again.

The content of both these political developments, however, enabled their progressive elements quickly to be contained: first by organizational procedures, then through the economy axe (the Geddes cuts in 1921 and the cuts associated with the sterling convertibility crisis in 1947). Council housing was fought for because it involved principles of collective ownership and provision on the basis of need. However, time and again, once acceptance had been won for an increase in government support for the sector little attempt was made to implement necessary far-reaching reforms of its mode of operation. The result is that council housing has not been able to break out of its paternalistic origins towards genuinely collective, democratic provision on the basis of need. The politics of tenure and the forms of representation of working-class political demands, via Labourism, helped to produce these difficulties. Centrally subsidized council housing was seen as the end: beyond an interest in minimum housing standards the way in which it was provided was of little concern. (Aneurin Bevan's attitudes to council housing during his years as Minister of Housing, 1945–7, illustrate the limits imposed by the approach; see Foot 1975.)

The continued failure to transform the social relations of council housing provision has robbed it of sustained political strength. It has had to rely for political strength on beliefs about council housing generated within the politics of tenure, ones which it has not been able to justify. The National Health Service, whatever its faults, did create a set of social relations that have sustained its strength and political significance despite attempts at drastic cutting in recent years. Furthermore, it is not simply being primarily state capital account expenditure that has weakened the position of council housing. Defence expenditure contains similar large capital account components, yet the social forces supporting it have managed to increase it throughout the recent years of economic crisis. Council housing succumbed to economic pressures instead because it had lost much of its political strength at the time it mattered.

Owner occupation and recent developments in housing policy

In trying to understand the broad developments in British housing policy since the late 1960s, this section will bring together the points that have been made earlier in the chapter about the structures of council and owner-occupied housing provision, the politics of tenure and wider economic class relations. The argument must inevitably be schematic and tentative for brevity and because of the nature of the historical material being dealt with.

It is important not to treat economic forces as the 'real' underlying determinant of any change in housing policy. Whilst in the late 1970s and early 1980s it may be the case that state expenditure on council housing has been cut because of general economic crisis, that action is still a specific political response by the state to that crisis (one that is part of a wider back-to-the-private-market, monetarist programme that makes the crisis worse). Moreover, if there had not been earlier political pressure for council housing there would be no state expenditure on council housing to cut as part of a specific economic programme. The ideological forms in which political debate and action take place nevertheless can reinforce the dominance of particular economic interests in the formulation of state policies. In terms of the economy as a whole, the interests of capital frequently appear as inevitable, quasi-natural economic necessities. Economic growth, for example, is equated with capital accumulation; new investment generally is posed as the problem of encouraging new private investment (via inducements, not directives); improvements in labour productivity have to be on capital's terms, for instance. The result is that there is enormous pressure on any government to introduce policies that place the emphasis on reducing working-class living standards and conditions of work during economic crisis, with mass unemployment used as a stick to weaken resistance to such measures. As a significant component in the cost of maintaining working-class living standards, housing conditions will come under severe attack.

Housing policies, in other words, do have clear links to specific ideologies about the economy so it is not surprising to find developments in housing policy related in part to economic policy. The state-financed component of housing costs is particularly important in this respect because successive governments in post-war Britain have been dominated by economic ideologies which see high state expenditure as a cause of inflationary economic crises. (This is true for mainstream Keynesian theory as well as its monetarist variant, albeit for slightly different reasons.) Governments have also since 1945 tried to keep down the rate of increase in money wages. Incomes policies and unemployment have been used as direct means of influencing wage settlements. There has been an attempt to influence housing costs as an indirect means of moderating inflationary pressures. On occasions the link has been explicitly announced: for instance, council rent rises have been moderated or temporarily frozen and strong pressure brought to bear on building societies over mortgage rates at various times. Even when the link is not explicit the relationship remains to influence governmental thinking on housing policy.

The key components of shifts in housing policy since the end of the 1960s were described in chapter 1. To summarize, there has been a move away from subsidies directly associated with housebuilding towards subsidies which offset housing consumption costs. At the same time the growing importance of owner occupation shifted the tenure to which consumption subsidies were directed. The problem that now needs to be addressed is how can these changes be assessed in terms of the general state of class relations, to what specific political strategies do they relate, and have they been successful?

The most obvious thing about the increasing emphasis on owner occupation is that it is a movement back towards the dominance of private market exchange in housing provision. This corresponds to the broader economic strategy of the post-1979 Conservative government which Gamble (1981) has called the social market strategy. It is also a rolling back of the welfare state in so far as housing provision is now even less related to need and more to the ability-to-pay. In these terms it is worth exploring which social groups have benefited economically from these changes in housing policy. The most obvious candidates are the capitalist interests which dominate British society.

The period of expansion and labour shortage during the long boom of the 1950s and 1960s, given the poor state of the existing housing stock, created a need for large-scale housebuilding and urban renewal to satisfy the workforce requirements of employers as well as the housing needs of the population. State expenditure directly generated new building (with council housing and urban renewal) or helped to prime it (with suburban owner occupation). State expenditure also helped to keep down the pressure of housing costs on wages. In general terms, therefore, there was a broad correspondence of interests over the expansion of housebuilding during the long boom. When the rate of new housebuilding faltered as it did at the end of the 1950s, owing to government policies, it quickly created a political reaction and a recommitment to state involvement because a wide range of economic interests were affected by the ensuing housing crisis. The end of the long boom changed that. Rising unemployment ended general labour shortages and the restructuring of British industry slowly altered the characteristics of the labour force still required by capital in terms of the work tasks needed and their spatial location. This broke the economic advantages to capital of the contemporary housebuilding programme with its high demands on state expenditure. So shifts in housing policy since the early 1970s can be said to have a certain capital logic to them, in the sense that they are related to the needs of general capital accumulation in the British economy. It is useful to present the case for this interpretation of subsequent developments in housing policy before making some important qualifications to it.

For most of the 1970s state housing policies were closely geared to attempts to hold down wages and cut public expenditure. Emphasis oscillated between which aspect was more important, but neither depended on an expansion of housing output (with the partial exception of the 1974 Labour government's social contract with the TUC which promised an increase in council housing). The growing emphasis on housing consumption cost subsidies reflected the concern over the impact of housing costs on wage demands. In this respect it is interesting to note the pattern of rates of increase in council rents and average earnings. Between 1955 and 1967, Merrett (1979) suggests that rents rose much faster than manual earnings. Yet after 1966 this no longer was the general rule. Between 1967 and 1971, years of prices and incomes policies and their aftermath, council rents rose less fast than average earnings (although still slightly faster than inflation in general). In 1972 and 1973, the years of the attempted transition to market-related 'fair rents' of the 1972 Housing Finance Act, rents rose faster than average earnings, but when incomes policies again became fashionable they rose much slower than earnings. Between 1974 and 1979 they barely kept pace with general inflation (HPR, figure 8 and *Social Trends* 1982). Rent rises after 1979 broke the incomes policy links of council rents. Rents rose so rapidly that left-wing Labour councils in their manifestos could put forward the 'radical' demand of limiting rent rises to the same level as earnings.

With the movement in the late 1960s away from subsidizing new building, attempts were made to make up the housebuilding shortfall through programmes of renovation and improvement of the existing stock. Renovation may be cheaper than development (although the financial calculation is rarely done). The life of the existing housing stock is lengthened so renovation is a cheap way of putting off a housing crisis, particularly if the incumbent household can be induced to do much of the work out of their own unpaid labour.

Tenure divisions have been an important component of these changes. Urban rebuilding has been associated with council housing, the switch to a policy of renovation principally with owner occupation. The improvement policy needed little direct state subsidy because of the tenure switch and the DIY propensities of some home-owners (although 'improved' owner-occupied houses will generate large amounts of mortgage interest tax relief). Other differences between the two tenure forms also have helped. The inner city orientation of council housing places it at a location which is increasingly unattractive for industrial capital. The housing requirements of the working class in the inner city are no longer a major component of the reproduction of labour power. The suburbs and freestanding towns have grown in relative importance as employment centres (Fothergill and Gudgin 1982) and this is the domain of owner occupation. Moreover, as chapter 9 pointed out, the shift of housing subsidies away from council housing has been a movement from a tenure whose households have become increasingly economically marginalized towards the tenure of those with greater economic power to undermine the profitability of capital. The incomes policy effect of housing costs, in other words, relates to owner occupation as well as to council housing and the former is now the crucial tenure for its implementation.

Owner-occupied housing may also weaken the link between housing costs and the general level of wages. A high cost has to be paid to enter the tenure and to move within it, yet most purchasers expect their housing costs to decline relatively over time. In this way a comparative acquiescence to high housing costs is generated as most owners accept they are on to a 'good thing'; rising house prices paradoxically reinforce that belief. None the less, the discussion of incomes and house prices in chapter 9 showed that there is still a close relationship between incomes and housing costs within owner occupation. Costs might decline for individual owners or, alternatively, they might get higher housing standards through trading up. But for many owners their housing costs are still high. The distribution across households in owner occupation of these housing costs, however, produces important effects which need to be explored.

Recent entrants pay the highest costs which then decline over time. Outright owners face the lowest costs, as long as they do not incur large repair and maintenance bills. The social composition of outright owners was discussed in chapter 9 where it was shown that over three-quarters were retired or approaching retirement and that the class location of outright owners was broad (for example, 37 per cent were manual workers). In terms of household characteristics, therefore, there are two general dimensions across which housing costs vary: they vary regressively across income groups with those on higher incomes incurring lower equivalent housing costs, and they also vary across age groups, declining with age. Money gains heighten the differentials.

The regressive income effect is obvious as higher priced houses produce a greater absolute amount of money gain and the mortgage interest tax relief is of greater benefit for those on higher than basic marginal tax rates. Yet, the older the owners the more likely they are to have low or zero mortgage outgoings and to have acquired a substantial money gain as well. Older owners are also in a better position to realize their potential money gain by trading down or moving to another tenure. The age distribution of realized money gain helps to explain the age composition of large savers with building societies. Over half the funds invested in societies are held by people over the age of 54 and nearly 30 per cent by those over 65. In terms of housing tenure, half of societies' funds are held by investors who own their houses outright (Stow Report 1979). Not all of this money comes from house sales, nor is it an accurate indicator of the destination of realized money gain; a lot, for instance, goes in legacies. Yet the high costs paid by younger owners do enable this redistributive effect to occur; owners unintentionally seem to be contributing to a bizarre and rather inefficient form of pension scheme. It does not necessarily make older households rich, as for most elderly owners or ex-owners the money realized from house sale provides only a small annual income.

Pressures for better state pensions or for higher wages to finance better private pension schemes may consequently be weakened by the existence of owner occupation to the benefit of capital. There is less need to save out of wages for retirement for the strata of the workforce where owner occupation is prevalent. By the age of retirement housing costs are reduced to repairs, maintenance and insurance and the retired household can 'trade down' and use the encashed money gain as an ongoing source of income. 40 per cent or more of married men and single women and men aged over 60 were outright owners in 1977–9 (*Social Trends* 1981), so this conclusion could be significant for the economy as a whole. Whether the relation between home ownership and the economic position of the elderly is a benefit for the latter or for the rate of profit on capital (or for reduced state expenditure) depends on its impact on struggles over wages.

Even if individual owner occupiers do actually get financial benefits from home ownership, this line of reasoning suggests their distribution is towards older owners. Younger owners, on the other hand, will experience mainly the costs. Yet they are the key group that can influence the profits of capital through wages struggle. If housing costs rise they bear the burden which tax relief subsidies only ameliorate. A series of conflicting pressures, therefore, can be seen to exist in the relationship between owner-occupied housing costs, wage levels and state subsidies.

The high cost of adding new housing to the nineteenth-century rented housing stock forced up the housing costs of all tenants and benefited only existing landlords who profited from the higher rents. With owner occupation, on the other hand, the high cost of new housing may force up house prices but the effects are diffuse. The agencies profiting from the existing structure of provision undoubtedly gain but the incidence of costs amongst consuming households is diverse. New owners bear the brunt of them, existing owners are partially shielded from the effect as the price of their houses rise as well, and those who are able to realize proceeds from house sales may actually gain. The age distribution of these effects may contribute to political acquiescence to their existence. But even so this effect is historically contingent on wider events.

The benefits to non-housing capital of the switch to owner occupation are great. But many of the effects described also create contradictory consequences. If, for example, owner occupiers are firmly enmeshed in the structure of owner-occupied housing provision it is not possible politically to isolate the key agents in that form of provision as it was possible with private landlordism in the First World War and with council housing in the late 1970s/early 1980s. But what happens if the structure of owner-occupied housing provision starts to generate effects against the interests of capital which outweigh its benefits? There is an impasse. The politics of tenure has no further opinions. A housing crisis then generates a general political crisis over housing, not based on class antagonism but on a stalemate created through a lack of alternatives.

Earlier chapters of this book have argued that the structure of owner-occupied housing provision is beginning to create serious social and economic problems. And they are problems for capital as well as for people who need housing. The crisis of production in owner-occupied housing provision means that new output is not being created, whilst housing costs rise. In the short run, variations in the level of owner-occupied housing costs may be determined by contemporary shifts in interest rates but over the longer term house price rises are the determining factor. The upward pressures of housing costs on wages and state expenditure, therefore, still generally exist and it can be reasonably predicted that they will get worse.

Such problems are exacerbated when the position of households is considered. 44 per cent of households after all are not owner occupiers. Their housing position can only get worse in the light of current developments in housing policy. The position of owner occupiers also is not bright. The financial benefits of owner occupation are only benefits relative to the initial starting cost. Rising house prices mean more outgoings on mortgage repayments to finance the price rises; some of this might be offset by increases in money gain but that effect is long term and uncertain. Moreover, the instability of the housing market may make it difficult to move during downturns without financial loss. Large house-price falls during particular periods also cannot be ruled out, given the instability of the housing market (it has happened in other European countries, for instance). An increasing number of households' housing needs cannot be met under current forms of provision. If political agitation grows for more housing and for a reversal of the current decline in housing standards, current policies and political practices will not be able to deal with them.

These statements are deductions and predictions from the analysis of contemporary structures of housing provision. To an extent, such results are already occurring although this rarely seems to be understood. Contemporary housing problems are usually seen as the product of immediate government policy, higher interest rates or temporary economic downturns. But structural problems with the two current major tenures are at the root of existing difficulties. It is unlikely that the problems will go away with more state expenditure or an economic upturn. One indicator of the severity of the longer-term problems of owner occupation, for instance, is the ratio of mortgage costs to incomes.

Because of the wide dispersion of housing costs in this tenure, it is not possible to make simple statements about housing costs and incomes. But one indicator is the initial average level from which a household's costs will decline (with inflation) over time (i.e. initial mortgage repayments), and comparison of that with average earnings. The figures do not take account of the taxation of income or mortgage tax relief, so they indicate directions of change rather than actual proportions of income absorbed by housing costs which are overstated. The data, nevertheless, do show a sharp rise in housing costs between the 1960s and 1970s. In the late 1960s, the ratio of initial mortgage repayments to average earnings was stable at 25 per cent, during the 1970s it fluctuated between approximately 30 and 40 per cent, in 1980 it rose to an incredible 45 per cent, whereas in 1981 it fell back to 38 per cent. In such circumstances it is unlikely that owner occupation has been much of a downward force on wages over the past decade, yet it has increased state expenditure. So it does seem that tax relief has kept the cost of owner occupiers' housing down only for a given level of costs. Because of the present structure of owner-occupied housing provision the existence of tax relief has simply enabled house prices to rise higher than otherwise, increasing the level of total mortgage costs and offsetting the initial tax deduction.

Difficulties are also being generated by the reaction of agents in owner-occupied housing provision to the crisis of housing production. The threat to the planning system was noted earlier in chapter 8, bringing with it problems of the adequate and efficient spatial reproduction of British capitalism. Other potential divisions are also beginning to emerge: the conflict between banks and building societies, for example, was highlighted in chapter 10. Divisions of interest between types of capitalist enterprise over owner occupation are beginning to appear.

Far from being a back-to-the-market capitalist solution to the housing provision of key sectors of the working population in an economic crisis, this argument would suggest that the present structure of owner-occupied housing provision is increasingly coming into contradiction with the needs of capital. The economic failure of owner occupation as a form of mass housing provision consequently goes far beyond affecting solely those households excluded from living in the tenure. It affects households in the tenure and also capitalism itself. This means the growing housing crisis in Britain cannot simply be put down to an onslaught by capital and its functionaries on the living standards of the working class as part of a capitalist way out of the current economic crisis. The economic benefits to capital needed to justify such an argument are not there; increasingly only the private agencies dominating this structure of provision are economically benefiting from them.

There are a number of parallels with Edwardian Britain in the current housing situation. Economic stagnation and decline then were exacerbated by an urban crisis in which inadequate and expensive housing provision played an important part. The private housing market in the form of rented housing was failing then and again it is doing so now with owner occupation. The implication is that a political space for significant changes in the nature of housing provision has opened up in Britain of a magnitude that has not existed since the end of the First World War. This space for change has arisen because a wide variety of interests stand to gain from such change, in a similar way to the coalition which squeezed private landlords from their dominant position sixty years ago. To talk of a 'political space', however, is only to suggest that pressures are growing for fundamental reform. It does not specify what those changes will be, or when they will occur. The parallel with the Edwardian period, however, ends beyond the specification of a discrepancy between economic interests and housing provision.

What is particularly different about the present situation and that prior to the First World War is the political effects of ideological allegiances to tenures and the consequences of the politics of tenure. Households living in private renting had little love or sympathy for the tenure form. This is not true of the two main current tenures. Council housing, despite its failings, still has widespread support amongst sections of the population. It does, after all, house almost a third of households. Similarly, owner occupation has widespread ideological support, not least because of the financial commitment many households have made to it. These ideologies have not been considered in detail in this chapter because, as chapter 9 pointed out, they operate at the level of the individual and are historically contingent. This does not mean that they can be ignored, however. They suggest, in particular, that reforms of these two existing tenures must be the main thrust of new policy initiatives rather than replacement of them by new tenure forms. Otherwise such ideologically based tenure support will act as a barrier to change rather than a means by which it can be implemented.

The effects of the politics of tenure reinforce this point. Prior to 1914 alternatives to private renting did exist in embryonic form. Council housing and owner occupation did already exist so the attack on private landlordism from diverse perspectives had practical alternatives to turn to. This is not the case now. The politics of tenure, in fact, acts as a barrier to change because it obscures the issues. The failings of structures of provision are not clearly understood because of it and the theoretical discourses that help to sustain it. The need to reform relations of production in housing is currently not an important area of political struggle. Because the real problems are not recognized, however, does not mean they do not exist. Inaction just exacerbates the housing crisis.

Future housing policy depends on the political options on offer. State expenditure to owner occupation may be cut in the name of financial expediency or fiscal fairness. Those measures will only exacerbate the housing crisis by choking off yet more output. Alternatively, restoring council housing cuts, bolstering demand in the owner-occupied sector, or directly subsidizing private housebuilding will not create much new housing, only sharpen inflation in construction and land costs whose impact will rapidly spread through to affect rehabilitation and modernization programmes and the costs of existing housing. Political programmes that alter the structures of provision in the two tenures, if fought for, however, would provide a viable and progressive alternative.

13

Towards new forms of
housing provision

Introduction

Earlier chapters have argued that present forms of housing provision are
increasingly unable to satisfy basic housing needs without generating
unreasonably high housing costs. Moreover, they impose poor employment and working conditions on building workers. And, furthermore,
the physical built environment generated is one of poorly planned urban
sprawl, inadequate building design and standards and increasing urban
decay. None of these characteristics is inevitable; instead they are products of the social organization of housing provision. The only way to
change them, therefore, is to change the social relations involved. This
chapter puts forward proposals for what are felt to be viable and feasible
reforms.

It is tempting when suggesting reforms of housing provision to present
detailed plans of a new scheme with all its intricacies to demonstrate how
changes would work in practice. Howard's Garden Cities or Unwin's
plans of economical working-class cottages and sketches of street layouts
to demonstrate that nothing is gained by overcrowding spring to mind as
respectable, if failed, precedents in a long lineage of housing schemes.
(So does, less fashionably, Le Corbusier's Radiant City.) Presentation of
detailed schemes, however, seems wrong in principle and poor political
practice. It suggests that there are unique but general organizational or
technical solutions to social problems. This is wrong. Social change is a

product of complex and historically specific social struggles. Their precise outcome can never be predicted in detail beforehand. Adherence to a rigid and detailed plan could not take account of inevitable changes in the course of events that would occur. Social change, however, cannot move in any direction. What is important is that it adheres to certain principles which have been demonstrated to be necessary for success. Within the broad framework of those principles the context of political debate and the need to win broad support should determine the detailed content of any programme for reform. The principle that has been argued for in this book is the need to take housing provision out of the market mechanism, and so out of the control of the private agencies that dominate there, in order to give people a greater control over this vital aspect of their lives. There is, in other words, a need to bring life and meaning to that well-known slogan 'housing for people, not for profit!'

An implication of this argument is that no amount of proposals for change, or alternative strategies, can by themselves produce the desired end. They are at best a means by which to get there. They could be distorted or diverted during their apparently successful implementation into something their proponents never meant them to be. The history of council housing is a reminder of the dangers of being mystified by the ideological images of organizational forms rather than being critically aware of the potentially negative aspects of their content. Organizational schemes can be no substitute for political struggle but their successful implementation nevertheless may still alter the content and balance of future political action.

Another characteristic of potential reforms is that they must have some rationale and links to the contemporary situation in which they are supposed to intervene. Pure ideal forms of social practice might be desirable, whilst anything less may be full of inconsistencies, yet those pure forms are utopian if they relate to no existing social practice. No one knows, furthermore, quite how perfection works. A wide variety of experiments with housing provision are required to see how better forms can evolve. But they must start off from situations that are both economically feasible and relate to people's contemporary consciousness and their existing housing situations. This implies, as the previous chapter suggested, that reforms need to tackle the problems of council housing and owner occupation rather than attempt to create a new mass housing tenure. In addition, and less palatably, it implies that housing inequalities cannot be removed overnight but rather that some of the worst aspects of the existing situation should be dealt with first.

It is from these perspectives that the issues of land and building industry nationalization are considered. They are argued to be part of the preconditions necessary to enable the transformation of housing provision rather than ends in themselves. Only with their existence can the control of housing provision by private capital and landownership be questioned. Obviously they are major reforms which to many would seem politically impossible; revolutionary or transitional demands rather than practicable reforms. But the arguments of the earlier chapters, which showed the extent of the underlying housing crisis and its cause in the failures of existing structures of provision, suggest that anything less would have little effect. Moreover, any radical reform is always seen as politically impossible by many until it happens. The history of council housing prior to 1914 or of land-use planning prior to the 1940s, two issues discussed in chapter 7, are littered with political impossibilities swept aside by the dynamic of the respective crises to which those reforms were related.

Why the reform of housing finance is not enough

Most expert opinion on housing argues for the reform of housing finance to solve the housing crisis. In particular, it argues for the taxation of owner occupiers' money gains but sometimes also for changes in council housing finance. The arguments are often technical and complex but essentially they miss the point. Housing finance is a set of accounting conventions that reflect the current state of social relations in housing provision. Trying to alter the accounting conventions without changing those social relations is fraught with difficulties and unforeseen consequences. Changes in housing finance may well be required but they should follow changes in social relations rather than try to avoid them. Unfortunately the plausibility of financial reform alone has widespread credence amongst housing activists and the Left in general, not least because one of the principal housing lobbies, Shelter, has been expounding this approach for a number of years in its magazine, *Roof*, and in a series of pamphlets such as a recent one entitled *Housing and the Economy* by Kilroy and McIntosh (1982). As this 'housing finance' view is so prevalent, it is necessary to highlight some of the theoretical inadequacies of its particular consumption-orientated approach.

The problem with the Shelter literature is that it accepts much of the jargon of expertise in housing finance as inevitable, even adding its own. The result is arguments which few people are technically equipped to understand, but as they come from experts blessed by Shelter most are prepared to accept the conclusions on trust or resign themselves to leaving housing finance to the experts. The approach, in other words, reinforces the politics of tenure described in the previous chapter. The Kilroy and McIntosh (1982) pamphlet is a confused mixture of Keynesian economics and monetarist jargon (like 'crowding out'). The content of the argument is of less interest than the conclusions: raise the general level of council rents (through national rent pooling) and reintroduce schedule A tax for owner occupiers 'to reduce the spur to demand for higher standards in housing which crowds out crucial housing priorities and may affect production in the rest of the economy'. Or 'Rising housing standards cannot continue as in the past without causing shortages and deterioration as well as a competition for resources with the rest of the economy' (Kilroy and McIntosh 1982, p. 13). Put another way, most people's housing situation has to be made worse to make some others' better. But why think in such narrow distributional terms? Why not 'crowd out' the landowner, speculative housebuilder or contractor, and the money-dealing capitalist? To treat housing solely as a distributional issue is to impose unnecessary costs on households in the name of equity whilst ignoring the creators of the inequalities themselves. It is rather like discovering that Robin Hood worked for the Sheriff of Nottingham as an undercover agent after all.

Like all other consumption-orientated studies, distributional studies ignore many of the potential structural constraints on state policies. Housing costs and housing policy, in particular, are not seen in terms of the impact of specific structures of housing provision. This is shown clearly in King and Atkinson's recent study (1980) which attempts to compare the rate of return on capital for owner occupation and council housing and suggests remedies by which these rates of return could be made more equitable. Their key policy proposals again are national rent pooling for the council sector and additional taxation for owner occupiers. The tax reforms proposed for owner occupation are either the abolition of mortgage interest relief or preferably the reintroduction of a schedule A style tax on the imputed income derived from home ownership. The political 'carrot' of such proposals, King and Atkinson argue, would be the possibility of income tax cuts made feasible by the additional state revenue derived from housing.

These proposals have considerable political credence so it is worth considering their argument for them in detail. The usual intellectual justification for such policies is based on comparisons between households' housing costs over time. The time dimension, however, implies the need for a comparison of individuals' subjective evaluations of different temporal profiles of cost. As theoretically there are an infinite number of possible preferences for consumption now versus consumption in the future, some simplifying assumptions must be made for calculation and comparison to be feasible. Yet the simplifications can have no theoretical basis so they are both arbitrary and unreasonable.[1] King and Atkinson try to overcome this difficulty by adopting what they call a return-to-capital approach. The difficulty of comparing subjective preferences is said to be avoided by treating the housing stock as if it were capital on which a rate of profit is made.

One obvious objection to this approach is that the purpose of much political agitation, especially over council housing, has been to remove the social relation of capital in housing (moves encapsulated in the slogan 'housing for people, not for profit!'), so to treat council housing as capital is to ignore one of the major justifications for its existence. And the fallacy of treating owner occupiers as capitalists was criticized earlier in chapter 9. But, even on its own terms, this capital approach faces severe difficulties.

A rate of profit is obviously a ratio, and to make comparisons between rates of return the numerators and denominators of the ratios have themselves to be comparable. This is not the case with owner-occupied and council housing. They are organized, priced and financed in totally different ways, reflecting the fact that they are distinct systems of housing provision. Any attempt to produce a surrogate measure of a unitary entity called 'capital' across these two tenures therefore is invalid (calling two different things by the same name does not make them the same).

The impossibility of making comparisons of the real incidence of costs to individual households in different tenures means that there can be no justification for schemes that claim to provide a unified approach to housing consumption costs across tenures. Such unified approaches turn out to be just an amalgamation of their authors' pet housing finance schemes for different tenures. A comparison within tenures, however, has more justification. It is easy to illustrate that all owner occupiers derive considerable benefits from inflation, and that the longer the household has been an owner and the richer the household the greater the financial benefits of ownership. The precise benefits to households still cannot be evaluated, nevertheless, as that would involve attempts to compare individual preferences. But even so distributional equity studies are on firmer ground when they discuss taxation reforms for owner-occupied housing.

Here again, however, problems arise over just looking at demand. It is commonly argued that one benefit of taxing owner's imputed rental income would be a fall in house prices, which would improve the situation of potential new owner occupiers. The price fall would occur, it is argued, because households would face higher costs for any level of housing price and, therefore, would reduce their demand. King and Atkinson, for example, provide an estimate with prices falling by 20 per cent. But more would change than one demand parameter.

First, falling house prices might create difficulties for mortgagors if they no longer had adequate security for their mortgage at the new valuation of their home. Foreclosures and disruptions to the whole system of mortgage finance could easily result, affecting especially lower-income households and those with minimal money gains from earlier house price rises.

Second, if prices fall substantially the supply of new houses could easily dry up so that in the long run house prices may actually rise. The fall in house prices and the disruption to the new housing market that would result, even if the reform was gradually introduced, would lead to the collapse of the housebuilding industry. It cannot be assumed that the only effect on housebuilding would be a reduction in land costs. Land development, after all, is where most of the speculative builder's profit comes from. Depending on the extent of any subsequent price rise, existing owners might still end up with higher money gains (after paying the tax) than now. New entrants and poorer owner occupiers would incur the cost through reduced housing opportunities, higher prices and additional taxation.

The effects of reforms on the wider structure of housing provision consequently cannot be ignored. Housing policy changes must not be seen solely in terms of altering one parameter determining demand. This is not to argue that reform is impossible but it does cast doubt on the feasibility of simple fiscal adjustment.

There is one further problem with tax reforms associated with the reintroduction of schedule A tax or any similar measure. This sort of tax is ongoing and paid out of current income, like rates. Yet the tax is supposed to be on the wealth tied up in housing. This is unrealizable until the house is sold. The financial burden on households, therefore, would be considerable, because schedule A would have to be paid out of current income with no offsetting financial benefit from the house being taxed. The tax cannot be paid out of money gains as they are purely notional until the house is actually sold. The problem did not arise when schedule A was directed at private landlords as their incomes from housing were actual rents rather than the imputed ones of owner occupiers. It was suggested in chapter 11, furthermore, that money gains may not be an unqualified benefit to individuals anyway because of the possible counteracting effects on other components of income.

The political difficulties of introducing taxation measures on owner occupation consequently are exacerbated by the consequences of adopting a tax on income. A tax on owner occupation, if needed, is likely to be more politically feasible and equitable if it is a direct tax on realizable wealth: for instance a sales tax or a wealth tax on the realized money gain component of the selling price.

The consequences of land and building industry nationalization

The argument for nationalization is that it transforms the structures of provision of both owner occupation and council housing. This section will examine the mechanism of nationalization including the question of compensation and the principles under which a successful nationalized building industry should operate. Obviously when talking about the nationalization of land and the building industry many spheres of economic and social life are being affected, as there would be a transformation of the whole of the built environment. Here, however, as in earlier chapters, only housing provision is dealt with.

With respect to land the area of interest here is building land: either redevelopment or greenfield sites. The state takeover of building land could either be part of a general transfer of ownership of all land as in the scheme prepared by Massey, Barras and Broadbent (1973). Or it could simply be of land as it came ripe for development or redevelopment, as with Uthwatt's 1942 proposals or the Community Land Act of 1975. Under the total nationalization proposals all land would be used on leaseholds from the state. A change of use would require a new or renegotiated lease on the part of users whilst the conditions of the lease would enable the state to redevelop the land when required. As Uthwatt and others have argued, total nationalization is preferable because it creates unified land ownership, whilst partial nationalization would be subject to all the vagaries of political or legal sabotage, as the Community Land Act so aptly demonstrated (Barrett, Boddy and Stewart 1978).

Nationalization of the building industry would avoid the situation where the struggle over the appropriation of development gain was transferred from private landowner/private builder to public land-owner/private builder. It could, moreover, create preconditions whereby the organization of the production process in building was no longer dependent on the attempt to appropriate speculative development gain or to maximize profits under the contracting system.

Nationalization of an industry concerns the takeover by the state of the productive potential of the industry: the fixed plant and equipment, stocks, and technologies required for subsequent production. With the building industry the nationalization proposals would have to be drafted with care as in many cases ownership of building firms does not necess-arily lead to the ownership of productive assets because of the complex networks of subcontracting and plant hire in the industry. The takeover of development land, given the nature of the speculative housebuilding industry, is the acquisition of most of the physically productively useful assets of speculative housebuilders. Similarly many of the assets of build-ing contractors are investment ones, like office blocks, shopping centres and industrial estates, or cash reserves or the paraphernalia associated with tendering and client disputes. Few are of direct usefulness to the physical act of building and so of little consequence for the new state building enterprise.

Compensation

Compensation for the nationalization of building firms and land is essen-tially a political matter of minimizing opposition to the nationalization. Most previous nationalizations of industries in Britain have been on exceedingly generous terms to shareholders, given the declining nature of those industries. Because of the nature of most building firms' assets and the social consequences and costs of their failings as productive enterprises, it should be possible to conduct political campaigns that do not lead to such generosity with the building industry. The acquisition of speculative housebuilders, in particular, raises problems because to compensate shareholders on the basis of contemporary share prices is to pay for the expected development profit to be made on firms' land banks. This could easily be in contradiction with the lower compensation paid to other landowners.

Most previous land nationalization schemes in Britain have argued for compensation to be given to landowners on the basis of the current-use value of their land: for example, the value of farming land in that activity rather than at a value which includes some expectation of future urban development (see the survey in Lichfield and Darin-Drabkin 1980). The issue of compensation, however, is not clear cut. The loss of development rights under the 1947 Act, for example, was not completely compensated for. Nor have successive rent control measures on private housing to rent, a similar form of real property, been linked with compensation schemes. Another means by which the state reduces the economic benefits of landownership is through land taxation. It would obviously be absurd to compensate landowners for the taxes they pay. Nationalization involves the loss of economic benefits of ownership by the previous owners, which is why compensation schemes are usually proposed. The necessity of compensation, however, is a political matter rather than one of social justice, as was concluded in chapter 7.

It could be argued that for landowners as a whole the years of zero taxation since 1947, despite the state's already existing ownership rights in land via the right to develop, have more than compensated them for any future loss of ownership. Compensation consequently could be reduced to a matter of individual landowner hardship rather than a general handout to landowners. One method by which a no-compensation policy could be implemented whilst maintaining the niceties of the rights of property would be to introduce a 100 per cent tax on revenues from land first and then compulsorily purchase all land at its new, zero price.

Surprisingly, it tends to be governments which profess socialist principles that have been most respectful of the rights of private property. As was noted earlier, the Conservatives abolished Labour's 1947 £300m. compensation fund for the nationalization of development rights, whilst Neville Chamberlain, hardly a revolutionary firebrand, suggested to the Cabinet in 1935 a *de facto* compulsory acquisition of strips of land on either side of main roads *without compensation* to solve the problem of ribbon development (Sheail 1979).

One possible approach to the problem is to direct compensation principally to land users rather than to landowners. This would separate off owner occupiers (households, farmers, shopkeepers and industrial and commercial capitalists) from other landowners. The latter could be compensated on individual merit if at all, whereas the potentially disruptive effect on, say, mortgage finance for owner occupiers could be taken account of by loan guarantees whilst the financial means to move elsewhere could be provided when the owner occupier wanted to move or when farmland was required for new building.

Operating criteria for the new building enterprise

Once the building industry and land have been nationalized the way in which the process of urban development subsequently operates depends on the precise forms of organization that arise out of the contemporary political situation, especially on the political alliances and compromises necessary to get the programme into operation. There are however certain broad features which any new forms of building production and land control must adhere to if they are going to break with current practices. Nationalization by itself does not mean very much. The Morrisonian style of public corporations adopted for nationalized industries in Britain to date are widely recognized to reproduce many features of pre-existing private ownership. They have reproduced with only minor modification capitalist relations of production with the imperative of accumulation and its necessary oppressive control of the workforce. Little possibility of production for need or of giving the workforce greater control over their work tasks has been possible as the capitalist criterion of efficiency, profit maximization, has not been questioned. If anything it has been reinforced as the state has had the resources and power derived from the centralization of capital to force through widespread rationalizations of industries against the opposition of the workforces in a way in which the fragmented capitals of pre-nationalization could not achieve. The examples of coal, railways, shipbuilding and steel highlight the difficulty of seeing nationalization as an end in itself.

From the discussion of speculative housebuilding in earlier chapters it can be seen that the crisis of production in housebuilding is not simply a problem of ownership but of the relations of production implied by that ownership. It has become fashionable, for example, to highlight problems of finance for productive industry in Britain as an explanation of low rates of growth of productivity (Minns 1982). This clearly is not the case with speculative housebuilding; if anything the industry's access to long-term capital for investment has increased over the past decade, as chapter 3 argued. If when nationalized the housebuilding industry tried to maximize profits through land banking and land development it would, by the rationale of the market in which it was operating, be forced to adopt very similar production and employment practices to those currently used by speculative housebuilders. The present crisis of production can only be resolved by going against the criteria of the market, not by changes of ownership alone.

The mistaken belief that profit-maximization leads to the most efficient forms of production is so well entrenched that it is worth repeating the conclusion of earlier chapters that the crisis of housing production is a product of organizing it on profit-maximizing lines. A change to organizational forms which emphasize social need and improved forms of production and employment, therefore, does not just lead to benefits in terms of greater social justice but also of greater productive efficiency. The current housebuilding industry is extremely wasteful of the labour power, building materials and land it uses as inputs and it produces comparatively poor-quality dwellings. Once a break with the overwhelming drive for profit is made considerable improvements in production are possible.

Quite how changes in production techniques would take place is difficult to predict because, as chapter 6 argued, the preconditions for their existence have never been put into practice. Yet the analysis of that chapter concerning the nature of the current crisis of production does suggest what those preconditions are, with implications for the organization of the industry. It was suggested that speculative housebuilding could not achieve continuity of production nor economies of scale and that the poor conditions of employment made it difficult to sustain a viable workforce. The implications are that production should be organized with those criteria in mind. A large-scale planned building programme could generate continuity of production, with phased shifts of production tasks between sites, and working conditions could be improved through the employment of workers permanently on union-negotiated wage rates and terms of employment.

Decasualization of the building industry is a precondition for any improvement in the productive capability of housebuilding. No longer would it be possible to use a casual, self-employed workforce to keep wage rates and working conditions down. Similarly, wage rates could be less directed towards piecework and other payment-by-results systems. Training programmes for existing workers in new tasks and for new workers must also become a priority for the long-term viability of the industry.

Decasualization would considerably increase the collective power of the workforce economically, in terms of wages and conditions, and also politically, in terms of an ability to exert power over the form and content of building programmes both nationally and locally. Decasualization only creates a precondition for the exercise of that power. It also depends on the collective political will of the workforce to use that power and create organizational forms through which it can be exerted. New trade union practices would be required in the industry for that to be possible. A break would have to be made with the internecine jealousies of present-day building trades unionism (possibly through the final achievement of one union for the industry), whilst, if that new-found power was to be a genuine reflection of the collective will of the workforce, union leaderships would have to be clearly accountable to ordinary union members. This would necessitate the existence of an active branch/site based local level of trades unionism to ensure the continued existence of genuine democracy within union structures. It would be important to organize the nationalized building industry in a way which made such active local-based unionism possible.

One object of nationalization, in other words, should be to shift economic power to the workforce. Far more resources would have to be devoted to better wages, conditions and training than currently is the case. To many people brought up on an ideological diet of anti-unionism and wage-rises-cause-inflation this might seem like a recipe for disaster. However, as chapter 6 argued, it has been the systematic over-exploitation of the workforce that has been a principal cause of the current crisis of housing production. Investment is currently unproductively centred on land to get development profit at the expense of efficient production and the workforce. With nationalization the whole emphasis of investment can shift from speculative profits to production. Labour is the central productive force. Improvement of the position of building workers, therefore, can be seen as an investment in productive capacity. Transforming the employment conditions in the industry is an improvement in the industry's productive potential. Improvements in the economic situation of building workers, therefore, are not simply gains to one sectional interest within the working class, important though those gains are. Much broader groups in society would also benefit from the improved productive capability of the housebuilding, and wider construction, industry.

There are important implications for costing, pricing and financial objectives of a nationalized building industry that follow from the arguments about the need to change the content and methods of production. Speculative housebuilders place emphasis on land development because of the drive for profits. The same pressures would be on a nationalized building industry if it were subject to similar profit criteria. Conflicts between landownership, land development and housing production would arise again within the new state bodies, reproducing the previous contradictions of private ownership in a new supra-state form. New financial criteria and social objectives have to be devised to stop the evolution of such regressive practices.

The fundamental principle required is that the difference between revenues and costs (i.e. gross profit) should not be the principal operating criterion for the state enterprise. One way of doing this is to plan building programmes on the basis of need, whilst making the operating criteria of the enterprise relate to meeting the objectives of the plan, subject to special constraints on its own practices. There are three components of this process which need to be elaborated: the formulation of the building programmes and the concept of need, the role of revenue and the operating criteria of the enterprise. They are all complex issues which can only be briefly dealt with here. Because of the importance of wider issues in the first two components, the last one will be dealt with first.

The operating criteria of the building enterprise have to relate to specified planning goals, to the efficient use of building resources and to its own internal working practices. This implies that there should be an amalgam of physical and financial goals rather than an attempt to reduce economic performance to one unitary measure. There should, in other words, be a social audit rather than the conventional private company style financial statement. The success of the productive unit could be measured in physical terms, such as number of dwellings built, or in the often more meaningful criteria which try to capture the qualitative aspects of building products, such as square metres of living space or the quality standard described in chapter 5 (table 5.2). Another dimension of a social audit would concern the enterprise's success in increasing the productive potential of the building industry during a given time period in terms of, say, additional plant and equipment, the development of new techniques, design formats or the adoption of new materials, plus those associated with the workforce such as training programmes and the health and safety record. Similarly productive efficiency could be measured in physical terms such as number of labour hours per dwelling, bricks used and so on. There is no reason, however, why money costings also should not be used as elements within the accounting processes of building, as long as they do not relegate other measures to a secondary role whilst they, in the guise of profit-maximizing or cost-minimizing, become paramount.

What is at stake in the content of the operating criteria of enterprises are the issues of accountability and control. In capitalist enterprises accountability is only to senior management and to shareholders, and internal firm control is a coercive form of management supervision of the workforce. Accountability needs to be far more democratic and broad based if building industry nationalization is to be successful. It must relate to the final product users, to the built environment planning process to which it relates and to the workforce of the enterprise. This is why some form of social audit is vital. Accountability must include some form of political representation at local levels and trade union representation if the needs of users and workers are to have effective forms of expression. The organization of the industry would have to be designed to accommodate those features. This point will be raised again later.

Within the enterprise itself management obviously is still required. The aim, however, should be to reduce its coercive content yet enable organizational failings to be identified and rectified. Given the heterogeneous nature of building work, in particular, it is important to have detailed profiles and costings of individual sites so that remedial action can be taken if things start to go wrong. Unlike the present-day building industry, with its closely guarded commercial secrets, this information should also be part of the wider accountability process so that clients and other interested bodies can be aware of the detailed operations and problems faced by the building enterprise. Without the existence of such details, public accountability could become a sham as problems and failings could be lost or 'laundered' in aggregate or average data.

It would be naive to expect that once the profit motive had been removed as the sole operating criterion of the building industry all social conflict over the aims and practices of the state building enterprise would cease. It is easy to imagine how the interests of external groupings, like users, may diverge from those of groups inside building production, or that management objectives may conflict with the interests of particular groups of workers. The existence of such conflicts may actually be fruitful as they help to ensure the existence of continual, broad-based vigilance over the conduct of the building enterprise, and help to keep in focus the debate over its objectives. A problem with the profit-based capitalist forms of organization is that conflict is seen as something to be overcome in the interests of a small group of managers and owners alone at the expense of everyone else. The new form of organization, on the other hand, could bring potential conflict into the open and the opposing interests would be more evenly balanced in their economic power. Conflict, in other words, would be politicized in the broad sense of the term instead of distorted economically in the name of profit.

Acceptance of the necessity of conflict and the encouragement of organizational forms which enable its expression to become a progressive force must be a central concern in the state takeover of building. It is another way of stressing the importance of mass, democratic accountability and control over the building enterprise. Only with the existence of such democratic forms can the twin problems of state ownership and bureaucratization and corporatism be confronted. As the state is a site of class struggle, particular state organizational forms have developed in the past that reflect the unevenness of that struggle and the dominance of capital. These effects of state intervention, which have been classically reproduced in council housing provision (see chapter 12), are not inevitable products of present-day society but tend to arise from the particular ways in which reforms and policies have been politically enacted. Social change has essentially been treated as a process of administrative reform from above. There has been no real possibility for the mass of people to be themselves involved in implementing programmes of reform, reflecting their interests rather than the reproduction of the present social structure. Hence, social reforms have tended to fail, leading to disappointed aspirations and apathy on the part of most users of state-provided facilities and the workers in them, whilst those state activities have been pushed more and more towards functioning as a means of social control.[2]

Decentralization must, as a result, be an important component of the organization of a nationalized building industry. So far, for expositional purposes, discussion of nationalization has been posed in terms of a new single state enterprise. In practice, it might be preferable to have a series of enterprises divided by function (civil engineering, repair and maintenance, housebuilding and so on) and by locality. Again a whole series of compromises have to be made, so a blueprint cannot be handed down. A key compromise in terms of decentralization is between the enhanced likelihood of popular control with greater decentralization and the potential loss of continuity of building work and economies of scale with small local units of production. To an extent these conflicting principles operate at different scales for each type of construction work. The operation of motorway building at the level of district councils is obviously absurd but housing repair and maintenance is viable at that scale. With new housebuilding, however, the smallest feasible level is likely to be that of a metropolitan area.[3]

A further advantage of decentralization is the variety of organizational forms of the industry that could develop. Comparison of their respective experiences would provide a useful means of learning lessons and high-lighting the changes needed to overcome problems as they arise.

Land development, land-use planning and building programmes

Public ownership of land and the building industry would transform the development process and land-use planning. Positive schemes for land-use change could take the place of the essentially negative planning control functions of today. Landowners' personal whims could no longer influence the pattern of urban development nor would speculative land development gain be the driving force behind new building. Land-use planning instead would become part of the wider process of planning the built environment, in the three interrelated spheres of new building, preservation and improvement. Land-use planning, therefore, comple-ments the drafting of planned building programmes discussed earlier.

For the transformation of planning to be successful the present practice of planning would have to be altered radically. In particular, the role of the planning bureaucracy would have to change considerably. The professionalization of planning practice was discussed earlier in chapters 7 and 8, where the attempt to treat planning as an essentially technical exercise to be performed by a professional planning élite was suggested to be a major contributory factor in the depoliticization of planning and its excessive concern with the needs of the developer. Plan-ning again needs to be seen as a political process subject to conflicting pressures in order to break down the false view of planning as the discovery and creation of ideal solutions to the process of land-use change.

Central to genuine changes in planning practice is the reformulation of the notion of need in planning objectives. Earlier, in chapter 7, it was pointed out that needs are currently identified in vague general terms like 'full employment', 'better housing' and so on. More detailed needs might subsequently be identified in the 'survey' part of the planning process via investigations of road traffic flows, shopping trips or housing conditions. But the formulation of such needs depends on the techniques of planning (e.g. traffic flows are amenable to computer modelling), whereas which of them the planning process tries to respond to depends on the feasibility of their satisfaction given the existing dominance of private development interests. Needs that the planners do not 'see' or cannot influence fail to get a response. With the advent of public ownership what is feasible in planning terms is altered and with it the role of planners as identifiers and interpreters of social needs would have to change.

The problem of formulating needs raises the whole question of democratization again. The interests of the mass of the population have to be given effective forms of expression if planning is to be more than the product of the will of a bureaucratic élite. This means that creating the conditions under which effective mass expression can take place must be central to the reform of the planning process. Political channels and organizational forms need to exist to make that possible. General interest in planning anyway should be greater as the process of planning will be more meaningful in terms of meeting its objectives and, hence, more directly political. Similar points could be made in relation to building design, and also to the means by which building programmes are formulated.

Arguments about the need to reform the content of planning, or of any other professional activity associated with the process of land development, should not be interpreted as denying the importance of technical specialists. Rather it is an issue of in whose interests they operate. A division of labour in the production of knowledge, and so individual specialism, is as important as in the production of physical objects. Plans need to be formulated, information gathered, economic constraints understood, spatial functional linkages explored, and co-ordination with national spatial and economic planning undertaken, for instance. Similarly knowledge is required to design buildings. Nor should technical specialists merely have to respond to predetermined instructions, as that would deny the possibility of progressive innovations in technical practices. A necessary tension must exist between what seems technically preferable and what satisfies articulated needs. As was argued earlier the point of changing present forms of housing provision is not to remove the existence of conflict but to channel it in progressive forms, and to be continually aware of whose interests ultimately are being served.

Implications of new structures of housing provision

Again principles rather than detailed proposals must be considered when looking at the effects of land and building industry nationalization on council housing and owner occupation. Other reforms will also be required, particularly of housing finance, to reflect and reinforce the proposed changes in housing provision.

The key element of the reform of council housing is the removal of the private agencies dominating its structure of provision. Land and building industry nationalization remove the two most important but socialization of design would still be required and so particularly would the removal of loan capital. The obvious implication of the latter is the reform of council housing finance. The archaic system of building houses by borrowing over a sixty-year period and then subsidizing the interest costs of borrowing must be replaced by a fairer system related to the direct costs of building, improving, maintaining and managing the existing stock. Central government, in other words, should allocate funds for housebuilding to councils on a direct basis, in a similar way to the present-day funding of central government sponsored construction work, such as motorways or hospital building.

Once council housing was no longer dependent on loan funding the pressures of financial necessity on rent setting, housing allocation and housebuilding would be considerably diminished. Council housing finance would no longer be uniquely susceptible to fluctuations in interest rates. Nor would the pressure for rent pooling be so great, as direct funding of new building reduces considerably the knock-on effect of new building on a council's housing revenue account. Current subsidies or additional ones for building undoubtedly would still be required as the housing provided must be cheap and good if the new organizational forms are to be successful rather than dogged by financial crises and disenchantment with the new housing built. But the questions of rent levels and subsidies should be a secondary consequence of prior plans related to needs and physical resources, rather than paramount as at present.

It is often argued that loan financing of council housing is a good thing as borrowing spreads the costs of new building over a long time period, but such arguments confuse accounting practices with economic principles. What is being suggested here is that council housing accounting costs should no longer include large, recurrent interest payment items. How the money allocated to council housing is funded, through taxation or borrowing, is an entirely different matter of macroeconomic policy. What needs to change is the accounting appearance of council housing costs which force rents up to astronomical levels yet still seem to show council housing as a massive drain on the Exchequer in comparison to other state activities.

With owner occupation the changes obviously would have to be somewhat different. A central concern must be to break the links between need and the ability-to-pay. Housing, in other words, should be built on the former criteria in this tenure, on lines suggested in an earlier section, whilst pricing policy must be kept separate. This would mean that particular types of housing could be built or emphasis concentrated on areas of current shortage on the basis of planned building programmes. Two consequences would be the possibility of reversing the dramatic decline in new housing standards and of consistently concentrating on house types of greatest need rather than reproducing the emphasis on the luxury sector generated by the criteria of ability-to-pay.

Although the building industry would be a major supplier of houses to the owner-occupied market, and hence a major influence on the operation of that market, second-hand sales would still outnumber its output. Co-ordination of buying and selling in this market by a new-style, publicly accountable exchange agency offers untold information and cost benefits over present practices. This could be done through local authorities or distinct housing exchange bodies, nationally or via regionally/locally linked networks. Computerization of exchange information and conveyancing procedures would be made feasible, avoiding the current multi-duplication of tasks to the cost of households and the benefit of the exchange professionals. Similarly the conditions under which exchanges took place could be improved. Houses could be assessed on nationally recognized quality standards in place of the estate agents' double-speak of 'studio' (i.e. one room only), 'garden flat' (i.e. in the basement) or 'executive style' (i.e. no first-time buyer can afford it). Offers of houses for sale could also be made dependent on a comprehensible house-condition survey being made available to prospective purchasers.

The basic principles of mortgage finance would not necessarily have to be altered in any major way. They are well understood and accepted by most owners and do not have the cost pooling effects of council housing loan finance. As one of the major beneficiaries of the current structure of provision, however, building societies, unless reformed, are likely to try to sabotage any new structure of provision. Various solutions could be used to tackle this problem. State takeover of mortgage finance is one option (or, in fact, a series, as it could be done in a number of ways). Alternatively, as Minns (1982) has suggested, the existing friendly society status of building societies could be taken advantage of by modifying their statutes to change the method of appointment of their directors and/or the conditions under which they can borrow and lend in order to force conformity with the new housing market relations.

Central to the operation of the transformed owner-occupied housing market, of course, is the determination of prices. The long-term influence on the level of market prices is the prices at which new dwellings are sold, so the pricing strategy of the state housebuilding enterprise is crucial. Before dealing with that issue, however, it is worth noting the variety of strategies that can be used to deal with short-term demand and supply imbalances and second-hand sales. There are two basic possible approaches.

The first approach is the use of administered prices, at which sales have to be made, such as so many £s per m² for housing of a particular quality level in an area or region. The determination of such prices could be an amalgam of demand, building cost and political factors. This procedure has advantages and disadvantages. It would necessitate a degree of administrative allocation where demand was greater than supply in the form of first come, first served queues or some other such principle; without such administrative control black markets with 'key' money would develop.

The second approach would be to let the market find its own clearing price level. Ability-to-pay comes to the fore here and this method could lead to speculation in second-hand sales and have a destabilizing effect on the sales of new housing. Yet taxation measures could minimize the price inflation effects. An example is a sales tax related to the money gain component of the selling price that varies with the individual rate of house price inflation compared to some norm, and the tax could rise to 100 per cent in cases of severe inflation.

Black markets are always likely to arise with shortages; the point is to minimize them and avoid creating the conditions where sustained profiteering can occur. Public registration of transactions and their prices (as already occurs in Scotland) would be a necessary complement of either of the two pricing approaches discussed above. This would make it more difficult for the black market payments to be reproduced at every subsequent sale. But most importantly the existence of black markets depends on the size of the housing shortage and the pricing of new housing output.

One approach to the pricing of new housing is to set it very low and fund housebuilding out of large-scale state subsidies. This method however does not seem appropriate. It contradicts the principle that absolute price falls should not be encouraged in owner occupation because of the wealth existing owners have tied up in their houses and it also implies the need for large-scale taxation to fund it. A better procedure would seem to be setting output prices near to existing levels. This would not only fund the owner-occupied proportion of housing output but also provide large surpluses for use in other programmes. Those surpluses could be channelled toward council housing or other forms of building, to central government, or elsewhere via the administrative prices charged for housebuilding land. Subsequent price changes could be geared to the rate of change of money housebuilding costs. If they were rising because of, say, general price inflation a sales tax could be imposed on second-hand sales to remove the potential for further money gains out of owner-occupied housing. The rate of tax could again be made dependent on the rate of house price inflation.

The essential point is that once housing supply is controlled considerable leverage is given over the operation of the owner-occupied housing market that no amount of playing around with 'housing finance' can ever achieve. A wide variety of pricing and output combinations are feasible, each with different effects. What necessarily changes for the consumer are the financial aspects of owner occupation, which change as a consequence of removing the tenure from the private market and the speculative interests that dominate it. A gradual blurring of the distinction between ownership and renting could be envisaged once market criteria are no longer overwhelming. Tenure and how housing is paid for could become of less importance to households than styles of living and choices of built form.

Conclusion

Only a broad outline of principles has been given in this chapter for reasons stated earlier. Elements are missing and it is quite impossible to highlight problems with any detailed proposal or to raise problems of the transition towards new forms of housing provision. The principles discussed have been highlighted to point out the changes in the social relations of housing provision necessary for improvements actually to occur. Anything less would lead to yet another experience of the disappointments of half-hearted reforms that cannot match the rhetoric used in creating them. The disappointments of reformism have been the unfortunate outcome of British housing policy to date. This book has tried to suggest why that was the case and that there is an alternative.

Even though only the bare outlines have been given of possible changes in housing provision, hopefully they are enough to convince most people that it is not beyond the wit of human beings to devise far more satisfactory systems of housing people and building their homes than currently exist. No attempt has been made to design a perfect housing system for the simple reason that none could exist. The point instead is to start devising forms of housing provision that break out of the mesmerizing and debilitating effects of the politics of tenure, so that new political groupings can coalesce around them to push for substantial change. Only in this way is it possible to start giving people greater control over that vital aspect of their lives.

There are grounds to think that significant mass support could be won for radical changes in current forms of housing provision, because the way in which both major tenures have developed in Britain has ended up producing housing that is more expensive and less physically satisfactory than it need be for virtually every household in the country. The wider social implications are also enormous, affecting the physical environment in which people have to live and the life-styles which they can adopt. An important issue not covered in this book, for example, is the effect on household structures. The nature of housing provision has widespread repercussions on personal life, acting as a severe restriction for instance on attempts to break down the dominance of patriarchal, nuclear family structures.

The proposals about changing the nature of structures of provision are ways of maintaining the formal definitions of house rental and ownership whilst transforming their content. Political opposition to change will obviously be enormous but only through land and building industry nationalization can the preconditions exist for the progressive transformation of housing provision. In this way it is possible to get a unity across housing tenures of progressive forces amongst consumers of housing and amongst building workers as producers of housing. To many this might sound like pie-in-the-sky idealism. But the collapse of the current structures of housing provision has no parallel since the demise of the structure of private rented provision in the early years of this century. That earlier collapse created the political opportunity for mass council housing; the current one is doing the same for dreams like mine.

Appendix

The survey of housebuilding firms and planning authorities: research method

Parts of the discussion in chapters 3, 5, 6 and 8 were based on data derived from an interview survey of housebuilders and planning authorities undertaken by the author and Andrew Cullen in 1980 and 1981. The aim was to get information on the operating characteristics of different types of firm in the housebuilding industry, and to examine the relations between planning authorities and private housebuilders.

Firms differ both by size and by their spatial location; the latter was felt to be important as the literature suggested that there were widespread variations in the practices of firms in different regions of the country. In addition considerable variability was expected between housebuilding companies in their efficiency and behaviour, because of the volatility of the markets in which builders operate and the relationship between land development profit and the production process discussed in chapters 5 and 6. These features of the building industry had a strong influence on the sampling and survey techniques adopted and on the range of information collected.

The potential variability of company behaviour in private housebuilding suggests that any attempt to use statistical techniques relying on an underlying conception of average (or normalized) behaviour would conceal the differences which it was hoped to highlight. Instead a stratified sampling technique was adopted to maximize the potential differences in firm behaviour that could be investigated given the limited resources of the research project. Four regionally distinct local housing markets were selected within which a sample of active housebuilders were interviewed. The names of the firms approached were provided by the relevant local authorities and the sample selected on the basis of firm size and already known differences in company behaviour. The purpose of the interviews was to understand the operations of housebuilders in a local market and relate that to the broader operations of the companies who build houses over a wider geographic area.

The types of data that can be collected on housebuilding also influenced the approach of the survey. Some of the difficulties were known beforehand, others became apparent after an initial few pilot interviews. Because of the heterogeneity of building work, it is difficult to estimate and to generalize from specific physical production data collected from one or a few building sites. Measurements of productivity and variations in productivity, which are central to the analysis of developments in production, consequently are fraught with theoretical and statistical difficulties (Fleming 1966). No attempt was made, in the light of these problems, to undertake detailed investigations of site production methods. Originally it had been hoped however to get detailed financial information from the companies interviewed about their development and production processes, their land banks and sources of finance. Virtually none were prepared to give such detailed information. Speculative housebuilders are very secretive about their financial calculations for fear of giving competitors an advantage. Yet most were happy to talk about the general history and nature of their company's operations over a wide range of issues. Specific questions were asked on topics such as the history of the company, types of work undertaken, land holdings and purchase, labour force, speed of building under different demand conditions, competition, marketing, the advantages of traditional and timber-frame methods, plant-hire, materials supply and attitudes to the planning system. Some firms, however, still expressed a desire for their company's name to be kept confidential. As a result, it was decided that none of the information collected in the surveys would be identified with individual firms. All the company histories described in this book instead have been derived from secondary source data.

Although the sampling problems discussed above would impose severe difficulties on an econometric study of housebuilding, they did not present insurmountable difficulties for this particular project. Diverse responses by firms to changing market conditions can be explained only by certain industrial structures. Theoretical and empirical consideration is consequently required of the industrial structure that enables such responses by the firms interviewed. Examination of individual firms' behaviour, in other words, helps to elaborate the nature of the housebuilding industry in ways described in the relevant chapters of this book. Over forty-five housebuilders were interviewed; in some cases separate interviews were held with managers of different parts of the company structure. The survey covered subsidiaries of most of the volume housebuilders and also companies covering the whole spectrum of firm sizes.

The four areas selected as local housing markets were Southampton, Swindon, Nottingham and Warrington. Initially a short list of ten towns was compiled from which the final four were drawn after detailed discussions with local planners about local housing and land markets. The four towns range from 100,000 to 500,000 in population size and exhibit different regional characteristics. Each, moreover, had different conditions of housing demand, land supply and local policy towards private housebuilding. The large national housebuilders are represented in all these areas. This latter consideration led to the exclusion of certain areas. The planners in Norwich, for example, explained that landownership patterns and the housebuilders themselves were highly localized. Choice of that area would consequently have minimized the possibility of contacting the larger national builders.

An important distinctive feature of each of the towns is the relationship of the local authority to the supply of land for private housing. Southampton is a classic area in Southern England where strong restrictions are placed on the overall availability of land and its specific location. Swindon, on the other hand, is an expanding town where there is close co-operation between the local authority and builders on the rapid expansion of the town, especially in terms of land assembly and release by the authority. Nottingham had recently switched from large-scale council developments to the encouragement of private housebuilding, in particular through build-for-sale schemes. Warrington, finally, is a new town in which the development corporation is the landowner. It does however form part of the much larger South Lancashire/North Cheshire housing market, representing one of the few potentially attractive areas of large-scale land release in North Cheshire.

Finally, I should like to thank all the respondents to this interview survey who kindly gave up a number of hours to explain their understanding of national, local and firm level issues associated with private housing. No amount of secondary source reading or data gathering could have replaced those hours of invaluable discussion.

Notes

1 Housing problems and owner occupation

1 *National Income and Expenditure 1981*, table 9.4 and *Housing Policy*, Cmnd 6851, figure 3.

2 See the annual housing survey in *Social Trends* for these and other data on physical housing conditions.

3 The *Housing Policy Review* with its technical appendices is a major, if deliberately obscurantist, reference text for the analysis of housing provision in Britain. Data from it will be used many times in this book. Given the long titles of its published components abbreviations will be used. The Green Paper itself, *Housing Policy: a Consultative Document*, Cmnd 6851, will be referred to as HPCD and its three technical volumes as HPTV I to III respectively, whilst the *Review* overall will be called HPR.

4 Sources: HCEC 1981 and *The Times*, 17 September 1981.

5 1981 housing expenditure was £5.7m. plus £2m. mortgage interest tax relief whereas expenditure on the National Health Service was £13.4m. and on education £13.7m. *National Income and Expenditure 1982* and *BSA Bulletin* 27.

6 Gross domestic fixed capital formation on housing by local authorities was £1952m. in 1976 and at £305m. in 1981 at 1975 prices. *National Income and Expenditure 1982*, table 9.4 deflated.

7 *The Government's Expenditure Plans 1982–83 to 1984–85*, vol II, Cmnd 8494 II.

8 Official data for the UK tabulated in *BSA Bulletin* 27 deflated by retail price index.

9 *National Income and Expenditure 1982*, table 8.5 (deflated by retail price index) and *Housing and Construction Statistics 1970–80*, table 107.

10 The phrase 'structure of housing provision' is rather unwieldy but it is difficult to think of a better replacement. The phrase is much shorter than its own definition and so it has been adopted here. Occasionally, to relieve the monotony caused by repetition, the rather vague expression 'form' will be substituted for 'structure' and also the full phrase 'structure of housing provision' will be shortened where the meaning is clear. In many ways it would have been preferable to call this conception of housing provision modes of housing provision rather than structures; not least because of the Pavlovian reaction against the notion of a structure among some sections of the British Left. That alternative expression, however, does have the overriding disadvantage of creating confusion with the central Marxist concept of the mode of production. As particular types of housing provision can involve interrelations between modes of production, it was felt that the confusion and ambiguity might get out of hand, so the word 'structure' has been used instead.

2 The origins of mass home ownership

1 The sharp rise in real land prices during the early 1930s shown in figure 2.2 was as much a product of the fall in the general price level as it was of a rise in the actual money price of land. The data shown, furthermore, overstate considerably the price large housebuilders had to pay. The data are based on auction sales which represent one of the most expensive ways to buy land and so is avoided by large firms. The data are also not weighted to take account of the upward bias to prices caused by a numerical preponderance of small sites. Small sites sell at a large premium over large sites, especially during booms, yet large sites are the domain of the large builder. By way of comparison the later post-1965 DoE series are based on all transactions rather than just auctions and are weighted to avoid the small site problem, and Vallis's figure for 1967–9 is 40 per cent higher than the latter's acreage figure for 1969 alone. The DoE figure moreover is still a market average, and the land market like others is one where size generates substantial purchasing economies. There is

little reason to expect, therefore, that large builders were actually facing sharply rising land costs during the peak years of the 1930s boom.

2 Although the conditions described below also helped to generate one consisting of office, shops and rented housing, the real take-off of the property market had to wait until the end of the 1950s (Marriott 1967).

3 The historical analysis of housing policy unfortunately has still to develop beyond the kings and queens approach to history. Changes in national housing legislation and contemporary data on national housing conditions are the focus of most studies. The political conflicts, many of a local nature or between local and national government, that must have formed a central element in housing politics are unfortunately as a result virtually unknown. Melling (1980) Benwell CDP (1978) and Wohl (1977), however, provide some useful information.

4 See Jackson's (1973) superb study of inter-war suburban London, from where the advertising quotes were taken.

3 The modern speculative housebuilding industry

1 This invaluable reference source ceased publication in 1980; only small parts of it are now published occasionally in *Housing and Construction Statistics*. Attempts have been made by the government to persuade a reluctant construction employers' organization, the NFBTE, to take over data collection on the construction industry.

2 On the concept of a portfolio of contracts and its usefulness in analysing the construction industry, see Ball 1983.

3 Direct Labour Collective 1978, ch. 6, provides a detailed description of the methods and pitfalls of tendering mechanisms.

4 The discussion of types of firm in speculative housebuilding draws on information from the interview survey of senior management in housebuilding firms. Most of the information on the differences between petty capitalist and small family capital categories was derived from a particularly helpful interview with the owner of a small speculative housebuilding firm.

5 One builder interviewed told of an unfortunate recent entrant to speculative building whose first small development of six or so 'executive' houses brought financial ruin as the facing bricks cracked badly during their first winter and had to be replaced. Negligence by

the brick company could only be proved after lengthy litigation, and possibly not even then, so the firm was forced to cease trading.

6 Optimistic accounts of housing output give the image of a dynamic enterprise which can benefit housebuilders. The firm's share price may strengthen. Financiers are more likely to extend credit if the firm is expanding. Landowners will feel more assured of payment for their land and are more likely to feel that the purchaser will still be around at the end of protracted negotiations and legal settlement. House buyers and building societies will be impressed by the marketability of the firm's product. Finally, materials suppliers and subcontractors are likely to be more responsive in price negotiations if they believe more work is on its way.

7 In December 1978 Grove Charity Management owned 49.9 per cent of the Wimpey building company, George Wimpey Ltd. Mitchell's death in December 1982 created uncertainty over Wimpey's future activities. For a description of Wimpey and its history see the centennial review in *Building*, 14 March 1980.

8 Publicly quoted building contractors were consistently able to earn higher rates of return than manufacturing and service companies throughout the 1970s, despite the long-term decline in construction work during the decade and the dramatic slump of 1973–4. See Ball (1983).

9 Two important changes that have occurred during the 1970s are the switch to the imputation system of taxing dividends in 1973, which removed double taxation on distributed profits, and tax relief on profits from stock appreciation introduced in 1974. Variations in tax relief for investment and the existence of investment grants also influence the attractiveness of different activities; see Kay and King (1980).

10 On the economic crisis of British capitalism, see Aaronovitch and Smith (1981).

11 This description of Christian Salveson is derived primarily from a profile of the company given in the *Financial Times*, 22 February 1978. Anecdotally, the Managing Director of Whelmar, Tom Baron, has been the Secretary of the Environment's housing adviser since the Conservative government came to power in 1979 until at least the date of writing this note.

12 Broackes outlines his version of the takeover by Trafalgar House in his autobiography (Broackes 1979). An appendix to that autobiography explains his 1967 views on the new corporation tax system for

property companies. Further information can also be found in articles in *The Sunday Times*, 21 December 1971 and 29 November 1981.

13 The involvement of P&O in housebuilding enables some interesting anecdotes about interlocking directorships to be noted. Inchcape is also on the board of Guardian Royal Exchange Insurance, the parent of Broseley. Bovis is an ex-firm of the Samuel family, a member of which is Sir Keith Joseph, in 1982 the Secretary of State for Education and a director of Bovis until 1978. Nigel Broackes, chairman of Trafalgar House, was in the 1960s on the board of the Bovis property subsidiary, and Sir Keith Joseph asked him to become chairman of Bovis in 1966, a post which Broackes refused (Broackes 1979). Broackes has since been appointed as chairman of the London Docklands Development Corporation, a public body set up in 1980 by the Conservative government to take over redevelopment of London's docklands from the relevant local authorities. It's a small world for some!

14 The exact extent of these takeovers is difficult to quantify as most of the acquired firms are small private companies. They were however a recurrent theme of the interview survey.

4 Building for the new housing market

1 Since this book was written the DoE has started to produce a weighted index of house prices that takes account of a number of the problems of the unweighted index outlined in this appendix. See *Economic Trends*, October 1982. The new weighted index also indicates that much of the difference between the new and existing house price indices shown in figure 4.7 arises from changes in the relative compositions of output. The faster rate of increase of new house prices since the mid-1970s results mainly from the move up-market by builders and so disappears in the new weighted index. There, however, is still a slight variation in the rate of change of prices with new house prices showing a marginally less volatile rate of change.

5 Housing development and land dealing

1 This section has benefited considerably from the work of Ellen Leopold at University College London (cf. Leopold and Bishop 1981).

2 The communal aspect of council housing sometimes has had

disastrous consequences as architects have used it as a rational basis on which to live out their fantasies of how the modern worker should live. In combination with the state's desire to break the economic power of the craft building worker and building contractors' frantic peddling of new concrete technologies, this architectural ideology led to some very expensive, poor-quality council housing in the 1950s and 1960s. Such follies were possible because of the absence of a communal element in the design process as well as in the design form. That would require mass democratic participation in the design and construction process, something which is impossible with the current structure of council housing provision and the dominant political ideologies supporting it.

3 Housebuilders were found in the Private Enterprise Housebuilding Survey to own 9200 hectares of land without planning permission and a further 1900 hectares held on option. The estimate of the white land bank was derived by adding those two figures together, converting the total to dwelling plots (by assuming thirty dwellings per hectare) and applying the same ratio of starts to land banks as that for land with permission in June 1980. The size of the white land bank at this date was much lower than at the time of the previous survey five months earlier, when 16,700 hectares were recorded; using the same assumptions that represents a land bank of more than eight years' duration.

6 The housing production process and the crisis of production

1 Surprisingly a number of managers interviewed suggested that this seemingly elementary piece of site organizational practice had only recently become universal. Palletization and the bulk transhipment of building materials helped its general adoption. Prior to such developments, minimizing of working capital presumably outweighed even efficient movement around a site.

2 The contracting system, however, still does not produce a technically efficient approach to building for reasons that are explored in Ball (1983).

3 One interviewee suggested to us that one of the most spectacular large bankruptcies of the early 1970s occurred because the firm in question ignored the effect of such bonds on borrowing limits. The firm started on many sites to maximize sales turnover, but each new

site reduced its outstanding credit limit sharply. The banks finally called in the liquidator, to the surprise of the firm.

4 This description is paraphrased from Phelps-Brown (1968) which provides a detailed analysis of labour-only subcontracting (LOSC) in the building industry. This official government report came out strongly in favour of LOSC with minor legislative changes proposed. SHAC (1970), a government report on the high cost of Scottish housebuilding, also came out in favour of LOSC and provides some interesting comparisons of LOSC work practices with those of unionized Scottish craft labour. On the relationship between trade unionism and the Lump see the excellent article by Austrin (1980).

5 On attempts by management in manufacturing industry to break down the power of craft workers over the past century, see Braverman (1974) and Friedman (1977).

6 Many Lump workers are trade unionists, although frequently well behind on their dues. A number of active and militant shop stewards have been forced to work on the Lump as a result of employers' blacklisting them from direct employment.

7 The Health and Safety Act of 1975 had a similar result. Prior to its introduction bricklaying gangs used to provide and erect their own scaffolding. The legislation placed the responsibility for site safety with the housebuilder, and led to a legal requirement that skilled scaffolders had to erect scaffolding. Both the task and the responsibility consequently disappeared from the bricklaying trade.

8 A more comprehensive report on the growth of timber-frame housebuilding in Britain, arising from the speculative housebuilding project, can be found in Cullen (1982).

9 The volumetric approach to timber-frame does not have all these advantages as it consists of factory-assembled units (e.g. first floor and roof). Originally formulated for mass public housing projects, it is difficult to transport long distances because of its bulk and imposes more constraints on design through its unit configuration.

10 Building regulations have played a role in the development of timber-frame in private housing. Until 1963 model by-laws had to be waived before timber-frame methods could be used. Subsequent building regulations have gradually increased wall insulation requirements, and this has boosted the development of timber-frame as insulation material can easily be fixed in the panels during its fabrication.

11 An estimate by M. Lindsay, *The Housebuilder*, February 1981. The

same journal provides this profile of one timber-frame producer in the 1970s:

> In 1973 the RILEYFORM timber frame building method was introduced and John Laing Construction Ltd alone have built about 4000 dwellings in the public sector. As a result of this experience the Laing Group decided to switch from traditional to timber frame construction for their private housebuilding activity and, in so doing, achieved record profits and the fastest rate of growth of any housebuilding company in 1980 which however was a disappointing year because of public sector schemes which aborted just as they were about to go forward and because of the slow down in the private sector. This gave a breathing space for technical and organizational reappraisal and planning towards the still further improvement of the company's services to private housebuilders. A number of them, notably McLean's South West, began building in RILEYFORM, while Laing Homes Ltd were active. Components are manufactured throughout the country, including Northern Ireland.   (*The Housebuilder*, February 1981, p. 39)

7 Land-use planning and speculative housebuilding

1 See Hobsbawm (1969) and Gamble (1981).

2 For rents see Offer (1980), on housebuilding Parry Lewis (1965), on building costs Fleming (1966) and on municipal income and expenditure Peacock and Wiseman (1968).

3 New towns built by the state since 1946 have experienced a similar dilemma of the relationship between public investment, public purchase of land at agricultural value and subsequent appropriation of development gain and encouraging private development in the new town (shops, offices, houses, etc.). See Schaffer (1972).

4 As mentioned in chapter 2, the regional structure of industry also changed dramatically. The new industries primarily grew up in the suburbs of London and the Midlands, whilst the traditional export-based industries located in regions that constituted Britain's pre-1914 industrial base, slumped dramatically. See McCrone (1973).

5 A number of large property companies of today started in this way (Marriott 1967).

6 Development is, of course, a dynamic process so any land in practice

will not have a hope-value with respect to only one development but to the timing and profitability of a series of potential future developments. This makes the definition of total value in development quite complex as it has a temporal dimension, but in the ten-farm static case given earlier it is easier to calculate: total value in development will simply be the value of the land of nine farms in agricultural use plus one in housing use at the time of the housing development.

7 The 1951 Conservative government in effect set the precedent by abandoning the compensation procedures set up by the 1947 Act.

8 The argument here contrasts strongly with a number of other interpretations from within the Marxist perspective. Castells (1977), for example, manages to write a whole chapter on urban planning without once mentioning landownership or the processes by which buildings get created (called here structures of provision). His omission is far from unique. Virtually all of 'New Left' writings on the urban question that have blossomed over the past decade have studiously ignored the provision of the built environment, which (with tongue-in-cheek) can be said to be one of the most concrete of products of the relations they are trying to investigate (cf. the articles in Harloe and Lebas 1981). The state somehow manages to intervene into consumption (collective or otherwise) alone. This consumption-orientated perspective was criticized earlier in chapter 1.

8 The demise of planning control

1 This restriction enables certain issues to be highlighted but the significance of housing as a land use should not be underestimated. Half the land use of British urban areas is housing; in small settlements (under 10,000 population) the percentage rises to 80 per cent (Best 1981, table 12). It is consequently by far the most extensive urban land use and so constitutes an important concern of land-use planning, not to mention the roads, shops, etc. associated with it.

2 The poorly publicized, discrete nature of the actual designation of planning permission and the difficulty of quantifying in aggregate the strength of planning control over development make it impossible to be precise about the weakening of planning constraint. Anecdotal evidence of Ministerial statements, structure plan revisions, planning appeals and the policies of individual development control departments is all that is readily available. CPRE (1981) has brought

together an impressive array of such evidence to demonstrate its case that controls have been lowered.

3 Foster (1980) for the census decades 1951–61 and 1961–71 adds a note of caution on how such differences in population and household changes for Britain as a whole should be interpreted spatially:

> The proliferation of households is usually held to be commonest in city centres, because that is where the old, the young, childless couples, the separated and divorced tend to be disproportionately large segments of the population. While this may be true for central cities, it is not for conurbations as a whole. Households grew much faster outside the conurbations (16% in 1951–61; 18% in 1961–71) than within them (3%; 2%).   (Foster 1980, p. 126)

4 Underwood has summarized the debate well:

> In recent years there has been a growing level of criticism of the planning system, much of which focuses on the role and functioning of development control. The more prominent role of 'Ugly Sister' may be as unwelcome to control officers as that of the previous years of neglect when development control was cast as the 'Cinderella' of the planning system. The control task is a thankless one, ill-defined in statute and open to interpretation by the various conflicting interests concerned with land and development which it attempts to mediate on behalf of an equally ill-defined 'public interest'. The main burden of criticism has come from major applicant interests who complain of delays in the handling of planning applications at local authority and appeal level, of an unwarranted degree of intervention particularly on matters of design, of planning policy frameworks either too vague and uncertain or too constraining and rigidly applied, and of the development costs associated with delay, particularly in a period of inflation. The consumers of the 'planned' environment also complain of inadequate consultation, poor quality developments and lack of protection to the environment from both major and minor applications.   (*The Planner*, July/August 1981, p. 100)

5 The major piece of legislation in recent years was the Local Government, Planning and Land Act 1980, the planning measures of which included a repeal of the Community Land Act 1976, the introduction of charges for applying for planning permission, a redefinition of the development control functions of county and district

councils, and the relaxation of criteria requiring planning permission to be sought. As far as new development is concerned most of these measures are either cosmetic or an administrative tidying up. The 1974–9 Labour government, for example, put its own land reform legislation on ice; later Conservative repeal was essentially a formality affecting a few schemes only. Far more important changes have occurred in the process of planning through central government directives and advice, through local authority inclination, and public expenditure cuts.

6 Bassett and Short (1981) provide a survey of housing policy and the inner city in the 1970s.

7 Two quotations will show the extent to which this point is recognized and also the ideological contortions of language used to diminish its potential political effect:

> [Builders] were afraid that local authorities might attach unrealistically high values to sites they owned in areas of weak demand. Land should be sold for what it would fetch from private housebuilders on the open market (i.e. its 'market value') and authorities should take advantage where appropriate of the various subsidies available for purchasing, improving and servicing development land prior to sale. . . . Builders argued that this might even mean negative land values.
>
> (*Manchester Housing Land Availability Study*, HBF/DoE 1979, p. 6)

*

> The biggest-ever campaign to persuade councils to help build houses for sale is being launched by the government and the House Builders Federation. . . .
>
> The government says . . . that local authorities – as major landowners – hold the key to low cost housing.
>
> The options spelled out include local authorities retaining ownership of the land and ultimately selling the freehold to the purchaser at a considerable discount, the builder paying for land in kind by giving houses to the local authority, or the builder selling houses to purchasers on a shared ownership basis. Alternatively the authority could sell the land 'at a price well below market value', but retain the right to nominate purchasers from its waiting list.   (Article in *Construction News*, 3 September 1981)

10 Building societies and the changing mortgage market

1 Because building societies are capitalist enterprises does not, of course, make the investors with building societies capitalists themselves. Their class situation is not altered simply because they lend relatively small amounts of money to a financial institution (see Harris 1976).

2 Mayes (1979), for example, concludes at the end of his econometric study of the impact of building societies on house prices that:

> The particular influences are straightforward and can be expressed in terms of 1970 values: (1) a 1% rise in mortgage advances raises house prices by 0.58% in the long run: (2) a 1% fall in the mortgage rate of interest causes house prices to rise by a 0.65% immediately (a more sensitive effect than (1)): (3) a 1% fall in the savings rate causes a 0.2% rise in house prices (the mortgage rate of interest changes as a consequence and has more effect in the short run than the increase in the inflow of funds). All these changes assume that the exogenous variables remain at their actual values but that the endogenous variables in the model can all change.
>
> (Mayes 1979, p. 14)

13 Towards new forms of housing provision

1 The problem of trying to make subjective economic comparisons between individuals is a classic example of the dilemmas of cost-benefit analysis criticized in Ball (1979).

2 The parallel position over the industrial strategies of successive Labour governments has been well documented by a Trades Council Inquiry:

> It follows from the traditional approach which sees socialist industrial policies mainly in terms of the extension of the existing state that the development of political consciousness is seen primarily as a matter of winning people's votes for a programme which politicians will implement. Since, on this approach, there is no real sense in which people themselves are involved in implementing this programme, then there is little need for them to be actively involved in formulating it, reflecting on the experience of past programmes, or making their own general aspirations more specific. (CLNNTC 1980, p. 156)

3 A city of 2 million people typically has a housing stock of about 800,000. A probable building rate of 2 per cent of the existing stock per annum would give an annual housebuilding requirement of 16,000 dwellings.

References

Aaronovitch, S. and Smith, R. (1981) *The Political Economy of British Capitalism*, London, McGraw-Hill.

Austerberry, H. and Watson, S. (1981) 'A woman's place: a feminist approach to housing in Britain', *Feminist Review*, 8, 49–62.

Austrin, T. (1980) 'The "Lump" in the UK construction industry', in Nichols, T. (ed.), *Capital and Labour*, New York, Fontana.

Backwell, J. and Dickens, P. (1979) 'Town planning, mass loyalty and the restructuring of capital: the origins of the 1947 planning legislation revisited', *Urban and Regional Studies Working Paper 11*, University of Sussex.

Baldwin, R. and Ransom, W. (1978) *The Integrity of Trussed Rafter Roofs*, Watford, Building Research Establishment, Current Paper 83/78.

Ball, M. (1978) 'British housing policy and the housebuilding industry', *Capital and Class*, 4, 78–99.

Ball, M. (1979) 'Cost-benefit analysis: a critique', in Green, F. and Nore, P. (eds), *Issues in Political Economy*, London, Macmillan.

Ball, M. (1980) *The contracting system in the construction industry*, London, Birkbeck College Economics Discussion Paper 86.

Ball, M. (1981) 'The development of capitalism in housing provision', *International Journal of Urban and Regional Research*, 5, 145–77.

Ball, M. (1982) 'Housing provision and the economic crisis', *Capital and Class*, 17, 60–77.

Ball, M. (1983) *Capital in the Construction Industry*, mimeo.

Ball, M. and Cullen, A. (1980) *Mergers and Accumulation in the British Construction Industry, 1960–1979*, London, Birkbeck College Economics Discussion Paper 73.

Ball, M. and Kirwan, R. (1975) *The Economics of an Urban Housing Market, Bristol Area Study*, London, Centre for Environmental Studies Research Paper 15.

Barlow Commission (1940) *Report of the Royal Commission on the Distribution of the Industrial Population*, Cmd. 6153, London, HMSO.

Barrett, S., Boddy, M. and Stewart, M. (1978) *Implementation of the Community Land Scheme: Interim Report*, University of Bristol, School for Advanced and Urban Studies Occasional Paper 3.

Barrett, S. and Underwood, J. (1981) 'Planning at the crossroads', *Planner News*, January, 3.

Bassett, K. and Short, J. (1980) *Housing and Residential Structure*, London, Routledge & Kegan Paul.

Bassett, K. and Short, J. (1981) 'Housing policy and the inner city in the 1970s', *Transactions of the Institute of British Geographers*, 6, 293–312.

Benwell CDP (1980) *Private Housing and the Working Class*, Newcastle-upon-Tyne, Benwell Community Project.

Best, R. (1981) *Land Use and Living Space*, London, Methuen.

Boddy, M. (1980) *The Building Societies*, London, Macmillan.

Boleat, M. (1981) 'The housing market – an economic framework', in *The Determination and Control of House Prices*, London, Building Societies Association.

Bowley, M. (1944) *Housing and the State*, London, Allen & Unwin.

Bowley, M. (1960) *Innovations in Building Materials*, London, Macmillan.

Braverman, H. (1974) *Labour and Monopoly Capital*, New York, Monthly Review Press.

Briggs, A. (1968) *Victorian Cities*, Harmondsworth, Penguin.

Broackes, N. (1979) *A Growing Concern*, London, Weidenfeld & Nicolson.

BSA (1981a) *Building Societies in 1980*, London, Building Societies Association.

BSA (1981b) *The Determination and Control of House Prices*, London, Building Societies Association.

Building Statistical Services (1973) *Chains of Sales in Private Housing*, London.

Bundock, J. (1974) 'Speculative housebuilding and some aspects of the activities of the speculative housebuilder within the Greater London outer surburban area 1919–1939', University of Kent, M. Phil. thesis.

Burnett, J. (1978) *A Social History of Housing 1815–1970*, Newton Abbott, David & Charles.

Butler, A. (1978) 'New price indices for construction output statistics', *Economic Trends*, 297, 97–110.

Cairncross, A. (1955) *Home and Foreign Investment 1870–1913*, Clifton, Kelley.

Castells, M. (1977) *The Urban Question*, London, Arnold.

Clarke, M. (1977) 'Too much housing', *Lloyds Bank Review*, 126, 17–33.

Cleary, E. (1965) *The Building Society Movement*, London, Elek.

CLNNTC (1980) *State Intervention in Industry*, Newcastle-upon-Tyne, Coventry, Liverpool, Newcastle and North Tyneside Trades Councils.

Cowley, J. (1979) *Housing for People or for Profit?*, London, Stage 1.

CPRE (1981) *Planning – Friend or Foe?*, London, Council for the Protection of Rural England.

Cullen, A. (1982) 'Speculative housebuilding in Britain. Some notes on the switch to the timber-frame production method', in *The Production of the Built Environment 3*, London, University College, Bartlett School.

Cullingworth, J. B. (1960) *Housing Needs and Planning Policy*, London, Routledge & Kegan Paul.

Cullingworth, J. B. (1979a) *Essays in Housing Policy*, London, Allen & Unwin.

Cullingworth, J. B. (1979b) *Town and Country Planning in Britain*, 7th edn, London, Allen & Unwin.

Damer, S. and Hague, C. (1971) 'Public participation in planning: a review', *Town Planning Review*, 42, 212–32.

Davies, H. (1980) 'The relevance of development control', *Town Planning Review*, 51, 7–24.

Dickens, P. and Goodwin, M. (1981) 'The political economy of Sheffield's council houses', mimeo.

Direct Labour Collective (1978) *Building with Direct Labour: Local Authority Building and the Crisis in the Construction Industry*, London, Housing Workshop of the Conference of Socialist Economists.

Direct Labour Collective (1980) *Direct Labour under Attack*, London, Housing Workshop of the Conference of Socialist Economists.

Dobry Report (1974) *Review of the Development Control System. Interim Report*, London, HMSO.

Drewett, R. (1973) 'Land values and the suburban land market', ch. 7, vol. 2 in Hall, P., Gracey, H., Drewett, R. and Thomas, R. *The Containment of Urban England*, 2 vols, London, PEP and Allen & Unwin.

Duffy, M. (1975) 'On the short-term forecasting of private housing investment in the United Kingdom', *Applied Economics*, 7, 119–34.

Dunleavy, P. (1979) 'The urban bases of political alignment', *British Journal of Political Science*, 9, 409–43.

Dyos, H. J. (1961) *Victorian Suburb*, Leicester, Leicester University Press.

EIU (1975) *Housing Land Availability in the South-East*, Economist Intelligence Unit, London, HMSO.

Elphick, P. (1964) 'Cramlington. Some problems encountered in building a new town', *Town Planning Review*, 35, 59–75.

Engels, F. (1969) *The Condition of the Working Class in England*, London, Fontana.

Evans, A. W. (1974) 'Private sector housing land prices in England and Wales', *Economic Trends*, 244, xiv–xxxvii.

Eversley, D. (1973) *The Planner in Society*, London, Faber.

Feinstein, C. (1972) *National Income, Expenditure and Output of the United Kingdom 1855–1965*, Cambridge, Cambridge University Press.

Fishman, R. (1977) *Urban Utopias in the Twentieth Century: Ebenezer Howard, Frank Lloyd Wright, and Le Corbusier*, New York, Basic Books.

Fleming, M. (1966) 'The long-term measurement of construction costs in the UK', *Journal of the Royal Statistical Society*, series A, 129, 534–56.

Fleming, M. and Nellis, J. (1982) 'A new housing crisis?', *Lloyds Bank Review*, 144, 38–53.

Foley, D. (1960) 'British town planning: one ideology or three?', *British Journal of Sociology*, 11, 211–31.

Foot, M. (1975) *Aneurin Bevan*, vol. 2, London, Granada.

Forbes, W. (1969) *A Survey of Progress in Housebuilding*, Watford, Building Research Establishment, Current Paper 25/69.

Ford, J. (1975) 'The role of the Building Society manager in the urban stratification system; autonomy versus constraint', *Urban Studies*, 12, 295–302.

Foster, C. (1973) 'Planning and the Market', in Cowan, P. (ed.) *The Future of Planning*, London, Heinemann.

Foster, C. (1980) 'Housing in the conurbations', in Cameron, G. (ed.) *The Future of the British Conurbations*, London, Longman.

Fothergill, S. and Gudgin, G. (1982) *Unequal Growth: Urban and Regional Employment Change in the UK*, London, Heinemann.

Fox, W. (1934) *London Speculative Builders*, London, Labour Research.

Francis, H. and Smith, D. (1980) *The Fed: a history of the South Wales miners in the twentieth century*, London, Lawrence & Wishart.

Fraser, D. (1979) *Power and Authority in the Victorian City*, Oxford, Blackwell.

Friedman, A. (1977) *Industry and Labour*, London, Macmillan.

Gamble, A. (1981) *Britain in Decline*, London, Macmillan.

Ginsburg, N. (1979) *Class, Capital and Social Policy*, London, Macmillan.

Glass, R. (1959) 'The evaluation of planning: some sociological considerations', *International Social Science Journal*, 11, 393–409.

GMC (1979) *Greater Manchester County Structure Plan*, Manchester, Greater Manchester Council.

Gough, I. (1979) *The Political Economy of the Welfare State*, London, Macmillan.

Gough, T. (1975a) 'Forecasting private housing investment: a reply to Duffy', *Applied Economics*, 7, 135–8.

Gough, T. (1975b) 'Phases of British private housebuilding and the supply of mortgage credit', *Applied Economics*, 7, 213–22.

Gray, F. (1979) 'Council house management', ch. 8 in Merrett, S. *State Housing in Britain*, London, Routledge & Kegan Paul.

Grieveson, Grant & Co. (1981) *The UK Construction Industry*, London, Grieveson, Grant & Co. Investment Research.

Hadjimatheou, G. (1976) *Housing and Mortgage Markets, the UK Experience*, Farnborough, Saxon House.

Hall, P. (ed.) (1966) *Land Values: A symposium*, London, Sweet & Maxwell for Acton Society Trust.

Hall, P. (1975) *Urban and Regional Planning*, Harmondsworth, Penguin.

Hall, P., Gracey, H., Drewett, R. and Thomas, R. (1973) *The Containment of Urban England*, 2 vols, London, PEP and Allen & Unwin.

Harloe, M. and Lebas, E. (1981) *City, Class and Capital*, London, Edward Arnold.

Harris, L. (1976) 'On interest, credit and capital', *Economy and Society*, 5, 145–77.

Harrison, A. (1977) *Economics and Land-use Planning*, London, Croom Helm.

Hayek, F. (1960) *The Constitution of Liberty*, London, Routledge & Kegan Paul.

HBF/DoE (1979) *Study of the Availability of Private House-Building Land in Greater Manchester 1978–1981*, 2 vols, London, Department of the Environment.

HCEC (1980) *Enquiry into Implications of Government's Expenditure Plans 1980–81 to 1983–84 for the Housing Policies of the Department of the Environment*, First Report from House of Commons Environment Committee, HC714, London, HMSO.

HCEC (1981a) *DoE's Housing Policies*, Third Report from House of Commons Environment Committee, HC383, London, HMSO.

HCEC (1981b) *Council House Sales*, Second Report from the House of Commons Environment Committee, HC 366-I, London, HMSO.

Hird, C. (1975) *Your Employer's Profits*, London, Pluto.

Hobsbawm, E. (1969) *Industry and Empire*, Harmondsworth, Penguin.

Horsfall, T. (1900) 'Housing of the labouring classes', unpublished lecture, quoted in Sutcliffe, A. (1981) *Towards the Planned City*, Oxford, Blackwell.

Howard, E. (1898) *Garden Cities of Tomorrow* (1965 edn) London, Faber.

Issacharoff, R. (1977) 'The building boom of the interwar years: whose profits and at whose cost?', in Harloe, M. (ed.) *Proceedings of the Conference on Urban Change and Conflict*, London, Centre for Environmental Studies.

Jackson, A. (1973) *Semi-detached London*, London, Allen & Unwin.

Jessop, B. (1982) *The Capitalist State*, Oxford, Martin Robertson.

JURUE (1977) *Planning and Land Availability*, Joint Unit for Research on the Urban Environment, University of Aston, Birmingham.

Kay, J. A. and King, M. A. (1980) *The British Tax System*, Oxford, Oxford University Press.

Kellett, J. (1969) *The Impact of Railways on Victorian Cities*, London, Routledge & Kegan Paul.

Kemeny, J. (1980) *The Myth of Home-Ownership*, London, Routledge & Kegan Paul.

Kilroy, B. (1978) *Housing Finance – Organic Reform?*, London, Labour Economic Finance and Taxation Association.

Kilroy, B. and McIntosh, N. (1982) *Housing and the Economy*, London, Shelter.

King, M. and Atkinson, A. (1980) 'Housing policy, taxation and reform', *Midland Bank Review*, Spring, 7–15.

Kirk, G. (1980) *Urban Planning in a Capitalist Society*, London, Croom Helm.

Labour Party (1981) *A Future for Public Housing*, London, Labour Party.

Laing & Cruikshank (1980) *Private Housebuilding – an Investment Strategy*, London, Laing & Cruikshank.

Land Enquiry Committee (1914) *The Land: The Report of the Land Enquiry Committee, vol. II: Urban*, London, Hodder & Stoughton.

Lemessany, J. and Clapp, M. (1978) *Resource Inputs to Construction: the Labour Requirements of Housebuilding*, Watford, Building Research Establishment, Current Paper 76/78.

Leopold, E. and Bishop, D. (1981) *Design Philosophy and Practice in Speculative Housebuilding*, London, Building Economics Research Unit, University College.

Leung, H. (1979) *Redistribution of Land Values. A Re-examination of the 1947 Scheme*, Cambridge, Department of Land Economy, Occasional Paper 11.

Levin, P. (1976) *Government and the Planning Process*, London, Allen & Unwin.

Lichfield, N. and Darin-Drabkin, H. (1980) *Land Policy in Planning*, London, Allen & Unwin.

McAuslan, P. (1975) *Land, Law and Planning*, London, Weidenfield & Nicolson.

McCrone, G. (1973) *Regional Policy in Britain*, London, Allen & Unwin.

Mackay, D. and Cox, A. (1979) *The Politics of Urban Change*, London, Croom Helm.

Marriott, O. (1967) *The Property Boom*, London, Pan Books.

Massey, D., Barras, R. and Broadbent, A. (1973) 'Labour must take over land', *Socialist Commentary*.

Massey, D. and Catalano, A. (1978) *Capital and Land*, London, Edward Arnold.

Massey, D. and Meegan, R. (1982) *The Anatomy of Job Loss*, London, Methuen.

Mayes, D. (1979) *The Property Boom*, Oxford, Martin Robertson.

Maywald, K. (1954) 'An index of building costs in the United Kingdom 1845–1938', *Economic History Review*, 2nd series, 7, 189–203.

Melling, J. (ed.) (1980) *Housing, Social Policy and the State*, London, Croom Helm.

Merrett, S. (1979) *State Housing in Britain*, London, Routledge & Kegan Paul.

Mingione, E. (1981) *Social Conflict and the City*, Oxford, Blackwell.

Minns, R. (1982) *Take Over the City*, London, Pluto.

Neuberger, H. and Nichol, B. (1975) *The Recent Course of Land and Property Prices and the Factors Underlying It*, London, Department of the Environment.

Nevin, E. (1955) *The Mechanism of Cheap Money*, Cardiff, University of Wales Press.

Nevitt, A. (1966) *Housing, Taxation and Subsidies*, London, Nelson.

Nicholls, D., Turner, D., Kirby-Smith, R., and Cullen, J. (1980) *The Private Sector Housing Development Process in Inner City Areas*, University of Cambridge, Department of Land Economy.

Odling-Smee, J. (1975) 'The impact of the fiscal system on different tenure systems', in *Housing Finance*, Institute of Fiscal Studies Publication No. 12.

Offer, A. (1977) 'The origins of the Law of Property Acts, 1910–1925', *Modern Law Review*, 40, 505–22.

Offer, A. (1980) 'Ricardo's Paradox and the movement of rents in England, *c.* 1870–1910', *Economic History Review*, 2nd series, 33, 236–52.

Parker, H. (1966) 'The history of compensation and betterment since 1900', in Hall, P. (ed.), *Land Values*, London, Sweet & Maxwell for Acton Society Trust.

Parry Lewis, J. (1965) *Building Cycles and Britain's Growth*, London, Macmillan.

Peacock, A. and Wiseman, J. (1968) *The Growth of Public Expenditure in the UK*, London, Allen & Unwin.

PEHW (1975) *Political Economy and the Housing Question*, London, Housing Workshop of the Conference of Socialist Economists.

PEHW (1976) *Housing and Class in Britain*, London, Housing Workshop of the Conference of Socialist Economists.

Phelps-Brown, H. (1968) *Report of the Committee of Inquiry into Certain Matters Concerning Labour in Building and Civil Engineering*, Cmnd. 3714, London, HMSO.

Pollard, S. (1973) *The Development of the British Economy 1914–1967*, London, Arnold.

Poulantzas, N. (1975) *Classes in Contemporary Capitalism*, London, New Left Books.

Randolph, W. and Robert, S. (1981) 'Population redistribution in

Great Britain 1971–1981', *Town and Country Planning*, September, 227–31.

Ravetz, A. (1980) *Remaking Cities*, London, Croom Helm.

Reiners, W. and Broughton, H. (1953) *Productivity in Housebuilding*, London, National Building Studies Special Report No. 21.

Richardson, H. and Aldcroft, D. (1968) *Building in the British Economy Between the Wars*, London, Allen & Unwin.

Rose, H. and Rose, S. (1982) 'Moving right out of welfare – and the way back', *Critical Social Policy*, 2, 7–18.

Saunders, P. (1979) *Urban Politics*, London, Hutchinson.

Savory Milln (1981) *Savory Milln's Building Book 1981*, London.

Savory Milln (1982) *Savory Milln's Building Book 1982*, London.

Schaffer, F. (1972) *The New Town Story*, London, Paladin.

Self, P. and Storing, H. (1971) *The State and the Farmer*, London, Allen & Unwin.

SHAC (1970) Scottish Housing Advisory Committee, *The Cost of Private House Building in Scotland*, London, HMSO.

Sheail, J. (1979) 'The Restriction of Ribbon Development Act: the character and perception of land-use control in inter-war Britain', *Regional Studies*, 13, 501–12.

Simmie, J. (1974) *Citizens in Conflict*, London, Hutchinson.

Simon & Coates (1978) *The Leading Contractors*, London.

Skinner, D. and Langdon, J. (1974) *The Story of Clay Cross*, Nottingham, Spokesman Books.

Stedman-Jones, G. (1971) *Outcast London*, London, Oxford University Press.

Stow Report (1979) *Mortgage Finance in the 1980s*, London, Building Societies Association.

Sutcliffe, A. (1981) *Towards the Planned City*, Oxford, Blackwell.

Swenarton, M. (1981) *Homes for Heroes*, London, Heinemann.

Tarn, J. (1973) *Five Per Cent Philanthropy: An Account of Housing in Urban Areas Between 1840 and 1914*, London, Architectural Association.

Treasury (1981) *The Government's Expenditure Plans 1981–82 to 1983–84*, Cmnd. 8175, London, HMSO.

TRG (1981) *Training on Trail: Government Policy and the Building Industry*, Birmingham, Training Research Group.

Underwood, J. (1981) 'Development control: a review of research and current issues' *Progress in Planning*, 16, 179–242.

Uthwatt Report (1942) *Expert Committee on Compensation and*

Betterment, Final Report, Cmnd. 6386, London, HMSO.

Vallis, E. (1972) 'Urban land and building prices parts I–IV', *Estates Gazette*, 222, 1015–19, 1209–13, 1406–7 and 1604–5.

Vickery, D. (1978) 'Development control: a new sense of purpose', *The Planner*, January, 24–5.

Vipond, M. J. (1969) 'Fluctuations in private housebuilding in Great Britain, 1950–66', *Scottish Journal of Political Economy*, 16, 196–211.

Whitehead, C. (1974) *The United Kingdom Housing Market: an Econometric Model*, Farnborough, Saxon House.

Wiener, M. (1981) *English Culture and the Decline of the Industrial Spirit*, Cambridge, Cambridge University Press.

Wilson Committee (1980) *Report of Committee to Review the Functioning of Financial Institutions*, Cmnd. 7937, London, HMSO.

Wilson, E. (1977) *Women and the Welfare State*, London, Tavistock.

Wohl, A. (1977) *The Eternal Slum*, London, Edward Arnold.

Wright, E. (1978) *Class, Crisis and the State*, London, New Left Books.

Index